POLYMER BLENDS
AND COMPOSITES

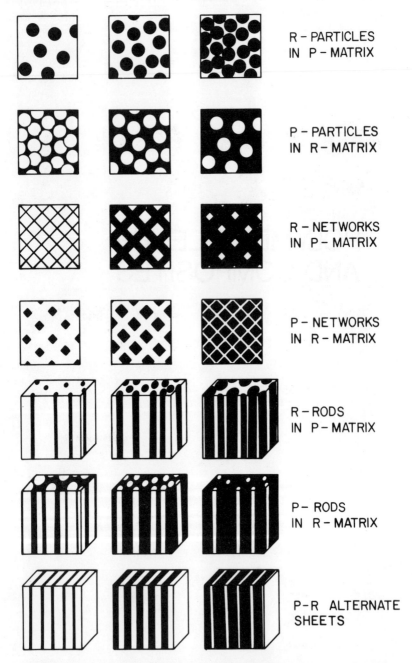

R – PARTICLES
IN P – MATRIX

P – PARTICLES
IN R – MATRIX

R – NETWORKS
IN P – MATRIX

P – NETWORKS
IN R – MATRIX

R – RODS
IN P – MATRIX

P – RODS
IN R – MATRIX

P–R ALTERNATE
SHEETS

Schematic illustration of some of the phase structures that can be composed from plastic (P) and rubber (R) components. Rubber content increases from left to right. M. Matsuo and S. Sagaye, in *Colloidal and Morphological Behavior of Block and Graft Copolymers*, G. E. Molau, ed., Plenum, New York (1971).

POLYMER BLENDS AND COMPOSITES

John A. Manson and Leslie H. Sperling
Lehigh University

PLENUM PRESS · NEW YORK AND LONDON

Library of Congress Cataloging in Publication Data

Manson, John A   1928-
  Polymer blends and composites.

  Bibliography: p.
  Includes index.
  1. Polymers and polymerization. I. Sperling, Leslie Howard, 1932-   joint
author. II. Title.
QD381.M373   547'.84   75-28174
ISBN 0-306-30831-2

© 1976 Plenum Press, New York
A Division of Plenum Publishing Corporation
227 West 17th Street, New York, N.Y. 10011

Printed in the United States of America

To our wives and children, who studied, played,
or sewed quietly for 500 consecutive nights that
this book might be written.

Preface

The need for writing a monograph on polymer blends and composites became apparent during presentation of material on this subject to our advanced polymers class. Although the flood of important research in this area in the past decade has resulted in many symposia, edited collections of papers, reviews, contributions to scientific journals, and patents, apparently no organized presentation in book form has been forthcoming. In a closely connected way, another strong impetus for writing this monograph arose out of our research programs in the Materials Research Center at Lehigh University. As part of this effort, we had naturally compiled hundreds of references and become acquainted with many leaders in the field of blend and composite research.

Perhaps the most important concept stressed over and over again is that engineering materials are useful because of their complexity, not in spite of it. Blends and composites are toughened because many modes of resistance to failure are available. Although such multimechanism processes are difficult to describe with a unified theory, we have presented available developments in juxtaposition with the experimental portions. The arguments somewhat resemble the classical discussion of resonance in organic chemistry, where molecular structures increase in stability as more electronic configurations become available.

For convenience, the monograph is divided into four parts. The first, Chapter 1, relates to homopolymers, and gives the reader a brief glimpse at the basic structures and properties of polymers. Readers who are already familiar with polymers through course work, research, or engineering will find this material to be an elementary review. Although detail has been kept to a minimum, we have attempted to describe the concepts of polymer science that are vital to an understanding of later sections.

The second part, Chapters 2–9, is concerned with polymer blends, including mechanical blends, graft copolymers, block copolymers, ionomers, and interpenetrating polymer networks. The development of most chapters proceeds from synthesis to morphology, and then shows how morphology affects or controls the physical and mechanical behavior of the finished material. The most exciting development of the past decade, the electron microscope studies of the details of phase separation, is emphasized.

The third part, Chapters 10–12, examines the properties of composites, where one component is usually nonpolymeric. Two broad classes of composites are treated: polymer-impregnated materials, such as wood or concrete, and fiber- and particle-reinforced plastics and elastomers. The approach throughout draws upon basic chemistry, materials science, and engineering. The phases discussed here tend to be large and well-defined, but the interactions at the phase interfaces remain crucial in determining properties.

Chapter 13, comprising part four, gives a brief view of very recent developments, together with some unsolved problems. Broadly speaking, differing physical behavior leads to three major classes of polymers: elastomers, plastics, and fibers. Based on applications, two subclasses, coatings and adhesives, are usually considered. It is interesting to note that significant proportions of each of the five materials are consumed in the form of polymer blends, composites, or combinations of both. Thus, the reader will find references to apparently diverse industries side by side on the same page.

During the course of writing this monograph, it occurred to the authors that a systematic classification of polymer blends and composites was sorely needed. In how many significantly different ways can polymers be mixed with other polymers, or with nonpolymers? What interrelationships exist among the known modes, and how may we go about uncovering yet undiscovered combinations? While the classification theme pervades the text, weaving in and out, the actual classifications are left to Chapter 13.

Every monograph is beamed at particular audiences. We hope that chemists, chemical engineers, and materials scientists working in, or newly entering, the field of polymers will constitute the prime audience. Although the book was not intended as a textbook, it may well serve as a source book for a second course, especially at the first year graduate level.

We wish to thank many people for their helpful discussions. Of special mention, Dr. D. A. Thomas of Lehigh University pointed out important analogies between metallic alloys and polymer blends, Dr. C. B. Bucknall reviewed Chapter 3, and Dr. M. Matsuo of Japanese Geon critically reviewed a larger part of the unfinished manuscript. On the secretarial side, a great deal of typing was done by Mrs. Marion Bray, Mrs. Joyce Davis, Mrs. Susan Grates, Mrs. Pamela Van Doren, and Mrs. Margaret Benzak. We appreciate the help and cooperation of the Departments of Chemistry and Chemical Engineering of the Materials Research Center and of our graduate students.

1976   J. A. Manson
Lehigh University   L. H. Sperling

Contents

8. Interpenetrating Polymer Networks............. 237

9. Miscellaneous Polymer Blends.................... 271

POLYMER BLENDS AND COMPOSITES

Homopolymer Structure and Behavior

Polymer blends and composites display a broad gamut of behavior, ranging from toughened elastomers through impact-resistant plastics to fiber-reinforced thermosets and polymer-impregnated concrete. Such materials are of practical importance because their unique two-phased structure often allows for nonlinear and synergistic behavior. In this monograph, polymer blends are defined as combinations of two kinds of polymers, and composites are defined as systems containing polymeric and nonpolymeric materials.

Before discussing the behavior of the blend and composite systems themselves, let us briefly examine those important properties of the base homopolymers that underlie the properties of their constituent materials. The purpose of this chapter will be to review the synthesis, structure, morphology, and mechanical behavior of homopolymers. Topics covered will include morphological properties, such as polymer structure, tacticity, crystallinity, and network formation. The relationships between morphology and such physical properties as glass–rubber transitions, rubber elasticity, viscoelastic flow, and failure phenomena will be briefly discussed. More detailed treatments are available in a number of textbooks and monographs, many of which are listed in the Bibliography at the end of this chapter along with the important polymer journals. Polymer synthesis will be briefly covered in Appendix A of this chapter.

An important secondary objective of this chapter will be to acquaint the reader with several important experimental techniques that will be discussed throughout. In the laboratory, similar experiments are often done on both homopolymers and more complex systems. However, their development in terms of homopolymers is simpler and will serve to set the stage for later chapters.

1

1.1. HIGH POLYMERS

High polymers are composed of many subunits or "mers" linked together to form long chains.* Indeed, the very high molecular weight implied in the term "high polymer" is responsible for the basis of the typical behavior of plastics and elastomers. With a few exceptions, such as some naturally occurring and structurally complex proteins, most polymer molecules are based on a relatively simple repeat structure involving only one or two kinds of mer, each derived from its own starting material, the "monomer." For example, the structure of polystyrene (based on styrene monomer) may be written (neglecting end groups) as a series of simple structural units chemically bonded together:

$$-CH_2-CH-CH_2-CH-CH_2-CH- \qquad (1.1)$$

The structure may be written more compactly by taking the mer or structural unit n times

$$\left(-CH_2-CH-\right)_n \qquad (1.2)$$

If, as in this case, the repeating unit is identical with the mer corresponding to the original monomer, the number of repeating units n is referred to as the degree of polymerization DP, and the molecular weight M, is given by the product of the DP and the mer molecular weight. For most high polymers, DP is of the order of hundreds or thousands, and M is tens of thousands to millions. Typical structures for polymers frequently mentioned in this monograph are illustrated in Table 1.1. If, on the other hand, the repeat unit contains two structural units (each derived from its own monomer), the number of repeat units n will equal one-half the DP; this is the case with polymers such as nylon 6,6 (Table 1.1).

1.2. MOLECULAR SIZE AND SHAPE

Important advances in the characterization of polymer chain properties were made in studies performed during the 1940's and 1950's. These studies

---

* Literally, the term "polymer" means many "mers" or parts.

Table 1.1
Common Polymer Repeat Structures

Mer	Polymer
$-CH_2-CH_2-$	Polyethylene
$-CH_2-CH-$ $\mid$ Cl	Poly(vinyl chloride)
$-CH_2-CH-$ (phenyl ring)	Polystyrene
$-CH_2-CH=CH-CH_2-$	Polybutadiene
CH_3 $\mid$ $-CH_2-C=CH-CH_2-$	Polyisoprene
CH_3 $\mid$ $-CH_2-C-$ $\mid$ $C=O$ $\mid$ O $\mid$ CH_3	Poly(methyl methacrylate)
(glucose ring structure) CH_2OH ... OH	Cellulose
$\overset{O}{\overset{\|}{-C}}(CH_2)_4\overset{O}{\overset{\|}{C}}-NH(CH_2)_6NH-$	Nylon 6,6
(2,6-dimethyl phenylene oxide ring) $CH_3 ... O ... CH_3$	Poly(2,6-dimethyl phenylene oxide) (PPO)
CH_3 $\mid$ $-CH_2-CH-$	Polypropylene
$-CH_2-CH-$ $\mid$ CN	Polyacrylonitrile
$-CH_2-CH_2-O-$	Poly(ethylene oxide)

involved examination of the polymers in dilute solution in which the individual molecules were separated from each other, and hence acted more or less independently (Flory, 1953, Chapters 7, 14; Billmeyer, 1971, Chapter 3). Considerable emphasis was placed on the determination of molecular weight and molecular weight distribution by osmometry, light scattering, and sometimes ultracentrifugation. A characteristic dimension of the chains, defined by the end-to-end distance r (see Figure 1.1), was also determined by light scattering instrumentation. For statistical reasons, the quantity r must be averaged over all molecular sizes present, and is commonly expressed as the root mean square end-to-end distance. In addition, important practical relationships between molecular weight and intrinsic viscosity $[\eta]$ were developed for routine characterization, since measurement of the intrinsic viscosity is especially convenient. More recently, gel permeation chromatography (GPC) techniques have made it possible to obtain information about molecular weights and distributions still more conveniently (Boni and Sliemes, 1968). The major findings of this body of literature will be summarized here, because properties in the bulk state, including polymer blends and composites, are influenced, and sometimes controlled, by chain properties such as molecular weight and size.

As mentioned above, molecular weights of high polymers are in fact exceptionally high; for example, commercial poly(vinyl chloride) usually has a molecular weight in the range of 70,000–100,000 g/mol. Typical polymer properties are observed when the chains are long enough so that extensive entangling occurs, and when the total bonding energy between the polymer chains (the sum of all the segment interactions) exceeds some critical value, such as the equivalent segment vaporization energy or the energy required to break the chain. Thus commercial waxes, whose molecules contain 25–50 carbons in the form of $-CH_2-$ groups, are not high polymers,

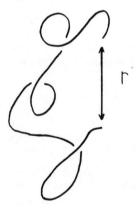

Figure 1.1. The end-to-end distance r of a polymer chain.

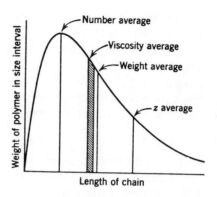

Figure 1.2. Illustration of molecular weight distribution in polymers, with averages. The quantity W_x is the weight fraction of x-mer, or the weight of polymer within a narrow interval of molecular weight. (McGrew, 1958.)

while polyethylene, whose molecules contain 2000–3000 carbon atoms of the same structure, is a typical high polymer.

With the exception of proteins and some specially synthesized polymers, most polymers exhibit a more or less broad distribution of molecular weights. For convenience, several average molecular weights (Flory, 1953) can readily be defined. For example, the number-average molecular weight $\overline{M}_n$ is given by

$$\overline{M}_n = \sum_i N_i M_i \Big/ \sum_i N_i \tag{1.3}$$

where N_i is the number of molecules having molecular weight M_i. This simple numerical average may be determined by such methods as osmometry or GPC. The weight-average molecular weight $\overline{M}_w$ may be written as follows:

$$\overline{M}_w = \sum_i N_i M_i^2 \Big/ \sum_i N_i M_i \tag{1.4}$$

and is usually determined by light scattering or GPC. For monodisperse systems $\overline{M}_w = \overline{M}_n$, but for many polymeric materials, including condensation polymers and addition polymers that terminate by simple disproportionation, $\overline{M}_w \cong 2\overline{M}_n$, at least at low degrees of conversion. A still higher molecular weight average, $\overline{M}_z$, is found from ultracentrifugation:

$$\overline{M}_z = \sum_i N_i M_i^3 \Big/ \sum_i N_i M_i^2 \tag{1.5}$$

The viscosity average molecular weight $\overline{M}_v$ is defined in a more complicated manner, but usually has a value slightly lower than $\overline{M}_w$. The several averages are shown schematically in Figure 1.2. The higher molecular weight averages $\overline{M}_w$ and $\overline{M}_z$ emphasize the longer molecules in a typical distribution, while $\overline{M}_n$ emphasizes the shorter molecules.

Different mechanical properties depend on different molecular weight (or size) averages (McGrew, 1958). For example, melt viscosity is usually recognized to be a function of $\overline{M}_w$, usually to the 3.4th power, because the longer chains are more entangled. Tensile strength in plastics, however, depends on $\overline{M}_n$, because $\overline{M}_n$ is sensitive to the presence of short chains, which do not appreciably contribute to the strength. Often the following empirical relationship is followed:

$$\sigma_B = A - B/\overline{M}_n \tag{1.6}$$

where σ_B is the ultimate tensile strength in the limit of high molecular weight, and A and B are constants that depend on the nature of the polymer. Although most mechanical properties improve with increasing molecular weight, the consequent increase in viscosity complicates processing. Commercial materials then usually have molecular weights high enough to give useful properties, but low enough to be reasonably processable.

1.2.1. Chain Conformation

A second property of interest is the shape or conformation of the chains in solution (Flory, 1953; McGrew, 1958; Nielsen, 1962). Light scattering and intrinsic viscosity studies have shown that the chains are best described as more or less random coils, usually with end-to-end distances of the order of several hundred angstroms. A complication arises because light scattering, which measures a weight-average molecular weight, measures a z-average end-to-end distance; however, corrections can be made from a knowledge of the molecular weight distribution, or through the use of sharply fractionated samples. Since the same polymer can usually be synthesized with a wide range of average molecular weights, it is often convenient to report values of

Table 1.2

Unperturbed End-to-End Distances[a]

Polymer	$(\overline{r_0^2}/M)^{1/2} \times 10^{11}$
Polyisobutylene	795
Polystyrene	735
Poly(methyl methacrylate)	680
Natural rubber	830

[a] Flory (1953); r_0 in cm and M in g/mol.

$(\overline{r_0^2}/M)^{1/2}$, a quantity that remains relatively invariant. The quantity r_0 has the special meaning of the "unperturbed" end-to-end distance and the bar signifies averaging of the square of r_0. This quantity is believed to be very close to the end-to-end distance in the bulk state (Krigbaum and Goodwin, 1965). The term "unperturbed" refers to the state of a molecule in a solution such that the entropic and energetic components of mixing are balanced, i.e., ΔF of mixing is zero for infinite molecular weights. Table 1.2 presents some typical values.

1.2.2. Chain Entanglement

Some types of high polymers, especially proteins, have rodlike or spherical conformations. However, the vast majority of all high polymers, natural and synthetic, exhibit some type of chain coiling and extensive entanglement at the molecular level. To an approximation, the entanglement behavior of a mass of polymer chains resembles the consequence of backlash in an old-time fishing reel.

At this point, we should emphasize that high polymers are capable of forming plastics, elastomers, and fibers because of chain entanglement, not in spite of it. The random coil—with its ability to entangle with neighboring chains—forms the basis for the high viscosity of polymer melts, the toughness of plastics, and the great extensibility of elastomers.

1.3. MOLECULAR STRUCTURE

For a given chemical composition, molecular weight, and conformation, polymer molecules may vary considerably in structure, depending on the number of ways available to join the structural units together and on the conditions of polymerization (Billmeyer, 1962, p. 117). This section discusses several common variations, including structural and stereochemical isomers, branching of several types, and crosslinking.

1.3.1. Configurations of Polymer Chains

The configurations of a polymer molecule may be defined as those arrangements of atoms in space that cannot be altered except by breaking and reforming chemical bonds. An example is the head-to-head vs. head-to-tail possibility introduced during synthesis:

$$\sim CH_2-CH\!-\!\!-\!\!-CH-CH_2\sim \qquad\qquad \sim CH-CH_2-CH-CH_2\sim \tag{1.7}$$

Head-to-head   Head-to-tail

The head-to-tail configuration is predominant in most polymeric materials, because of the greater degree of steric hindrance and polar repulsion encountered in head-to-head unions. The term configuration should be distinguished from the term conformation, which considers the arrangements of atoms in space, which can be altered by rotation about single bonds (Section 1.2.1). Conformational changes are important in rubber elasticity theory (Section 1.5.4); other configurational isomers and variants are discussed below.

1.3.2. Stereo and Geometrical Isomerism

A second, and often more important, type of configurational variation is stereoisomerism (Flory, 1953; Lenz, 1967, pp. 252–260; Schultz, 1974), which arises from differences in symmetry between substituted carbons along a polymer chain. In small molecules, carbon atoms that are attached to four different groups are clearly asymmetric, and mirror image isomers and optical activity are observed. For example, assuming tetrahedral bonding to the carbon atom,

$$\underset{\underset{\displaystyle CH_3}{|}}{\overset{\overset{\displaystyle H}{|}}{Br-C-Cl}}$$

has two optical isomers, which can be resolved experimentally, as in Figure 1.3.

If one carbon is arbitrarily termed "right-handed," or *dextro* (*d*), the other may be termed left-handed, or *laevo* (*l*). In high polymers containing vinylic carbons,

$$\underset{\underset{\displaystyle H\;\;X}{|\;\;\;|}}{\overset{\overset{\displaystyle H\;\;H}{|\;\;\;|}}{-C-C-}}$$

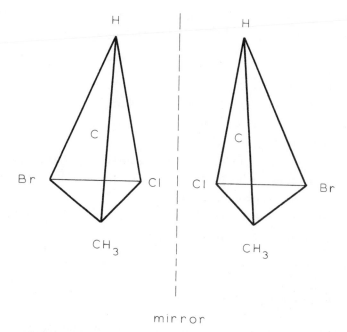

mirror

Figure 1.3. Two true optical isomers. There is no way to transform one configuration to the other without breaking bonds, and each isomer is a distinct and separable species.

three major types of stereoisomerism arise from analogous considerations*: isotactic, syndiotactic, and atactic. These are illustrated in Table 1.3, which makes use of two-dimensional projections of the tetrahedrally bonded carbon chains. The atactic polymer, usually prepared through free radical processes, has an essentially random structure. In terms of the classical *d, l* nomenclature mentioned above, the *d* and *l* forms occur to equal extents and in a random manner. Isotactic and syndiotactic sequences contain all-*d* (or all-*l*) and alternating *d* and *l* configurations, respectively. The length of a sequence of a given type in a molecule depends on the method of synthesis, and may vary from short to long runs comprising essentially the whole molecular length. The degree of such stereoregularity is very important, for it strongly affects the packing of chains, and hence properties such as crystallinity. Thus the atactic configuration is usually amorphous because the polymers lack sufficient regularity to form good crystals. Highly polar polymers with

---

* Strictly speaking, the analogy to the case of small molecules is not perfect, but it should be sufficient for present purposes; see Lenz (1967).

Table 1.3
Tacticity in Polymers

Structure	Designation	Morphology
(a)	Atactic[a]	Usually amorphous[b]
(b)	Isotactic	Usually crystallizable
(c)	Syndiotactic	Usually crystallizable

[a] Note that this particular arrangement shown is only one of many possible random arrangements along the polymer chain. The side group R may be either "up" or "down," in a random array.
[b] Polymers such as poly(vinyl alcohol) or poly(vinyl chloride) that possess small polar substituents can crystallize to some extent regardless of tacticity.

small substituents, such as poly(vinyl chloride) and polyacrylonitrile, however, do develop some crystallinity in the atactic state.

Briefly, other important types of isomerism include *cis–trans* geometrical isomerism (Lenz, 1967) about double bonds:

$$\tag{1.8}$$

and substitutional isomerism, as in 1,2 addition vs. 1,4 addition in polybutadiene:

$$\tag{1.9}$$

The 1,4-addition mode may involve either *cis* or *trans* configurations as in equation (1.8). Physical properties, such as degree of crystallinity, and associated mechanical properties, such as tensile strength, depend strongly on the detailed configuration, which affects the ease with which molecules can fit and pack together.

1.3.3. Random Branching

While strictly linear polymers are important conceptually and, as is the case with many condensation polymers, important industrially, many polymers, especially those made by addition polymerization, contain branches. Often such branching arises due to side reactions, though it may be introduced deliberately. Two types of branching may be distinguished:

1. Short-chain branching, presumably caused by intramolecular chain transfer between a terminal free radical and a hydrogen atom further back in the chain. For example, one reaction in polyethylene is postulated as follows (Billmeyer, 1962, pp. 364–375):

$$R-CH_2-CH_2-CH_2-CH_2-CH_2-CH_2 \cdot$$

transient
six-membered
ring formation

(1.10)

$$R-CH_2 \diagdown \quad CH_2 \diagdown$$
$$CH \qquad CH_2 \xrightarrow[\text{hydrogen transfer}]{\text{intramolecular}} R-CH_2-\overset{\cdot}{C}H-CH_2-CH_2-CH_2-CH_3$$
$$H \qquad CH_2$$
$$\cdot CH_2$$

where the dot indicates the site of the free radical. Further chain growth takes place from the new free radical site, thus yielding in this case a 4-carbon branch.

2. Long-chain branching, caused by intermolecular chain transfer or chain transfer to monomer. We again use polyethylene formation as an example:

$$R_1-CH_2-CH_2 \cdot + R_2-CH_2-CH_2-R_3$$

propagating
chain

dead polymer
molecule

(1.11)

$$\xrightarrow[\text{chain transfer}]{\text{intermolecular}} R_1-CH_2-CH_3 + R_2-\overset{\cdot}{C}H-CH_2-R_3$$

dead polymer
molecule

propagating chain

On the average, the chains formed as a result by addition to the new free radical site will be as long as the backbone chain itself. Similar branches may sometimes arise due to transfer of a growing polymeric free radical with a monomer molecule, and in termination by disproportionation. (See Appendix A.)

The product described in equations (1.10) and (1.11) is called "low-density" polyethylene, because the many branches in the molecular architecture lower its density from that found in the truly linear counterpart. When the polymerization reaction is carried out using special heterogeneous catalysts, a much more linear product forms; this so-called "high-density" polyethylene is higher melting and more crystalline because the fraction of branch-point imperfections is reduced.

A special case of branching, "crosslinking," which leads to the generation of three-dimensional networks is discussed in Section 1.3.5.

1.3.4. Nonrandom Branching

In the previous discussion, polymer molecules containing both short and long branches were examined; all these branches formed at random with the frequency determined by the ratio of rate constants for propagation and transfer. In contrast, Berry has developed a series of highly regular molecules resembling combs and stars (Berry, 1966, 1967, 1971; Berry and Fox, 1964); see Figure 1.4. While Berry's main interest has been related to differences in thermodynamic and hydrodynamic behavior between linear and branched polymers, his materials are of special interest to us because of their structural relationship to the graft copolymers to be discussed in subsequent chapters. Specifically, many types of graft copolymers have topologies similar to the comb-type molecules, except that the teeth and back of the comb are composed of two different polymers.

1.3.5. Crosslinking

Many important polymeric materials, such as elastomers and thermosetting resins, are crosslinked before use; for example, the vulcanization of

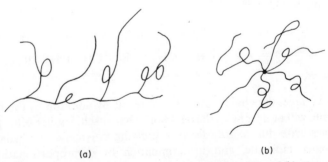

(a)   (b)

Figure 1.4. Idealized representations of (a) polymer combs and (b) stars.

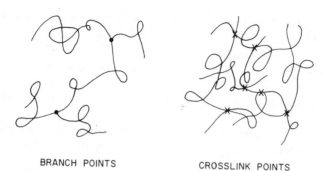

BRANCH POINTS   CROSSLINK POINTS

Figure 1.5. Branched and crosslinked polymer molecules. While a
branched polymer molecule retains finite size and an identifiable
molecular weight, the crosslinked polymer forms a three-dimensional
network of macroscopic proportions. (A rubber band is essentially
one molecule, since any two atoms are connected ultimately by
covalent bonds.)

rubber is a crosslinking process. When a polymer is crosslinked, all the individual polymer chains become joined chemically to each other, usually at
several points, eventually generating a three-dimensional network having
infinite molecular weight. Schematic structures of branched and crosslinked
polymers are compared in Figure 1.5.

Polymers may be crosslinked by a variety of reactions, such as the
following:

1. Curing reactions. In each case otherwise linear polymer molecules
are joined together by reaction with other molecules to form a crosslinked
network. A common example is the use of sulfur to vulcanize rubber; in this
case sulfur molecules react with double bonds in adjacent polymer chains,
and thus bridge them.

2. Use of multifunctional chain-growth, addition-type monomers. Use
of such monomers allows simultaneous polymerization and development of
a crosslinked network. For example, the copolymerization of small amounts
of divinyl benzene with styrene yields a crosslinked polymer:

$$R_1-CH_2-CH\cdot \ + \ CH_2=CH \longrightarrow R_1-CH_2-CH-CH_2-CH\cdot$$

reactions of
vinyl groups
independent
in time

$$R_2-CH_2-CH\cdot \ + \ CH_2=CH \longrightarrow R_2-CH_2-CH-CH_2-CH\cdot$$

(1.12)

Both free radicals then continue formation of their individual chains, and indeed each chain may form at a different interval of time.

3. Step growth condensation reactions involving multifunctional reactants. More or less densely crosslinked products such as epoxy and phenolformaldehyde resins may be prepared by the reaction of multifunctional monomers or prepolymers with each other or with other appropriate molecules.

In common parlance, linear polymers are termed thermoplastic, because they flow on heating. Crosslinked materials are called *thermoset*, because their structure is "set" on completion of polymerization (usually at an elevated temperature) and does not permit flow at elevated temperatures.

In elastomeric materials, the crosslink density is about 2×10^{19} crosslinks/cm^3, or about one crosslink per 110 mers, while in the thermoset plastics, crosslink densities are 10–50 times higher. Whatever the type of polymer concerned, crosslinking serves to reduce or prevent creep and flow. As we shall see in Section 1.5.4, the modulus of elastomeric materials is directly proportional to the crosslink density.

1.4. CRYSTALLINITY AND ORDER

In the previous section the effect of tacticity on polymer crystallizability was already mentioned. In general, high regularity of repeating units favors crystallinity, while branching (Section 1.3.3) and crosslinking (Section 1.3.5) decrease the ability of the chains to align. Polymers differ from small molecules in both extent and perfection of crystallization. Whereas small molecules, e.g., water, are normally* either totally crystalline or totally amorphous (liquid), polymers are usually only partly (if at all) crystalline. Because of the size and complexity of the polymer molecules, even highly crystalline regions normally contain many more defects than their lower-molecular-weight analogs.

In a very real sense, partly crystalline polymers are two-phase materials. While the major theme of this book is the behavior of two-phase materials, each phase being different chemically, the reader will observe a great deal of similarity between partly crystalline homopolymers, such as polyethylene, and the polyblends and composites. Important analogies or contrasts will be mentioned whenever appropriate. Indeed, the concept of highly oriented crystalline polymers as "molecular composites" has recently generated an exceptional degree of interest from both fundamental and engineering points

---

* Except, of course, under conditions corresponding to a phase change.

of view (Society of Plastics Engineers, 1975; Lindenmeyer, 1975). Let us now examine some models and experimental results related to polymer crystallinity.

1.4.1. Fringed Micelle Model

Dating back to the 1930's, the fringed micelle model represented the first modern attempt to account for the x-ray patterns of partly crystalline polymers. Polymer x-ray patterns are usually characterized by both considerable line broadening, indicating small or imperfect crystallites, and a broad halo or background of diffuse scattering, indicating the presence of an amorphous fraction (Billmeyer, 1962, Chapter 5; Bonard, 1969; Bryant, 1947; Hearle, 1963; Hosemann, 1962; Hosemann and Bonart, 1957; Lindenmeyer, 1965; Stein 1966). Early workers (Bryant, 1947) concluded from this x-ray pattern (Stein, 1966) (and general physical behavior of partly crystalline polymers) that there must be small crystallites embedded within an amorphous matrix, as shown in Figure 1.6. The crystallites were envisioned to behave as sites for a combination of crosslinking and reinforcement, while the amorphous portion allowed for the relatively high degree of rubberlike elasticity observed. A related model, the fringed fibrillar model, has been proposed for oriented fibers (Hearle, 1963; Lindenmeyer, 1965). Another model, which emphasizes lattice imperfections and distortions, is the paracrystalline model (Figure 1.7). This model includes some of the concepts of paracrystallinity, and also some aspects of the folded-chain single-crystal model to be discussed next (Bonard, 1969; Hosemann, 1962; Hosemann and Bonart, 1957). In any case, there are many variations of morphology in crystalline polymers, and, depending on the extent and nature of the crystallinity, one model or another may be more suitable for a given polymer.

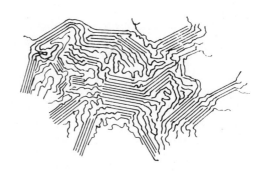

Figure 1.6. Fringed micelle model. Note alternating crystalline and amorphous regions. (Bryant, 1947.)

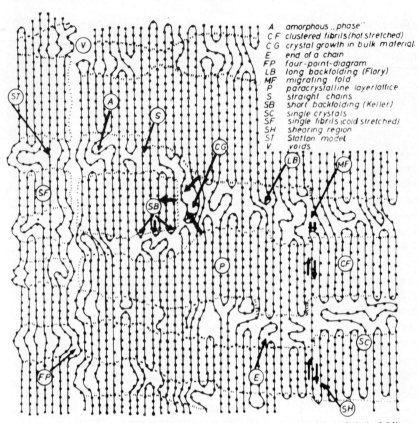

Figure 1.7. The Hosemann paracrystalline model, including the concept of chain folding. (Hosemann, 1962.)

1.4.2. Folded-Chain Single Crystals

Long after the existence of single crystals in polymers such as gutta percha had been reported, renewed interest in such relatively perfect forms was stimulated by the nearly simultaneous discovery by several investigators that single crystals could be easily formed by cooling dilute solutions of polyethylene (Anderson, 1964; Eppe *et al.*, 1959; Fava, 1969; Geil, 1963; Keller, 1957; Lindenmeyer, 1965; Odian, 1970, p. 25; St. John Manley, 1963). Electron microscopy showed these crystals to be regularly shaped, thin lamellae, about 100–200 Å thick, as shown by Figure 1.8, while x-ray diffraction measurements revealed the surprising fact that the polymer chain axis was perpendicular to the surface of the crystal. Since the molecular length was greater than the crystal thickness, the chains clearly had to be folded in order to be accommodated in the crystal (Keller, 1957). This discovery led to formulation of the folded-chain crystal model shown in Figure 1.9,

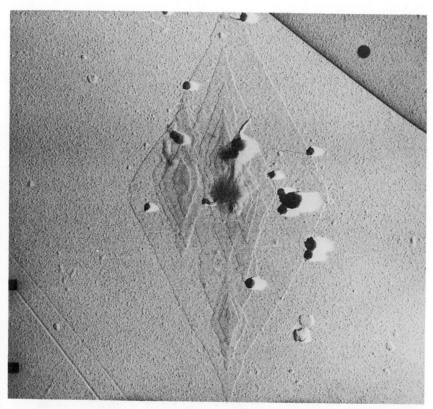

Figure 1.8. Single crystal of nylon 6 precipitated from a glycerin solution. The lamellae are about 60 Å thick. Black marks indicate one micron. (Geil, 1963.)

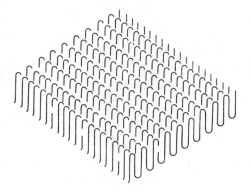

Figure 1.9. Idealized schematic showing chain folding in a crystalline lamella. (Fava, 1969.)

and called into question the accepted models for the structure of crystalline polymers.

Since the work with polyethylene, the phenomenon of single-crystal formation from dilute solution has been shown to be quite general (Anderson, 1964; Eppe *et al.*, 1959; Fischer, 1957; Geil, 1960; Keith, 1969; Keller, 1957; Price, 1958; Schlesinger and Leeper, 1953; Till, 1957) and lamellar crystals have been obtained with many different polymers (see, for example, Figure 1.8). Also, chain folding appears to occur in the bulk as well as in solution.

Reexamination of fracture surfaces of bulk-crystallized polymers also revealed the existence of folded-chain crystallites, often stacked up in lamellar form as shown for polyoxymethylene in Figure 1.10.

Figure 1.10. Surface replica of molded polyoxymethylene fractured at liquid nitrogen temperatures. While the lamellae to the lower left are oriented at an angle to the fracture surface, the lamellae elsewhere are nearly parallel to the fracture surface. The important point is that lamellae (resembling those in single crystals) are stacked up like cards or dishes in the bulk state. (Geil, 1963.)

We may now reinterpret the fringed micelle model in Figure 1.6 (Geil, 1960). The lamellae are presumed to be imperfect, with many chains failing to fold in the regular manner illustrated in Figure 1.9. This material, together with the uncrystallizable portion (if any), makes up the amorphous portion between the platelets. Chains run from one crystallite to another through this amorphous portion (see Figure 1.7). Thus the crystallites are held together in a manner similar to that depicted by the fringed micelle model, but the organization within them is now believed to be different in that chain folding to some degree is assumed. It must be emphasized that such crystalline polymers attain their great strength by the presence of amorphous, usually rubbery, chain segments connecting the lamellae in a cohesive mass.

At low degrees of crystallinity in the bulk state, a fringed micelle type of model (admitting the possibility of chain folding) may be appropriate, whereas at higher degrees of crystallinity, a paracrystalline type of model such as is depicted in Figure 1.7 may better reflect reality.

1.4.3. Extended-Chain Crystals

By use of special techniques (see, e.g., Porter *et al.*, 1975), it is possible to develop highly crystalline polymers containing extended-chain crystals, which act as highly reinforcing fibers (Chapter 12), in a matrix of coiled or folded chains (Society of Plastics Engineers, 1975; Halpin, 1975; Kardos and Raisoni, 1975; Lindenmeyer, 1975). Such materials exhibit remarkable strength and stiffness, because deformation of the extended-chain crystals in the chain direction requires the bending or stretching of bonds.

1.4.4. Spherulites

There is yet a larger scale of organization in many crystalline polymers, known as spherulites. These spherical structures are composed of many crystalline lamellae, which have grown radially in three dimensions and which are connected by amorphous molecular segments (Keith, 1969). Spherulites are easily seen with an optical microscope between crossed polarizers, and under these conditions they exhibit a characteristic pattern with circular birefringent areas possessing a Maltese cross pattern, as shown in Figure 1.11.

1.5. MECHANICAL RESPONSE: ELASTICITY AND VISCOELASTICITY

The response of a polymer depends upon temperature, time scale of the experiment, and molecular structure, morphology, and composition. In

Figure 1.11. Spherulites of low-density polyethylene as observed between crossed polaroids, revealing characteristic Maltese cross pattern. (Geil, 1963.)

this section, the role of molecular (and segmental) mobility and other factors in determining the several states in which polymers may exist is reviewed, and characterization of small-strain viscoelastic behavior by dynamic mechanical spectroscopy, stress relaxation, and creep is briefly described. The use of time–temperature superposition techniques to correlate a wide range of experimental data and predict behavior at different times and temperatures is also introduced. More detailed treatments are, of course, available in the references provided, which also describe additional techniques, such as dilatometry and dielectric and nuclear magnetic resonance relaxation.

1.5.1. Molecular and Segmental Motion

At low temperatures, the atoms in a polymer chain are restricted to isolated vibrational motions, and the bulk polymer is stiff and glassy in behavior (Ferry, 1970; Meares, 1965; Nielsen, 1962, Chapter 7; Tobolsky, 1960). As the temperature is raised, vibrations became more violent and of greater amplitude, and eventually coordinated rotational and translational motions involving from 10 to 50 carbon atoms become possible. Over a relatively narrow temperature range, usually 15–20°C, an amorphous polymer becomes much softer and rubberlike. This transition, usually called the glass–rubber transition, may also be observed in amorphous portions of semicrystalline polymers and corresponds to the glass–liquid transition in low-molecular-weight inorganic or organic glasses. At still higher temperatures, an uncrosslinked polymer will flow in a viscous manner as whole molecules begin to undergo translational motions. In crystalline regions, on the other hand, the ordered molecular segments pass more or less directly from the crystalline to the liquid state at higher temperatures, though there may also be additional transitions corresponding to changes of packing within the crystallite.

While the melting of crystallites is known as a thermodynamic first-order transition (since it involves a discontinuity, at a characteristic temperature, in a property, such as volume, that is a first derivative of free energy), the glass–rubber transition is often considered as a second-order transition (since it involves a discontinuity at a characteristic temperature in a property, such as coefficient of expansion, that is a second derivative of energy).*

The temperature of the glass–rubber transition T_g is time dependent, occurring at higher temperatures if the experiment is conducted more rapidly. This anomaly arises because the transitions may be called "observed" in a given experiment only after an appreciable fraction (say $1/2$ or $1 - 1/e$) of the chain segments have had sufficient time to execute the required motion or relaxation (Tobolsky, 1960); the higher the rate of testing, the more difficult such response becomes. Usually T_g rises from 3 to 7°C with each decade of increase of frequency in dynamic experiments (Ferry, 1970; Nielsen, 1962, Chapter 7). Relaxation behavior is discussed in more detail in the following sections.

1.5.2. Modulus–Temperature Behavior

It has been mentioned that an amorphous high polymer behaves like a glass at low temperatures, like a rubber at higher temperatures, and like a viscous liquid at still higher temperatures. In other words, depending on the temperature, behavior may be elastic (i.e., in conformity to Hooke's law),

---

* The view that the glass–rubber transition is truly thermodynamic in nature, and not just a kinetic anomaly, is not universally accepted, though a strong case for a thermodynamic origin can be made (Meares, 1965).

viscous, or in a state that exhibits aspects of both extremes, namely, visco-elastic. While thermodynamic measurements, such as that of volume as a function of temperature, serve to define the melting and glass–rubber transitions, they do not illuminate the rubbery state, which, as shown below, is in fact a true state of matter.

The response to mechanical stress, on the other hand, not only serves to define the rubbery state (see Section 1.5.4), but often differentiates in a very sensitive way between the various states in a polymer and indicates structural and compositional effects. In addition, the modulus itself is of inherent interest as a property of empirical importance. Hence mechanical measurements of, for example, modulus not only complement other measurements of state parameters, but yield additional information about polymer behavior.

The development of modulus–temperature relationships (so-called thermomechanical techniques) as a characterization tool has been due in large part to Tobolsky and his co-workers (Tobolsky, 1960, pp. 71–83), and has been used to considerable advantage by others (American Chemical Society, 1972). Consideration of this approach is especially appropriate to the subject of this book, since the modulus is of both fundamental and em-pirical importance.

1.5.3. Five Regions of Viscoelastic Behavior

According to Tobolsky (1960, pp. 71–83), five major regions of visco-elastic behavior may be defined for amorphous materials. Referring in the following to Figure 1.12, glassy behavior is always observed at the lowest temperatures, in region 1. Young's modulus E is commonly about 3×10^{10} dyn/cm^2, and is nearly independent of polymer structure.* The second region is the glass–rubber transition region, where a drop in modulus of 10^3 within a 20°C temperature span is common. The glass–rubber transition is generally associated with the onset of long-range coordinated segmental motion (see Section 1.5.1).

Above the glass transition lies the rubbery plateau region, region 3. The equations of state for rubber elasticity (see Section 1.5.4) apply here; if the material is crosslinked, these equations may apply up to the decomposition temperature (see dashed line, Figure 1.12).

At still higher temperatures, the rubbery flow and liquid flow regions are encountered, regions 4 and 5. In the former, flow is hindered by physical entanglements. At higher temperatures molecular motion is sufficiently rapid that molecules behave more nearly independently.

* Mechanical terms are defined in Appendix B.

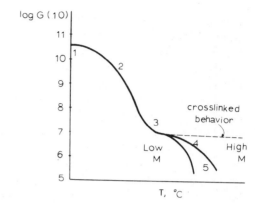

Figure 1.12. The logarithm of the shear modulus (measured at 10 sec) vs. temperature. Effects of increased molecular weight and crosslinking are also illustrated.

1.5.4. Rubberlike Elasticity

An elastomer may be defined as a crosslinked polymer network whose temperature is above its glass transition temperature. The molecular mechanism responsible for rubber elasticity is based on changes in chain conformation brought about by the overall strain (see Figure 1.13). Clearly, the number of possible chain conformations must be fewer in case (c) than in case (a), resulting in a reduction of entropy (Flory, 1953, Chapter 11). Statistically all possible chain conformations are equally likely, assuming negligible

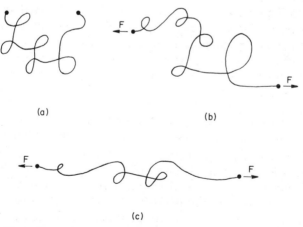

Figure 1.13. Illustration of entropy reduction on application of force F to network chain: (a) chain at equilibrium; (b) chain as force is being applied; (c) final state with force applied.

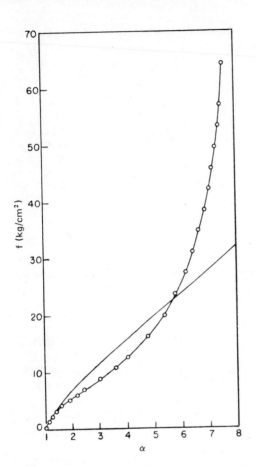

Figure 1.14. Theoretical (—) and experimental (-o-) stress–strain curves for simple elongation of gum-vulcanized rubber. The increase in modulus beyond $\alpha = 4$ or 5 is caused by the onset of crystallinity, and also by the finite extensibility of the chains (Shen et al., 1968.)

energy differences between *trans* and *gauche* rotational states.* However, most of these equally likely chain conformations exist when the distance between chain ends is relatively small. The decreased conformational entropy results in a positive free energy for the stretched state, providing a driving force that causes the material to snap back when the stretching force is released.

In this respect elastomers differ fundamentally from metal springs. The latter develop retractive forces primarily because of increased interatomic distances, which correspond primarily to a change in enthalpy rather than

---

* In a "*trans*" conformation, the chain carbons are rotated so that two successive carbon–carbon bonds lie in a plane; in a "*gauche*" conformation, the second bond is 60° out of the plane.

entropy. Thus, in terms of free energy, the entropy term ΔS is primarily responsible for the free energy change ΔF in elastomers

$$\Delta F = \Delta H - T\,\Delta S \tag{1.13}$$

while the enthalpy term ΔH is of prime importance in metal springs. The fundamental difference between elastomers and metal springs results in a quite different dependence of mechanical behavior on temperature. For elastomers, the equation governing the temperature dependence of the shear modulus may be written

$$G = nRT \tag{1.14}$$

where n represents the number of network chains per unit volume. Equation (1.14) applies to the modulus of the dashed-line portion of Figure 1.12. Equation (1.14) is analogous to the equation of state for ideal gases, $PV = nRT$, which is also based solely on entropic considerations. Neither theory says anything about details of chemical structure, but rather involves the modes of molecular motion.

The retractive force f developed by stretching an elastomer in tension may be written

$$f = nRT(\alpha - 1/\alpha^2) \tag{1.15}$$

where $\alpha = L/L_0$, the ratio of the final to the initial length. An experimental stress–strain curve is compared to values obtained from equation (1.15) in Figure 1.14 (Flory, 1953; Treloar, 1944).

Equation (1.15) suffers from the same defects as afflict the ideal gas equation, which makes no allowance for such factors as molecular attractive forces or spatial considerations.

An important semiempirical improvement is given by the Mooney–Rivlin equation (Mooney, 1948; Rivlin, 1948a, b):

$$f = 2C_1\left(\alpha - \frac{1}{\alpha^2}\right) + 2C_2\left(1 - \frac{1}{\alpha^3}\right) \tag{1.16}$$

where $2C_1$ may be equated with the nRT term of equation (1.15). The interpretation of $2C_2$ is, however, still uncertain (Gee, 1966). A newer improvement in equation (1.15) involves the so-called "front factor" (Tobolsky and Shen, 1966; Shen et al., 1968, 1971), which allows for energy differences between trans–gauche conformations. One form of the equation may be written (Tobolsky and Shen, 1966; Shen et al., 1968, 1971):

$$f = nRT\left[\frac{\overline{r_{i,V_0}^2}}{\overline{r_{f,V_0}^2}}\left(\frac{V_0}{V}\right)^{\gamma}\right]\left(\alpha - \frac{V}{V_0}\frac{1}{\alpha^2}\right) \tag{1.17}$$

where the quantity $\overline{r_{i,V_0}^2}$ represents the mean square length of the vector connecting the chain junctures, and $\overline{r_{f,V_0}^2}$ represents the mean square length that the chains would have if their ends were free. Each quantity is taken at the volume V_0 of the elastomer in the absence of an applied force. The quantity γ is a semiempirical parameter; although a value of $\gamma = 0$ is predicted for an ideal elastomer, a value of 0.33 was found experimentally for silicone rubber (Tobolsky and Sperling, 1968).

Active research in the field of rubber elasticity is currently centered on the calculation of polymer chain end-to-end distances and elastic response behavior from knowledge of individual bond properties. The reader is referred to studies by J. E. Mark (1972a,b, 1973) for details.

1.5.5. Dynamic Mechanical Spectroscopy

Considerable information about elastic and viscoelastic parameters may be derived by measuring the response of a polymer to a small-amplitude cyclic deformation. Molecules perturbed in this way store a portion of the imparted energy elastically, and dissipate a portion in the form of heat (Ferry, 1970; Meares, 1965; Miller, M. L., 1966, pp. 243–253; Nielsen, 1962, Chapter 7; Rosen, 1971; Schultz, 1974, pp. 67–71; Williams, D. J., 1971), the ratio of dissipation to storage depending on the temperature and frequency. In dynamic mechanical spectroscopy experiments, a cyclic stress is applied to a specimen, and two fundamental parameters are measured: the storage modulus E', a measure of the energy stored elastically, and the loss modulus E'', a measure of the energy dissipated. The loss modulus E'' may be calculated as follows:

$$E'' = E' \tan \delta \qquad (1.18)$$

where δ is the phase angle between the applied stress and resulting strain. The factor $\tan \delta$ is analogous to the power factor associated with an alternating electric field. The quantities E' and E'' compose the complex modulus E^* through the relation

$$E^* = E' + iE'' \qquad (1.19)$$

where $i = \sqrt{-1}$. With mechanical spectroscopy, the frequency may be varied at constant temperature, or, often more conveniently, the temperature may be varied at constant frequency. Results of the two experiments may be interrelated by means of the well-established time–temperature superposition principle formulated by Williams, Landel, and Ferry (see Section 1.5.7). Figure 1.15 is a plot of E' and E'' vs. temperature at a single frequency, or as a function of log frequency at constant temperature. The

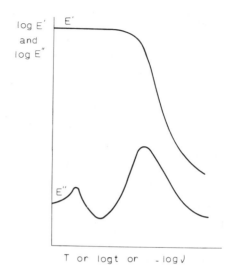

Figure 1.15. The storage and loss moduli plotted against absolute temperature. Through the equivalence of time and temperature (see Section 1.5.7), the abscissa may also be taken as log t or its inverse, which is equal to minus log frequency.

major maximum in E'' usually occurs at the glass–rubber transition, which corresponds to the onset of long-range (10–50 carbon atoms) coordinated motion as discussed in Section 1.5.1. Also shown in Figure 1.15 is a low-temperature E'' peak, which is usually attributed to the onset of motion of a smaller group of atoms, such as a short segment of a main chain or a side group (Williams, D. J., 1971, Chapter 11).

Dynamic mechanical spectroscopy has a direct analogy to other spectroscopic methods. For example, when an infrared beam of appropriate wavelength passes through a sample, the radiation will be absorbed in part and the molecules (or some group of atoms within them) raised to a higher energy state. At the frequencies corresponding to the different energy states, appropriate instrumentation will, of course, reveal maxima in absorption. Similar effects occur in dynamic mechanical spectroscopy, the quantity E'' being analogous to the spectroscopic optical density.

The analogy between mechanical and infrared spectroscopy is not exact, however. The energy absorption mechanism in dynamic viscoelasticity is of the relaxation type, while that of infrared spectroscopy belongs to the resonance absorption class. The two can be compared (see Figure 1.16) with the aid of models incorporating springs, dashpots, and mass elements.* The velocity of wave transmission is different for the two modes: Infrared waves are governed by the velocity of light, while mechanical wave transmission is governed by the velocity of sound in the medium. Amrhein (1967) has recently reviewed some aspects of the relaxation–resonance relationship.

* We are indebted to M. Takayanagi for pointing out this distinction.

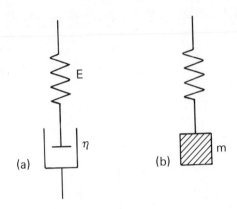

Figure 1.16. (a) Relaxation (mass is negligible) and (b) resonance (viscosity is negligible) absorption mechanisms (after Takayanagi, 1972).

1.5.6. Stress-Relaxation and Creep Behavior

In stress-relaxation experiments, a specimen is stretched rapidly from length L_0 to length L and the force required to maintain L is measured as a function of time (Tobolsky, 1960, Chapter 4) and, usually, of temperature. In creep experiments the extending force is held constant, and the length of the sample recorded as a function of time (Nielsen, 1962, Chapter 3). In general, stress relaxation and creep reflect the same phenomena, though in an inverse manner. Although creep studies probably are more important technologically, stress relaxation is more amenable to mathematical analysis, and so receives more scientific attention.

There are several simple, distinct mechanisms of stress relaxation and creep. The more important of these include:

1. Relaxation of the chains, especially in the glass–rubber transition region.
2. Molecular flow, which occurs in the rubbery flow and liquid flow regions (regions 4 and 5 in Figure 1.12).
3. Chain scission, a form of chemical degradation, which is often important in crosslinked elastomeric networks, especially at elevated temperatures.
4. Bond interchange (not a degradation) between molecules, for example, a molecular interchange between the Si–O bond interchange in silicone elastomers (Tobolsky, 1960, Chapter 4).

5. Thirion relaxation, which usually is reversible, involving complex adjustments of the chains in a network in response to a stress, and possibly involving chain motion of physical entanglements. The long-term relaxation of thermally stable elastomers illustrates this phenomenon (Nielsen, 1962, Chapter 3; Thirion and Chasset, 1962).

Briefly, if relaxation modes 3 or 4 are the result of single, isolated chemical reactions, a relaxation time τ may be defined according to

$$f(t) = f(0)\, e^{-t/\tau} \tag{1.20}$$

where the retractive force f is a function of time t.

Often it is convenient to represent the relaxation behavior of polymers by the behavior of simple combinations of springs and dashpots. As in real life, a spring is assumed to deform readily, and to retract fully on the release of applied load. A dashpot may be visualized as a plunger being pulled through a viscous medium; the motion is slow and irreversible, since there is no recovery force. It can be shown that equation (1.20) follows from the Maxwell model, in which a spring and dashpot are in series (Figure 1.17). More complicated models involving both parallel and series combinations of springs and dashpots are often invoked to analyze motion in creep and relaxation experiments; the reader is referred to Aklonis *et al.* (1972), Alfrey and Gurnee (1967), Nielsen (1962), Sperling and Tobolsky (1968), Thirion and Chasset (1962), and Tobolsky (1960) for further details.

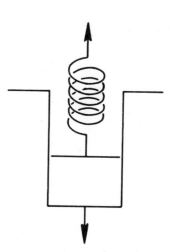

Figure 1.17. The Maxwell model for a viscoelastic body. The spring constant and dashpot viscosity are both variable.

1.5.7. Time–Temperature Relationship

The Williams–Landel–Ferry (WLF) equation (Williams, M. L., *et al.*, 1955) is probably the most powerful single relationship available for the correlation of viscoelastic behavior in amorphous polymers. Analogous relationships may often be used for semicrystalline and filled polymers. Based on the need of sufficient free volume for chain segments to undergo motion, it interrelates properties such as viscosity and modulus with time (or frequency) and temperature (see Tobolsky, 1960).

Let us briefly examine the relationships between free volume, molecular motion, and the glass transition (Ferry, 1961, pp. 218–224). Consider a close-packed array of amorphous high polymer molecules arranged on a quasi-lattice, with all the sites filled. The molecules are incapable of long-range, coordinated motion, because in order to do so, their neighbors must be pushed out of the way, and these, in turn, have no place to go. (A simple analogy is to the toy that arrays 15 numbers and an empty space on a square plate. Only the presence of the empty space, which corresponds to an element of free volume, allows the numbers to be rearranged.) Free and occupied volume are depicted qualitatively in Figure 1.18; note that the free volume rises rapidly above T_g. The rapid increase in free volume is due to the increased force of segmental motion, and subsequently allows motions to take place more freely. The occupied volume also increases slowly with temperature, in response to an increase in the violence and amplitude of vibrations.

The most common form of the WLF equation is given by

$$\log A_T = \frac{17.44(T - T_g)}{51.6 + (T - T_g)} \tag{1.21}$$

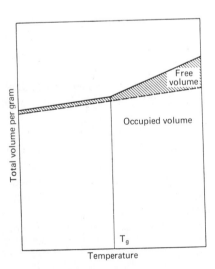

Figure 1.18. Free and occupied volume in a polymer (Ferry, 1961). The occupied volume corresponds to the volume associated with the molecules themselves.

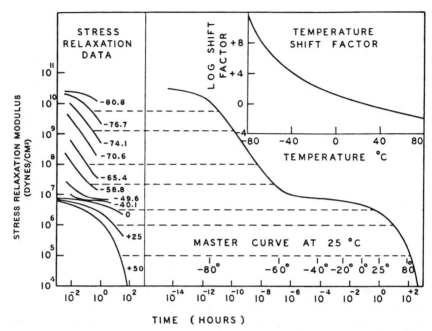

Figure 1.19. Time–temperature superposition principle illustrated with polyisobutylene data. The reference temperature of the master curve is 25°C. The inset graph gives the extent of curve-shifting required at the different temperatures. (Catsiff and Tobolsky, 1955, 1956.)

where A_T represents a reduced variables shift factor. The simplest application involves viscous flow, where $A_T = \eta/\eta_{T_g}$ for linear polymers, and is reasonably accurate up to a temperature of $T_g + 50°K$.

Application to a viscoelastic material involves the construction of a so-called master curve. Here experimental data such as log modulus are plotted vs. log time for a series of temperatures to form a family of curves. By horizontal shifts of the curves (the shifts corresponding to A_T), physical properties for very short or extended time ranges at a single temperature may be obtained. This is illustrated schematically in Figure 1.19 (Catsiff and Tobolsky, 1955, 1956; Nielsen, 1962, pp. 89–92).

The power of the WLF equation resides in the principle that the higher the temperature of an activated process, the shorter the time to reach an equivalent state. In a sense, time may be considered as a measure of molecular motion. At higher temperatures (with more energy), molecules (or segments of them) move more frequently, and time (for them) appears to go faster.*

---

* Consider a Maxwell demon watching an idealized molecular segment jump back and forth, and using these jumps as a clock. At higher temperatures, the higher time-average number of jumps will make him think time is elapsing faster.

The WLF equation permits us to estimate mechanical variables at times, frequencies, or temperatures difficultly accessible otherwise.

1.6. ENERGETICS AND MECHANICS OF FRACTURE

Because of the importance of failure phenomena in any consideration of multicomponent polymer systems, a brief review of the energetics and mechanics of fracture follows (Andrews, 1968, 1972; Berry, 1961; Broutman and Kobayashi, 1972; Eirich, 1965; Griffith, 1921; Halpin and Polley, 1967; Hertzberg *et al.*, 1973; Johnson and Radon, 1972; Kambour and Robertson, 1972; Krauss, 1963; Lannon, 1967; Manson and Hertzberg, 1973*a*; Mark, H., 1943, 1971; Orowan, 1948; Prevorsek, 1971; Prevorsek and Lyons, 1964; Radon, 1972; Riddell *et al.*, 1966; Rivlin and Thomas, 1953; Rosen, 1964; Tobolsky and Mark, 1971; Williams, M. L., and DeVries, 1970; Zhurkhov and Tomashevskii, 1966). Specific cases are discussed in Sections 3.2 and 12.1.2.

In general, catastrophic failure of a material such as a polymer may be considered to involve two stages: initiation and propagation of a crack or flaw (see also Section 3.2).

1.6.1. General Approach to Fracture

A useful approach to the question of defining conditions under which failure will occur may be based upon the reasonable supposition that all real materials contain flaws such as small cracks, imperfections in crystals, or heterogeneities. Such preexistent flaws will result in local stresses higher than whatever nominal stress may be applied, and if the stress (or energy input) is sufficient to induce propagation of the flaw (or of a crack associated with it), catastrophic failure will occur. With certain modifications (see references for details), this simple concept has proved to be exceedingly useful in describing and interpreting fracture phenomena.

In terms of molecular phenomena, polymeric materials fail by some combination of two distinct mechanisms: bond breakage and chain slippage (Figure 1.20). Chain slippage, or shear yielding, is also frequently accompanied by cavitation and the coalescence of voids, giving rise to the inhomogeneous response known as crazing (Section 3.2.3). Since the breakage of a van der Waals or hydrogen bond requires a much lower stress than the breakage of a primary bond [about 2×10^{-6}, 1×10^{-5}, and 5×10^{-4} dyn/cm^2, respec-

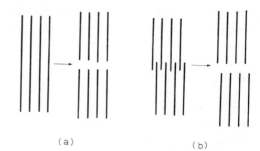

Figure 1.20. Schematic showing mechanisms of failure in polymeric materials. (a) Bond breakage, (b) chain slippage.

(a)

(b)

tively (Mark, H., 1943; 1971)], it is easy to see that initially chain slippage may occur at relatively low stresses. Stress may then be concentrated on a few primary bonds to a degree sufficient for rupture, thus forming a flaw, which is available for catastrophic crack propagation during which additional yielding and bond breakage* occur. Failure takes place progressively and at tensile strengths below those theoretically predicted on the basis of simultaneous rupture of all covalent bonds.

Thus the tensile strength of, for example, most commercial fibers such as rayon or polyamide range between 3 and 6 g per denier (4×10^4 to 8×10^4 psi), depending on the molecular weight, degree of orientation, and degree of crystallinity. While strengths as high as 15–20 g per denier can be obtained with especially well-oriented, high-molecular-weight fibers, the maximum theoretical strength is approached more closely, but not actually obtained. As pointed out elsewhere (Section 1.43 and Chapter 12), however, it has recently become possible to achieve strengths and moduli much closer to theoretical values by the use of special extrusion techniques (Lindenmeyer, 1975; Porter et al., 1975; Halpin, 1975).

The mode of fracture, which also reflects the balance between bond breakage and slippage, is also important, since it has a profound effect on the total *energy* (not just maximum stress) required to break a specimen. Even for a given stress to break, the shape of the stress–strain curve may vary widely from polymer to polymer, as shown in Figure 1.21. (For a given polymer, the shape of the curve may also vary, depending in particular on the temperature and rate of testing.) Behavior may be essentially brittle–elastic in nature (curve *a*), rubbery–elastic (curve *b*), or rather ductile (curve *c*), with many variations possible. Polymeric materials may thus exhibit many combinations

---

* Dramatic evidence for the breaking of bonds during the deformation of polymers, prior to actual fracture, has been provided by the use of electron spin resonance techniques, which detect the free radicals produced by chain rupture (Williams, M. L., and DeVries, 1970; Zhurkhov and Tomashevskii, 1966).

of strength and toughness (as measured by the area under the stress–strain curve). Frequently with polymers, high strength per se is not as important as the ability to combine a useful value of modulus with the ability to yield and dissipate considerable energy in an inelastic manner prior to fracture, as shown by stress–strain curve *c* in Figure 1.21.

Thus, while a brittle polymer could in principle require considerable energy to break if one could attain the maximum theoretical strength, in practice the inability to attain such an idealized value poses a severe limitation on engineering applications of a polymer. However, by modifying the energy absorption and dissipation characteristics by introducing ductility, it is often possible to obtain good levels of toughness without unduly decreasing modulus or strength. Of course, there may be times when, for example, modulus per se is paramount; frequently compromises are necessary in the selection of a polymer for a particular application.

In any case, the energy (or stress) required to induce failure under a given set of loading and environmental factors is clearly important. Different test methods may be appropriate, depending on the nature of the application—for example, on whether the polymer may be subjected to a slowly applied stress, to a sudden impact, to a cycling load, and so forth (see Section 1.7).

A general approach to the quantitative characterization of fracture may now be given in terms of the energy balance required to propagate a crack or flaw.

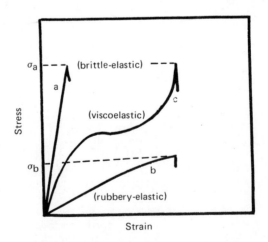

Figure 1.21. Typical types of stress–strain behavior for polymers. σ_a and σ_b represent tensile strengths for *a* and *b*, respectively.

1.6.2. Energy Balance in Fracture

Basically, some of the energy applied to a specimen may be stored elastically as bonds are strained; energy will also be dissipated if bond breakage or viscous flow (even on a segmental level) occurs. Assuming a glassy solid, in which energy is dissipated by the breakage of bonds and not in viscous processes, Griffith (1921) analyzed the balance between the energy applied and the energy released in bond breakage as a crack propagates, in order to define the critical condition for extension of a flaw to form a catastrophic crack. He suggested that when the release of strain energy per unit area of the crack surface exceeds the energy required to break the bonds associated with the unit area of surface (the latter being the intrinsic surface energy S), a crack would propagate. In other words, crack propagation will occur when the amount of strain energy available to a crack exceeds the energy required to break bonds and form new crack surface.

Griffith showed that this critical point is defined by the equation

$$\sigma = \left(\frac{2ES}{\pi a}\right)^{1/2} \tag{1.22}$$

where σ is the gross applied stress, E is Young's modulus, S is the surface energy, and a is the crack length. Although this analysis has been reasonably confirmed in inorganic glasses, it was shown by Berry (1961) (see also Rosen, 1964, Chapters IIa, IIb) and by others that experimentally determined values of S were 10^2–10^3 times values calculated on the basis of bond breakage alone. Since examination of the fracture surfaces of polymers reveals evidence of much plastic flow, the discrepancy mentioned presumably arises because of such viscous phenomena. Experiments on, for example, swollen rubbers that do not undergo plastic flow do yield values for S close to those predicted. Thus Griffith's expression needs to be generalized for polymers to allow for the characteristic occurrence of localized plastic deformation (viscous flow) even for apparently brittle fractures. A more general equation may be obtained by replacing S in equation (1.22) with a term $(S + P)$, where P accounts for the energy involved in plastic deformation (Orowan, 1948), or by using an all-inclusive term T to account for all energy-dissipating processes (Rivlin and Thomas, 1953). Experimentally, the term "fracture surface energy" γ is used to characterize the critical amount of energy required to induce fracture, and is defined by the equation

$$\gamma = \tfrac{1}{2}G_c = \tfrac{1}{2}(P + S)_c \tag{1.23}$$

where G_c and $(P + S)_c$ represent, respectively, the critical values of strain energy release rate and *effective* surface energy corresponding to the onset of fracture.

As is discussed in more detail later (Section 12.1.2), the energy balance equations have been shown to have their counterparts in terms of mechanical stress. The development of both energy and stress approaches—constituting the body of knowledge known as "fracture mechanics"—has been extremely fruitful in characterizing and interpreting failure in all kinds of materials subjected to static and cyclic loads (Andrews, 1968, 1972; Berry, 1961; Broutman and Kobayashi, 1972; Broutman and Krock, 1974; Eirich, 1965; Griffith, 1921; Hertzberg et al., 1973; Johnson and Radon, 1972; Kambour and Robertson, 1972; Manson and Hertzberg, 1973a; Orowan, 1948; Radon, 1972; Riddell et al., 1966; Rivlin and Thomas, 1953; Rosen, 1964). The fracture mechanics approach complements molecular and phenomenological theories of fracture based on the accumulation of damage due to the breakage of bonds (Halpin and Polley, 1967; Krauss, 1963; Lannon, 1967; Mark, H., 1943, 1971; Prevorsek, 1971; Prevorsek and Lyons, 1964; Tobolsky and Mark, 1971, Chapter 11; Williams, M. L., and DeVries, 1970; Zhurkhov and Tomashevskii, 1966).

While correlation of fracture behavior in glassy polymers with the time–temperature relationships is receiving increasing attention [see, for example, Broutman and Kobayashi (1972), Johnson and Radon (1972), Radon (1972)], the case of rupture in elastomers is rather well developed, and is discussed briefly below.

1.6.3. Viscoelastic Rupture of Elastomers

When simple crosslinked elastomers are stretched and held at high elongation, two consecutive phenomena occur: (1) a smooth relaxation period, and (2) sudden failure. This sequence is depicted in Figure 1.22 for poly-(styrene-co-butadiene) rubber (Rodriguez, 1970, Chapter 9; Scott, 1967; Smith, 1964; Smith and Stedry, 1960).

Our interest lies in the loci of failure points as a function of temperature. The family of curves of stress to break σ_b at various temperatures (of which the dashed-line portion of Figure 1.22 is representative) is plotted schematically in Figure 1.23 (Scott, 1967). Use may now be made of the time–temperature superposition principle and the WLF equation (Section 1.5.7) to construct a master curve, as shown in Figure 1.24.

The strain to break ε_b may be plotted against the stress to break in the master curve to yield a failure envelope, shown schematically in Figure 1.25 (Scott, 1967). The failure envelope is independent of temperature, time to break, and strain rate. It is a universal curve independent (at least ideally) of the type of rupture test. If σ_b is further divided by the crosslink density, the resulting failure envelope is also approximately independent of both the degree of crosslinking and the chemical structure of the elastomer. The latter

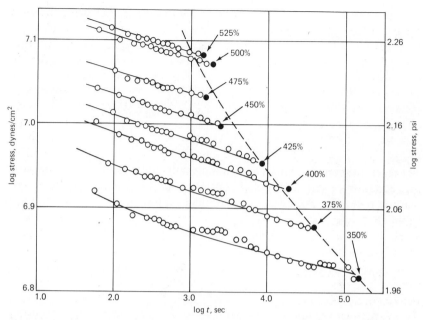

Figure 1.22. Stress relaxation of a crosslinked poly(styrene-co-butadiene) rubber (at 1.7°C) at elongations from 350 to 525%. Solid points indicate rupture, the dashed-line gives the ultimate stress. (Scott, 1967.)

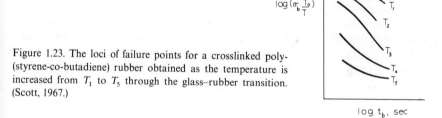

Figure 1.23. The loci of failure points for a crosslinked poly-(styrene-co-butadiene) rubber obtained as the temperature is increased from T_1 to T_5 through the glass–rubber transition. (Scott, 1967.)

Figure 1.24. Master curve of reduced stress to break vs. reduced time to break for a crosslinked poly(styrene-co-butadiene) rubber. (Scott, 1967.)

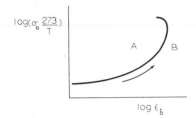

$\log(\sigma_b \frac{273}{T})$

A   B

$\log \epsilon_b$

Figure 1.25. The failure envelope for a poly-(styrene-co-butadiene) rubber. Area *A* to the left of the curve indicates stable levels of stress and strain. Values in region *B* will cause rupture. Arrow indicates direction of lower temperatures or higher strain rates.

statement, surprising but true for noncrystallizing elastomers, follows from the equation of state for rubber elasticity and the strength of the carbon–carbon bond. The failure envelope is indeed a powerful tool in dealing with ultimate properties.

1.7. MECHANICAL TESTING OF POLYMERS

A number of important mechanical tests of polymer strength properties at large strains culminating in fracture have evolved over the years. Examples include tests of such behavior as ultimate tensile strength, tearing or fracture energy, impact resistance, and fatigue. Although rigorous analysis of such tests is not always possible, correlations with molecular composition and behavior can often be made on an empirical basis. In some cases, such as fracture energy, it is possible to obtain characteristic parameters that are at least independent of specimen characteristics (see Section 1.6). Such parameters may be considered as material constants at a given temperature and load condition. In other cases, such as fatigue life, this is not the case, and only empirical comparisons can be made.

1.7.1. Stress–Strain and Fracture Behavior

The viscoelastic behavior of polymers is often sensitive to many chemical media, such as water, organic solvents, and oxygen. Measurements of stress–strain behavior are generally made using an instrument that permits operation over a range of strain rates and, with an appropriate chamber, over a range of temperature and environmental conditions.

In order to determine fundamental values for crack propagation rates, fracture energy, or the related critical stress intensity for fracture, it is necessary to use specimen geometries for which the stress distributions are known.

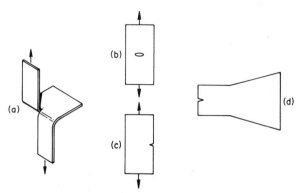

Figure 1.26. Some typical specimens for determination of characteristic energy and stress intensity parameters for fracture (after Rivlin and Thomas, 1953). (a) Trouser-leg design. (b) Center-notched panel. (c) Edge-notched panel. (d) Cantilever-beam type. Type (a) has been commonly used with elastomers, and types (b)–(d) with plastics.

Typical examples are given in Figure 1.26; both static and cyclic measurements may be made. Various experimental techniques may be used (Andrews, 1968; Kambour and Robertson, 1972; Rosen, 1964); the reader is referred to the literature for further details.

1.7.2. Impact Strength

Resistance to impact loading is of major concern in plastics for many engineering applications, and is an important attribute of many of the polymer systems discussed in this book (Lannon, 1967). Values obtained for impact strength depend on the specimen geometry, the presence of natural or artificial flaws, and the conditions of testing. A rigorous and fundamental interpretation of impact strength is not yet possible and values cannot be expected to be always consistent with values of other parameters related to toughness. However, in some cases correlation is possible with factors such as the presence of low-temperature loss peaks or an abnormally high free volume, which could permit localized relaxation of even rapidly applied stresses (Boyer, 1968; Eirich, 1965; Heijboer, 1968; Kambour and Robertson, 1972; Litt and Tobolsky, 1967).

Measurements of impact strength may be made (Lannon, 1967) using specially designed stress–strain testers to permit very high rates of loading, or, more commonly, using one of two types of instruments: the Izod and the Charpy impact test machines, each involving the striking of a specimen

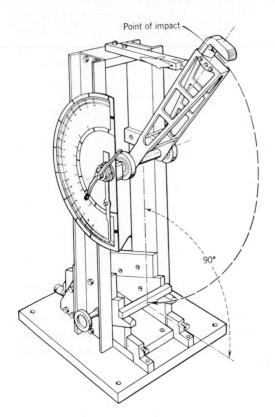

Figure 1.27. Simple beam (Charpy-type) impact machine (ASTM D 256-56).

with a calibrated pendulum. The Charpy instrument is shown in Figures 1.27 and 1.28; the Izod instrument differs principally in that the sample is supported at one end only, cantilever style. In both cases the sample is often notched to provide a standardized weak point for the initiation of fracture. With appropriate instrumentation, impact testers can be adapted to measure force-time behavior, as well as the impact energy itself (Bucknall, 1967a, b). Special high-strain-rate tensile testers may also be used.

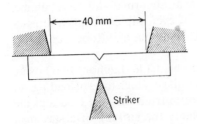

Figure 1.28. Charpy test piece and support. In some forms of the Charpy test, the notch has a rectangular cross section.

1.7.3. Fatigue

The case of creep in a polymer subjected to a constant stress has already been mentioned in Section 1.5.6. Cyclic or repeated stresses are also important in actual service, and give rise to the phenomenon of fatigue (Andrews, 1968, 1969, 1972; Manson and Hertzberg, 1973a; Riddell et al., 1966).

Thus, under cyclic stress, a polymer may fail at stresses well below the ultimate failure stress measured in simple extension. Often thousands or tens of thousands of cycles may elapse with little appearing to happen, followed by catastrophic failure; in some cases a so-called "endurance limit" is observed below which stress failure never occurs (see Figure 1.29). In general, the higher the stress, or the higher the temperature, the shorter the lifetime. In terms of molecular processes, fatigue failure involves two mechanisms: weakening by adiabatic heating due to the presence of energy-dissipating modes of relaxation, and mechanical initiation and propagation of a crack. The former mechanism is dominant at high (Riddell et al., 1966) and the latter at low frequencies (Andrews, 1972; Manson and Hertzberg, 1973a). As shown in Figure 1.30, chemical structure and composition exert a profound influence on mechanically induced crack propagation (Hertzberg et al., 1973). Experimental details are reviewed by Andrews (1969) and Manson and Hertzberg (1973a).

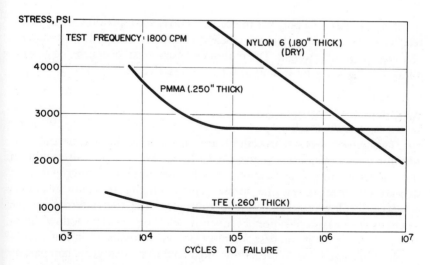

Figure 1.29. Fatigue life curves (lifetime vs. number of cycles) for nylon 6, poly(methyl methacrylate) (PMMA), and polytetrafluoroethylene (TFE). (Riddell et al., 1966.)

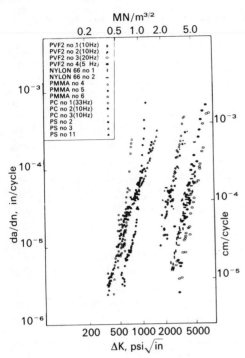

Figure 1.30. Relationship of fatigue crack growth rate per cycle in several polymers as a function of range of stress intensity factor ΔK, which is a measure of the range in stress concentrated at the crack tip during cyclic deformation (Hertzberg *et al.*, 1973). PVF, poly(vinylidene fluoride); PMMA, poly(methyl methacrylate); PC, polycarbonate; PS, polystyrene.

APPENDIX A. POLYMER SYNTHESIS

The synthesis of a polymer molecule involves the repetition of one of the relatively few reactions that are efficient enough to permit the attainment of a high molecular weight. The monomers used must, of course, be at least difunctional, that is, must have at least two reactive sites.

Addition Polymerization

Two major types of reactions are encountered in polymerizations: addition and condensation (Lenz, 1967; Ravve, 1967). Kinetically, addition and condensation polymerizations may follow either a chain-growth or a step-growth mechanism, the former corresponding to chain reactions observed for low-molecular-weight molecules, and the latter to simple reactions such as esterification. Frequently, addition reactions follow chain-growth kinetics (as is the case for most polymers discussed in this monograph), while condensation reactions often tend to be step-growth in nature. Addition reactions that exhibit a chain growth process may involve either free radical or ionic intermediates, and are characterized by several stages: initiation, propagation, termination, and transfer. For example, the photopolymerization of ethyl acrylate using benzoin as an initiator may be described as follows:

Initiation:

$$\text{benzoin} \xrightarrow[k_d]{h\nu} \quad + \quad \text{radicals} \tag{1.24}$$

or

$$R-R' \rightarrow R\cdot + R'\cdot$$

Compounds, such as peroxides, that can undergo a similar but thermal dissociation into free radicals are more commonly used to initiate polymerization ($R-O-O-R' \rightarrow R-O\cdot + R'-O\cdot$). Addition of the first monomer unit is usually considered as part of the initiation step:

$$R\cdot \text{ (or } R'\cdot) + \underset{\substack{C=O\\|\\O-C_2H_5}}{CH_2=CH} \xrightarrow{k_i} R-CH_2-\underset{\substack{C=O\\|\\O-C_2H_5}}{CH}\cdot \tag{1.25}$$

Propagation:

$$R-CH_2-\underset{\substack{|\\C=O\\|\\O-C_2H_5}}{CH}\cdot + \underset{\substack{|\\C=O\\|\\O-C_2H_5}}{CH_2=CH} \xrightarrow{k_p} R-CH_2-\underset{\substack{|\\C=O\\|\\O-C_2H_5}}{CH}-CH_2-\underset{\substack{|\\C=O\\|\\O-C_2H_5}}{CH}\cdot \tag{1.26}$$

addition to another
monomer molecule

The propagation reaction (1.26) may proceed rapidly hundreds or thousands of times, each adding another unit, before two growing chains (each having a free radical at one end), which react together, terminate the kinetic chain.

Termination:

$$R_1-CH_2-\underset{\substack{|\\C=O\\|\\O-C_2H_5}}{CH}\cdot$$

$$+$$

$$R_2-CH_2-\underset{\substack{|\\C=O\\|\\O-C_2H_5}}{CH}\cdot$$

$$\xrightarrow{k_{tc}} R_1-CH_2-\underset{\substack{|\\C=O\\|\\O-C_2H_5}}{CH}-\underset{\substack{|\\C=O\\|\\O-C_2H_5}}{CH}-CH_2-R_2 \tag{1.27}$$

$$\xrightarrow{k_{td}} R_2-\underset{\substack{|\\H\ C=O\\|\\O-C_2H_5}}{C=CH} + R_1-CH_2-\underset{\substack{|\\C=O\\|\\O-C_2H_5}}{CH_2} \tag{1.28}$$

where R_1 and R_2 represent long polymer chains. Equation (1.27) represents termination by combination, and (1.28) represents termination by disproportionation. Termination by combination yields twice the chain length and a narrower molecular weight distribution than termination by disproportionation.

Transfer:

Growing free radical chains may also undergo transfer reactions with monomer, existing polymer molecules, with solvent, or with deliberately added chain transfer agents. The last are often incorporated to control the molecular weight of the final product. For example, using *n*-dodecyl mercaptan as a chain transfer agent, a growing radical chain abstracts a hydrogen atom from the mercaptan, thus terminating the growing polymer molecule and leaving behind a mercaptyl radical capable of starting a new chain.

$$R_1-CH_2-\underset{\underset{O-C_2H_5}{\overset{\displaystyle |}{\underset{|}{C=O}}}}{\overset{\displaystyle |}{CH\cdot}} + C_{12}H_{25}SH \rightarrow R_1CH_2-\underset{\underset{O-C_2H_5}{\overset{\displaystyle |}{\underset{|}{C=O}}}}{\overset{\displaystyle |}{CH_2}} + C_{12}H_{25}S\cdot \qquad (1.29)$$

In general, transfer reactions affect only the molecular weight, not the polymerization rate. Transfer to monomer or polymer will, of course, lead to branching.

Condensation Polymerization

The second important polymerization reaction involves condensation; in this case a small molecule, such as water, splits out and must be removed. An example is the synthesis of nylon 66, a polyamide:

$$nH_2N-(CH_2)_6-NH_2 + nHO-\overset{\overset{\displaystyle O}{\|}}{C}-(CH_2)_4-\overset{\overset{\displaystyle O}{\|}}{C}-OH$$

$$\downarrow \text{heat} \quad \text{polymeric salt}$$

$$\qquad\qquad (1.30)$$

$$H-\left[NH-(CH_2)_6-\overset{\overset{\displaystyle H}{|}}{N}-\overset{\overset{\displaystyle O}{\|}}{C}-(CH_2)\right]_n OH + (2n-1)H_2O$$

In condensation reactions each polymer chain remains capable of further condensations with other chains, as opposed to addition reactions, where the chain is "dead" after the termination step.

APPENDIX B. BASIC MECHANICAL PROPERTIES AND RELATIONSHIPS*

Since the mechanical properties of polymer blends and composites are of major importance, some consideration of basic mechanical property terms is in order. In this appendix, such parameters are defined, and, where appropriate, interrelationships given. Common symbols are given in Table 1.4; it should be noted that in some cases several symbols are commonly cited for the same property.

Table 1.4
Some Common Mechanical Symbols

Symbol	Definition
T (or τ, f, or σ)[a]	Stress
E (or γ, s, or ε)[a]	Strain
G	Shear modulus
E	Young's (tensile) modulus
B	Bulk modulus
v	Poisson's ratio
η	Coefficient of viscosity
J	Tensile compliance

[a] More than one symbol in common use.

Hooke's law relates the strain s to the applied force f in a perfectly elastic system. In shear or in tension the relationships may be written, respectively, as

$$f = Gs \tag{1.31}$$

$$\sigma = Es \tag{1.32}$$

where G is the shear modulus and E is Young's modulus.

Newton's law relates the shear stress τ to the shear rate ds/dt, for simple viscous systems:

$$\tau = \eta(ds/dt) \tag{1.33}$$

Both Hooke's and Newton's laws represent limiting cases of ideal behavior. Few polymers ever obey these relationships exactly, but combinations of equations (1.31) or (1.32) and (1.33) form the basis for the important field of viscoelasticity.

* Ferry (1970); Nielsen (1962, Chapter 1); Rodriguez (1970); Tobolsky (1960).

Poisson's ratio v characterizes the volume change on deformation. If a tensile strain is applied in the x direction (s_x), then

$$v = -s_y/s_x = -s_z/s_x \qquad (1.34)$$

When $v = \frac{1}{2}$, there is no volume change on stretching, and when $v = 0$, there is no lateral contraction. Normally $v = 0.48$–0.49 for elastomers (indicating virtual incompressibility) and 0.20–0.40 for plastics.

Bulk modulus B is usually calculated from measurements of bulk compliance ($1/B$) or compressibility. The usual expression may be written

$$\frac{1}{B} = -\left(\frac{1}{V}\frac{\partial V}{\partial P}\right)_T \qquad (1.35)$$

where $(\partial V/\partial P)_T$ represents the change in volume with hydrostatic pressure at constant temperature.

As indicated above, the Young and shear moduli are simply given by the tensile or shear stress divided by tensile or shear strain. It is often useful, however, to define these moduli in terms of Poisson's ratio and the bulk modulus:

$$E = 3B(1 - 2v) = 2(1 + v)G \qquad (1.36)$$

which is a general relationship between the four basic mechanical properties. An important approximation often used in reporting data for polymers is $E \cong 3G$; the two moduli are exactly equal only for ideal elastomers ($v = 0.50$), but the relationship is sufficiently accurate even for most plastics, especially when a widely varying modulus is plotted on a logarithmic scale. Typical values of E are given in Table 1.5.

Tensile compliance is the inverse of Young's modulus for time-independent (perfectly elastic) materials:

$$J \cong 1/E \qquad (1.37)$$

For viscoelastic materials the relationship is more complex (Ferry, 1970); however, if the slope of $\log E$ vs. $\log t$ (in seconds) is shallow (closer to zero than -0.1 or -0.2), the correction is negligible for many purposes.

Table 1.5
Young's Modulus (dyn/cm²) of Selected
Materials[a]

Copper	1.2×10^{12}
Polystyrene plastic	3×10^{10}
Soft rubber	2×10^7

[a] Tobolsky (1960. p. 15).

BIBLIOGRAPHY OF POLYMER BOOKS AND JOURNALS

The following books and journals provide a broad background for polymer science and engineering.

Books

ALFREY, T., and GURNEE, E. F. (1967), *Organic Polymers*, Prentice-Hall.

ANDREWS, E. H. (1968), *Fracture in Polymers*, Elsevier.

BAER, E. (1964), *Engineering Design for Plastics*, Reinhold.

BATEMAN, L., ed. (1963), *The Chemistry and Physics of Rubber-Like Substances*, Wiley.

BAWN, C. E. H., ed. (1972), *Macromolecular Science* (MTP International Review of Science, Physical Chemistry Series One, Vol. 8), Butterworths and University Park Press.

BILLMEYER, JR., F. W. (1971), *Textbook of Polymer Science*, 2nd ed., Interscience.

DEANIN, R. D. (1972), *Polymer Structure, Properties, and Applications*, Cahners.

FERRY, J. D. (1970), *Viscoelastic Properties of Polymers*, 2nd ed., Wiley.

FLORY, P. J. (1953), *Principles of Polymer Chemistry*, Cornell.

GEIL, P. H. (1963), *Polymer Single Crystals*, Interscience.

HAM, G. E., ed. (1967), *Kinetics and Mechanisms of Polymerization*, Vol. I, *Vinyl Polymerization*, Marcel Dekker.

JENKINS, A. D., ed. (1972), *Polymer Science. A Materials Science Handbook*, North-Holland.

KAUFMAN, M. (1968), *Giant Molecules*, Doubleday.

KRAUS, G., ed. (1965), *Reinforcement of Elastomers*, Interscience.

LENZ, R. W. (1967), *Organic Chemistry of Synthetic High Polymers*, Interscience.

MARK, H. F. (1966), *Giant Molecules*, Time, Inc.

MARK, H. F., GAYLORD, N. G., and BIKALES, N. M., eds. (1967), *Encyclopedia of Polymer Science and Technology*, Interscience.

McCRUM, N. G., READ, B. E., and WILLIAMS, G. (1967), *Anelastic and Dielectric Effects in Polymeric Solids*, Wiley.

MEARES, P. (1965), *Polymers: Structure and Bulk Properties*, Van Nostrand.

MILLER, M. L. (1966), *The Structure of Polymers*, Reinhold.

NIELSEN, L. E. (1962), *Mechanical Properties of Polymers*, Reinhold.

NIELSEN, L. E. (1974), *Mechanical Properties of Polymers and Composites*, Vol. 1, Marcel Dekker.

ODIAN, G. (1970), *Principles of Polymerization*, McGraw-Hill.

PINNER, S. H. (1961), *A Practical Course in Polymer Chemistry*, Pergamon.

RAVVE, A. (1967), *Organic Chemistry of Macromolecules*, Marcel Dekker.

RITCHIE, P. D., ed. (1965), *Physics of Plastics*, Van Nostrand.

RODRIGUEZ, F. (1970), *Principles of Polymer Systems*, McGraw-Hill.

ROSEN, B., ed. (1964). *Fracture Processes in Polymeric Solids*, Interscience.

ROSEN, S. L. (1971), *Fundamental Principles of Polymeric Materials for Practicing Engineers*, Barnes and Noble.

SCHULTZ, J. (1974), *Polymer Materials Science*, Prentice-Hall.

SEYMOUR, R. B. (1971), *Introduction to Polymer Chemistry*, McGraw-Hill.

TOBOLSKY, A. V. (1960), *Properties and Structure of Polymers*, Wiley.

TOBOLSKY, A. V., and MACKNIGHT, W. J. (1965), *Polymeric Sulfur and Related Polymers*, Interscience.

TOBOLSKY, A. V., and MARK, H. F., eds. (1971), *Polymer Science and Materials*, Wiley–Inter-
science.
WARD, I. M. (1972), *Mechanical Properties of Solid Polymers*, Wiley.
WILLIAMS, D. J. (1971), *Polymer Science and Engineering*, Prentice-Hall.

Polymer Journals

Advances in Colloid and Interface Science
Advances in Polymer Science
Angewandte Makromolekulare Chemie
Biopolymers
British Plastics
British Polymer Journal
Colloid Journal (USSR)
European Polymer Journal
Fibre Chemistry (USSR)
Fracture
High Polymers Series, Wiley.
Inorganic Macromolecules Reviews
International Journal of Polymeric Materials
International Journal of Protein Research
Japan Plastics
Journal of Adhesion
Journal of Applied Polymer Science
Journal of Colloid and Interface Science
Journal of Composite Materials
Journal of Macromolecular Science,
    Part A Chemistry
    Part B Physics
    Part C Reviews in Macromolecular Chemistry
    Part D Reviews in Polymer Technology
Journal of Materials Science
Journal of Polymer Science, A-1, A-2, B, C; now
    Journal of Polymer Science, Macromolecular Reviews
    Journal of Polymer Science, Polymer Chemistry Edition,
    Journal of Polymer Science, Polymer Physics Edition
    Journal of Polymer Science, Polymer Letters Edition
    Journal of Polymer Science, Polymer Symposia
Journal of the International Rubber Institute (IRI)
Journal of the Rubber Research Institute of Malaysia
Kautschuk und Gummi
Kolloid Zeitschrift und Zeitschrift für Polymere
Kuntstoffe
Macromolecules
Makromolekulare Chemie
Modern Plastics
Modern Textiles

Polymer
Polymer Engineering and Science
Polymer Journal (Japan)
Polymer Mechanics (USSR)
Polymer News
Polymer Preprints
Polymer Science: USSR
Rubber Age
Rubber Chemistry and Technology
SPE Journal
Textile Research Journal
Transactions of the Society of Rheology

2

General Behavior of Polymer Mixtures

This part of the monograph will examine systems containing mixtures of two distinguishable kinds of polymer molecules. Such mixtures, known as polymer blends, polyblends, or simply blends, include mechanical blends, graft copolymers, block copolymers, and interpenetrating polymer networks.

Polymer blends containing one plastic phase and one rubbery phase will be emphasized in the next eight chapters. Depending on which phase predominates, such combinations yield impact-resistant plastics or reinforced elastomers. The briefer development of rubber–rubber blends given here belies the importance of the subject, since some 75% by volume of all rubber used is in blends. Also treated briefly are the plastic–plastic grafts, the best known of which are the castable polyesters.

In this chapter we will introduce the several types of polyblends. This chapter will be primarily concerned with a general description of the several types of polymer mixtures, their nomenclature and morphology, and of the physical properties and research techniques common to all such materials. A major theme developed throughout this monograph relates to the mixing of two polymers. Compatibility, miscibility, blending, and mutual solubility of two polymers, either on a molecular or supermolecular scale, are widely used terms that require definitions. The practical industrial chemist or engineer has tended to define two polymers as "compatible" or miscible if on mixing they blend together sufficiently well that his particular purpose is satisfied. A certain degree of clarity and/or adhesion between the components is often implied. In fact, most such materials do contain two phases, as has been shown by numerous recent research studies. (See also Section 13.4.)

While detailed descriptions of relevant terms are placed as appropriate throughout the text, brief definitions are in order at the outset. After blending together, polymer pairs may be qualitatively considered incompatible, semicompatible, or compatible, depending on whether two distinct or immiscible phases remain, partial mixing of the two polymers takes place at the molecular level, or a single thermodynamically stable phase is formed. True mutual solubility in polymer blends is rare, due to the small entropy gain on mixing. Important guidelines in determining compatibility often involve, besides optical clarity, a single, sharp glass–rubber transition and homogeneity at a scale of 50–100 Å.

2.1. METHODS OF MIXING POLYMER PAIRS

The principal methods of mixing two kinds of polymer molecules include mechanical blending, graft copolymerization, block copolymerization, and interpenetration of two networks. The last two are often considered as subgroups of the graft method.

2.1.1. Polymer Blends

Polymer blends may be defined as intimate mixtures of two kinds of polymers, with no covalent bonds between them. Historically, the oldest and simplest method involves mechanical blending, where a plastic and a noncrosslinked elastomer are blended either on open rolls or through extruders (Matsuo, 1968). Materials prepared in this manner usually contain several percent of elastomer dispersed in a plastic matrix, as shown schematically in Figure 2.1.

In simple mechanical blends the plastic component usually predominates, with the dispersed elastomer having dimensions of the order of several micrometers. The shear action of mechanical blending also generates free radicals through polymer degradation reactions. The free radicals thus induced by mechanochemical action subsequently react to form a small number of true chemical grafts between the two components. The quantity and importance of such grafted material obviously depend on the exact mode of blending (Casale and Porter, 1971). Significant improvements in impact

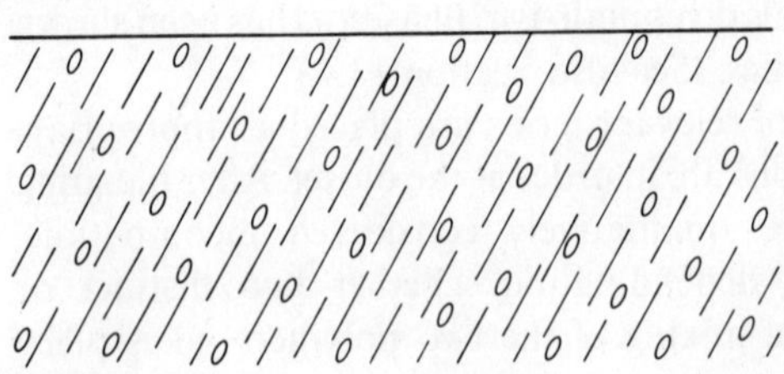

Figure 2.1. Schematic illustration of rubber droplets dispersed in a continuous plastic phase.

resistance and toughness are usually noted for such blends over the plain parent plastic, even in cases where no particular amount of grafting is noted.

2.1.2. Graft Copolymers

Further improvement in mechanical behavior can be obtained by graft copolymerization, which will be considered as a separate type of poly-blend in this monograph. In the graft copolymerization method (Battaerd and Tregear, 1967) the first polymer (usually the rubber) portion is dissolved in the plastic monomer, and polymerization is effected. During the poly-merization, some or all of the second polymer becomes joined to the first. Often the polymerization is carried out with stirring, to bring about phase inversion. This results in a much finer dispersion of the rubber phase, and a far more complex morphology. Polymer blends and grafts designed for impact resistance will be considered further in Chapter 3.

2.1.3. Block Copolymers

In block copolymers, the individual components are joined at their ends. Block copolymers have been synthesized by several methods, but perhaps the most elegant procedure follows the "living polymer" anionic polymerization process (Henderson and Szwarc, 1968). The unusual features of this reaction include simultaneous nucleation and uniform growth rates of all chains, and lack of termination reactions. After exhaustion of a first monomer the polymer chains remain alive, and addition of a second mono-mer results in a block copolymer of the form

$$A-(A)_{n-2}-A-B-(B)_{m-2}-B \qquad (2.1)$$

where n and m are the degrees of polymerization of the A and B mer units, respectively. More blocks can be added in the same manner, if desired. Block copolymerization results in extremely fine phases called domains, in the form of characteristic spheres, cylinders, or lamellae, depending on composition. An important class of block copolymers is made up of the triblock ABA thermoplastic elastomers (Holden et $al.$, 1969a) considered in Chapter 4.

2.1.4. Interpenetrating Polymer Networks (IPN's)

This novel class of polymers, together with simultaneous interpenetrating networks (SIN's) and interpenetrating elastomeric networks (IEN's)

(Klempner *et al.*, 1970) forms another important class of two-phase polymer systems. IPN's can be formed (Sperling and Friedman, 1969) by preparing a crosslinked polymer network, swelling in a second monomer together with activator and crosslinking agent, and polymerizing *in situ*. This second reaction forms a crosslinked polymeric network that interpenetrates the first network. A SIN differs from an IPN in that both networks are formed more or less simultaneously. An example might involve concurrent noninterfering addition and condensation polymerizations, each reaction having appropriate crosslinking agents (Sperling and Arnts, 1971). IPN's and SIN's also form distinct and characteristic domain-type structures, with the restriction that both components must be continuous throughout the macroscopic mass. IEN's are prepared by mixing and coagulating two different kinds of polymeric latexes, followed by a single crosslinking reaction. The result is a three-dimensional mosaic structure. The IPN's and IEN's will be considered further in Chapter 8.

2.2. INTERDIFFUSION

Although polymer pairs certainly do tend to be incompatible, it should not be supposed that interdiffusion does not occur. Indeed, two polymers in contact will exhibit varying degrees of segmental interdiffusion at temperatures that are high enough to permit a significant degree of segmental mobility. Such interactions have been studied extensively by Voyutskii and co-workers, whose book (Voyutskii, 1963) provides further detail; some aspects of interdiffusion will be considered in Section 13.4. Since interdiffusion of polymers has been shown to play an important role in the adhesion of one polymer to another, similar effects may be expected in polyblends that exhibit limited compatibility.

2.3. NOMENCLATURE

There are many distinguishable ways of mixing two kinds of polymer molecules. It is important that each mode of mixing or bonding be given a unique description. The present section will summarize the accepted nomenclature rules (Battaerd and Tregear, 1967, Chapter 2; Ceresa, 1962) and suggest some new ones.

Random copolymers may be indicated by use of the syllable -co- to join the two monomers involved, e.g., poly(butadiene-co-styrene), and have the general structure

$$A—A—B—A—B—B—B—A—B—B—A—A—B—A \qquad (2.2)$$

where the A and B mers or monomer units are added in random order. The relative numbers of A's and B's are, of course, determined by the relative quantities of monomer added and by relative reactivities.

Alternating copolymers have the structure (Furukawa, 1970; Furukawa *et al.*, 1971)

$$A—B—A—B—A—B—A—B—A—B \qquad (2.3)$$

and may be indicated by replacing the -co- by -alt-. Random and alternating copolymers usually form one phase, and as such will not be emphasized in this monograph.

A graft copolymer has the form

$$
\begin{array}{c}
A—A—A—A—A—A—A—A—A—A—A \\
| \\
B \\
| \\
B—B—B—B—B—B
\end{array}
\qquad (2.4)
$$

The symbol -g- may be employed, as in poly(butadiene-g-styrene). The first polymer mentioned forms the backbone chain, and the second the branches. Block copolymers of the type

$$A—A—A—A—A—A—A—A—B—B—B—B—B—B \qquad (2.5)$$

may be designated by the symbol -b- in place of the -g-. Block copolymers differ from graft copolymers in that the two homopolymers are always joined at the ends. Another type of nomenclature has recently become popular for block copolymers; in this terminology, diblock and triblock polymers have the general designation AB and ABA, respectively. Particular combinations are usually indicated by the monomer initials; thus poly(styrene-b-isoprene-b-styrene) becomes SIS.

More complicated structures can be considered. For instance, a cross-linked (cl) graft structure has been ideally described as

$$
\begin{array}{c}
A—A—A—A—A—A—A—A—A—A—A—A—A \\
|\qquad\qquad\qquad\qquad\quad | \\
B\qquad\qquad\qquad\qquad\quad B \\
|\qquad\qquad\qquad\qquad\quad | \\
(B)_n\qquad\qquad\qquad\quad (B)_m \\
|\qquad\qquad\qquad\qquad\quad | \\
B\qquad\qquad\qquad\qquad\quad B \\
|\qquad\qquad\qquad\qquad\quad | \\
A—A—A—A—A—A—A—A—A—A—A
\end{array}
\qquad (2.6)
$$

A commercially important class of materials, the crosslinked polyesters, are

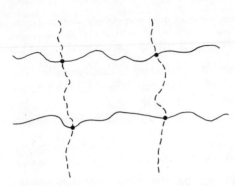

Figure 2.2. Schematic of a graft copolymer as described in the text.

synthesized by casting an unsaturated polyester with styrene monomer, which is subsequently polymerized. The hybrid condensation–vinyl polymer cross-linking results from reactions of the type (Sorenson and Cambell, 1961)

$$
\left[R-O-\overset{\overset{O}{\|}}{C}-CH=CH-\overset{\overset{O}{\|}}{C}-O\right]_n + CH_2=CH(C_6H_5)
$$

$$
\rightarrow \left[R-O-\overset{\overset{O}{\|}}{C}-\underset{\underset{CH(C_6H_5)-CH_2-}{|}}{CH}-\overset{\overset{CH_2-CH(C_6H_5)-}{|}}{CH}-\overset{\overset{O}{\|}}{C}-O\right]_n
$$

(2.7)

Such a structure could be written as a modification of structure (2.6), in which (Figure 2.2) the B's extend on both sides of the A chains, and the points of intersection indicate crosslinks. These polymers are members of the special class designated as "joined interpenetrating polymer networks," or AB-crosslinked polymers.

Since the general class of interpenetrating polymer networks (IPN's) have been developed largely since the publication of the work by Battaerd

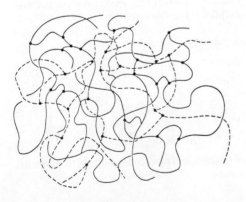

Figure 2.3. The idealized structure of the IPN's, in the limiting case of high compatibility. (After Frisch and Klempner, 1970a.)

and Tregear (1971), Ceresa (1962), and Sorenson and Cambell (1961), no standard nomenclature exists. These polymers contain two independent crosslinked networks (Frisch and Klempner, 1970a), as illustrated in Figure 2.3. The sequential type, where one network is formed before the other, may be conveniently written X/Y IPN, where X is the first polymer network and Y is the second network. An example is 50/50 PEA/PS, in which an equal weight of crosslinked polystyrene is formed within a network of poly(ethyl acrylate).

Other symbols occasionally used in the literature include (br) for branched materials and (iso), (syndio), and (a) for isotactic, syndiotactic, and atactic structures respectively. No symbol appears to exist for mechanical blends, although these materials are obviously important. Where necessary the symbol -m- will denote a mechanical blend, for example, poly(styrene-m-butadiene) for a mechanical blend of polystyrene with polybutadiene.

An important subclass of the IPN's are the semi-IPN's, where one polymer is linear and the other is locked in network form. Two subclasses may be described, semi-IPN's of the first or second kind, depending on whether the polymer synthesized first, polymer I (D'Agostino and Lee, 1972), or the polymer synthesized second, polymer II, is crosslinked.

In many places throughout the text, we shall refer to the first polymer synthesized as polymer I and the second polymer synthesized as polymer II. Their monomers, where appropriate, will also bear the designations I and II, respectively. Thus, in graft copolymers, the backbone chain is usually polymer I, and the side chains comprise polymer II.

2.4. ELECTRON MICROSCOPY

Scientists and engineers working in the fields of polyblends and block copolymers have realized for many years that phase separation of the two components takes place, and that this is indeed important to the development of the mechanical behavior characteristic of these materials. However, it was not until the development of the electron microscope that the structure of any but the coarsest mechanical blends could be discerned, and even then lack of contrast between the two phases remained serious. This problem was solved in 1965 by Kato (1966, 1968), who discovered that osmium tetroxide preferentially stains polymer molecules containing carbon–carbon double bonds, such as in polybutadiene and polyisoprene. The osmium tetroxide also hardens the rubbery phase, allowing convenient ultramicrotoming of specimens to ~ 500 Å thickness.

Osmium tetroxide staining can be accomplished by exposing a sample to osmium tetroxide vapor for a week, or by soaking overnight in a 1 %

aqueous solution. Both methods selectively stain and harden unsaturated rubber to a depth of a few micrometers, sufficient for the preparation of suitable sections.

An important result of such studies was the realization that almost all commercially important polymer blends, blocks, and grafts clearly exhibit phase separation, and that each has its own characteristic fine structure. An example of a typical morphology for an impact-resistant plastic is shown in Figure 2.4. Although the polymer contains only 6% rubber, much of the volume of polystyrene is occluded (Wagner and Robeson, 1970). Those interested in the details of electron microscope construction and operation, as well as experimental techniques, should consult Hall (1966) or Kay (1965).

Polymer blends, blocks, and grafts are sometimes called "polymer alloys" because of the similarity of microstructure that exists between these materials and their metallic counterparts; this analogy is developed briefly in Appendix A.

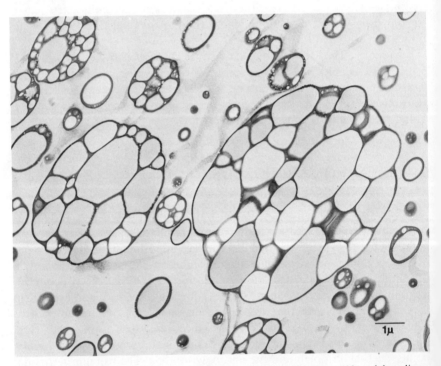

Figure 2.4. Rubber membrane structure of impact-resistant polystyrene (6% polybutadiene; 22% rubber phase volume). This graft-copolymer morphology, containing a fine structure within the discontinuous phase, is brought about by agitation during the early stages of polymerization. (Wagner and Robeson, 1970.)

2.5. THE INCOMPATIBILITY PROBLEM

It should be pointed out that in order to develop superior mechanical properties in a two-component polymer system, the components should not be so incompatible that they do not wet, nor so mutually soluble that they form one homogeneous phase (Tobolsky, 1960, pp. 81–82). Most of the presently important systems are compatible to the extent that a slight (but usually unknown!) degree of mixing takes place, or interfacial bonding is developed directly, as in grafts or blocks.

The experimental evidence for phase separation in polymer mixtures was briefly developed in the previous section. In the present section a thermodynamic explanation for mutual insolubility, or incompatibility, of polymer pairs will be outlined.

Polymer incompatibility arises from the very small entropy gained by mixing different kinds of long chains. In fact, it will be shown that in the limit of high molecular weight, only polymer pairs with zero or negative heats of mixing form one phase. Let us first develop the history of polymer incompatibility studies, since this offers insight into the status of our understanding and shortcomings at the present time.

Dobry and Boyer-Kawenoki (1947) investigated the phase relationships existing in ternary systems: polymer (1)–polymer (2)–mutual solvent (3). They prepared dilute solutions of polymers in common solvents, and then mixed the two solutions of interest. All of the polymer pairs studied were found to undergo phase separation at only 5–10% polymer concentration. For instance, cellulose acetate and polystyrene were immiscible at 5% concentration in toluene. These investigators concluded that incompatibility of two polymers even highly diluted is the normal situation.

2.5.1. Thermodynamics of Mixing

Shortly thereafter, Scott (1949)* offered a theoretical explanation of the Dobry and Boyer-Kawenoki results; specifically, he investigated the partial molal free energy of mixing $\Delta \bar{F}$ of polymer–polymer binary systems, and polymer (1)–polymer (2)–solvent (3) ternary systems. For binary polymer–polymer systems, $\Delta \bar{F}$ is given by

$$\Delta \bar{F}_1 = RT\left[\ln v_1 + \left(1 - \frac{m_1}{m_2}\right)v_2 + m_1 \chi_{12} v_2^2 \right] \tag{2.8}$$

---

* Note especially the section, "The Two Polymer System."

$$\Delta \bar{F}_2 = RT \left[\ln v_2 + \left(1 - \frac{m_2}{m_1} \right) v_1 + m_2 \chi_{12} v_1^2 \right] \qquad (2.9)$$

where v_1 and v_2 represent volume fractions of component one and component two, respectively, m_1 and m_2 are essentially degrees of polymerization,* and χ_{12} is the Flory heat of interaction term.

Scott set the first and second derivatives with respect to volume of $\Delta \bar{F}_1$ or $\Delta \bar{F}_2$ to zero, and found that when the two polymers are at their critical solution temperatures

$$(\chi_{12})_c = \tfrac{1}{2} \{ [1/(m_1)^{1/2}] + [1/(m_2)^{1/2}] \}^2 \qquad (2.10)$$

$$(v_1)_c = (m_2)^{1/2} / [(m_1)^{1/2} + (m_2)^{1/2}] \qquad (2.11)$$

$$(v_2)_c = (m_1)^{1/2} / [(m_1)^{1/2} + (m_2)^{1/2}] \qquad (2.12)$$

For $m_1 = 1$ and $m_2 = 1$ (mixing of normal fluids), $(\chi_{12})_c = 2$. For polymer–solvent systems, $(\chi_{12})_c = \tfrac{1}{2}$. However, for polymer–polymer mixtures where m_1 and m_2 are in the normal range $(\chi_{12})_c$ is of the order of 0.01. It can be seen from equations (2.10)–(2.12) that polymers of infinite molecular weight would be incompatible if there were any positive heat of mixing (per submolecule) at all. Similar considerations apply to the ternary system.

Since the heat of mixing usually is positive unless some specific interaction occurs, such as hydrogen bond formation, the above theory predicts two phases for most mixtures of interest.† Experiments by Slonimskii (1958) and by Tompa (1956) tended to confirm the above results.

These early works must be viewed in the context of the time during which they were performed. The results obtained are undoubtedly correct, yet their combined conclusion of rather total incompatibility of polymer mixtures tended to discourage efforts to find and study compatible or semicompatible polymer pairs. As we shall see below, the fact that many polymers could be usefully mixed and partial compatibility achieved was stumbled upon partly as a result of other lines of investigation. (See also Section 13.4.)

2.5.2. Polymer–Polymer Phase Diagrams

Most of the above-mentioned experiments were conducted on 50/50 binary mixtures, or on the corresponding dilute solutions. Detailed phase diagrams, which would permit estimation of the degree of mixing in any particular case, are much more difficult to obtain, primarily due to the very long times required for equilibrium to be reached.

---

* $m_1 = \hat{v}_1 / \hat{v}_0$, where $\hat{v}_1$ and $\hat{v}_0$ are the molar volumes of polymer one and monomer one, respectively.

† A crucial unsolved problem that will reappear several times in this book in different forms is the composition of the two phases, noting that total separation is quite unrealistic.

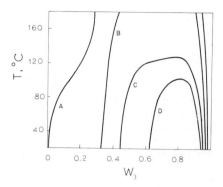

Figure 2.5. Temperature–composition co-existence curves for silicone (M.W. = 850, 1350, and 17,000) and polyisobutene (M.W. = 250 and 440) (Allen *et al.*, 1956). (A) 17,000/250; (B) 1350/440; (C) 1350/250; (D) 850/250. These phase diagrams show the basic incompatibility of even low-molecular-weight polymers.

In an attempt to solve this problem, Allen *et al.* (1956) studied phase diagrams of telomers as a function of molecular weight. Their results for the system polyisobutene–poly(dimethyl siloxane) are summarized in Figure 2.5.

Mixtures of telomers of molecular weights as low as ~ 1000 g/mol have critical temperatures above the decomposition points of the polymers, illustrating the high degree of incompatibility of this particular pair.

Recently R. Kuhn *et al.* (1968a, b; Kuhn, 1968) studied the phase relationships of polystyrene–poly(methyl methacrylate) by employing an ingenious dilute solution light-scattering technique. Equations derived from phase separation quantities could be extrapolated to the pure phases, yielding the conclusion that if both polymers had molecular weights less than about

Table 2.1
Some Compatible or Semicompatible Polymer Pairs

Component 1	Component 2	Reference
Poly(vinyl chloride)	Poly(butadiene-co-acrylonitrile)	Takayanagi (1963), Nielsen (1953), Breuers *et al.* (1954), Davenport *et al.* (1959)
Poly(vinyl acetate)	Poly(methyl acrylate)	Nielsen (1962), p. 177; Hughes and Brown (1961)
Poly(methyl methacrylate)	Poly(ethyl acrylate)	Hughes and Brown (1961), Sperling *et al.* (1970a)
Polystyrene	Poly(α-methyl styrene)	Baer (1964)
Polystyrene	Poly(2,6-dimethyl phenylene oxide)	Cizak (1968), Stoelting *et al.* (1969), MacKnight *et al.* (1971)
Polystyrene	Isotactic poly(vinyl methyl ether)	Bank *et al.* (1969)
Poly(vinyl chloride)	Poly(ε-caprolactone)	Koleski and Lundberg (1969)

Table 2.2
Selected Incompatible Polymer Pairs

Polymer 1	Polymer 2	Reference
Polyethylene	Polyisobutylene	Schmieder (1962)
Poly(methyl methacrylate)	Poly(vinyl acetate)	Jenckel and Herwig (1956)
Natural rubber	Poly(styrene-co-butadiene)	Dannis (1963)
Polystyrene	Polybutadiene	Turley (1963)
Polystyrene	Poly(vinyl chloride)	Buchdahl and Nielsen (1950)
Poly(methyl methacrylate)	Polystyrene	Baer (1964)
Poly(methyl methacrylate)	Cellulose triacetate	Movsum-Zade (1964)
Nylon 6	Poly(methyl methacrylate)	Shinohara (1959)
Nylon 6,6	Poly(ethylene terephthalate)	Papero *et al.* (1967)
Polystyrene	Poly(ethyl acrylate)	Sperling and Friedman (1969), Hughes and Brown (1963)
Polystyrene	Polyisoprene	Holden *et al.* (1963)
Polyurethane	Poly(methyl methacrylate)	Niederhauser and Bauer (1972)

30,000–40,000, the materials would form a single phase. Above this molecular weight, two phases would be formed. Kuhn *et al.* did not, however, systematically study the degree of separation of the two components as a function of molecular weight.

While most polymer pairs exhibit pronounced incompatibility, several important pairs are apparently either compatible or nearly so (Bohn, 1966, 1968; Krause, 1972); see Table 2.1. The experimental methods employed in such studies vary, but optical clarity and the presence of a single glass–rubber transition have been especially important in such judgements. Table 2.2 lists a few important pairs of incompatible polymers. It is interesting to note that nearly all materials are incompatible when one or both components crystallize.

2.6. BULK BEHAVIOR OF TWO-PHASE POLYMERIC MATERIALS

2.6.1. Glass Transitions

Simple homopolymers and random copolymers usually exhibit one principal glass transition, although one or more secondary transitions are common enough. When two incompatible polymers are mixed, the individual phase domains retain the glass transitions of their respective parent homopolymers. The result is that most blends, blocks, grafts, and IPN's exhibit *two* principal glass transitions. If significant molecular mixing takes place, the transitions will be broadened, and/or their temperatures will be closer together. Correspondingly, two maxima will be observed in the mechanical loss spectrum and two transitions will be observed dilatometrically.

Attention will now be turned to two important experimental aspects of glass transition behavior: the temperature dependence of the modulus, and stress-relaxation studies. This will be followed by a brief discussion of mathematical models that describe polyblend glass transition behavior.

2.6.2. Modulus–Temperature Behavior of Model Polyblends

A comparison of the properties of a polyblend with a random copolymer provides considerable insight into the differences in structure. For example, the exact position and breadth of the two transitions in polyblends reflect the degree of mixing obtained, while the random copolymer exhibits only one transition.

Figures 2.6 and 2.7 illustrate such differences in the cases of polystyrene/SBR polyblends and SBR random copolymers, respectively (Tobolsky, 1960, pp. 79–82); the overall S/B ratio is approximately the same in each. Figure 2.6 shows that both transitions depend on composition.

Observe that the higher transition of the SBR-rich blends is smeared, indicating that the polystyrene component may be incompletely phase-separated from the SBR. In contrast, the series of random copolymers in Figure 2.7 behave much like homopolymers having only one transition. The temperature of such a single transition usually depends upon the composition in some simple manner, often following an equation of the type

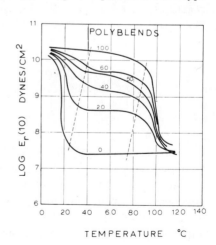

Figure 2.6. Modulus–temperature behavior of blends of polystyrene with a 30/70 butadiene/styrene copolymer. Dashed lines delineate change in T_g with composition. (Tobolsky, 1960.)

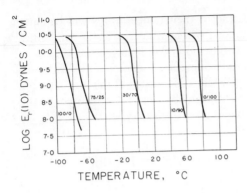

Figure 2.7. Modulus–temperature behavior of a series of random copolymers of butadiene and styrene with compositions (B/S) ranging from 100/0 to 0/100. Each composition exhibits one transition, indicating only one phase. (Tobolsky, 1960.)

(Tobolsky and Shen, 1965)

$$T_g = W_1 T_{g_1} + W_2 T_{g_2} + K W_1 W_2 \qquad (2.13)$$

where W_i and T_{g_i} are the weight fraction and glass transition temperature of the ith component, respectively, the constant K allowing for nonideality.

2.6.3. Stress-Relaxation Behavior

The relaxation behavior of individual chains depends on the simultaneous or cooperative relaxation of neighboring chains. If these are the same, different, or mixed in type, differences in stress-relaxation behavior may be expected. Thus, stress-relaxation experiments may be expected to reveal even the most intimate extents of chain mixing or segregation (Horino et al., 1965; Manabe et al., 1969, 1971; Soen et al., 1966; Takayanagi, 1972; Takayanagi et al., 1963; Tobolsky and Aklonis, 1964; Tobolsky and DuPre, 1968).

Let us examine the stress-relaxation behavior of an incompatible polymer pair (Figure 2.8) and compare it to the behavior of a semicompatible polymer pair (Figure 2.9). The blend poly(vinyl acetate)/poly(methyl methacrylate) illustrated in Figure 2.8 is seen to have two regions of rapid relaxation, the first near 35°C and the second near 109°C (Horino et al., 1965; Soen et al., 1966). These relaxations correspond to the glass temperatures of PVAc and PMMA, respectively. The authors remark that although the blend is a 50/50 mixture, the lower temperature transition appears less distinct. They point out that the mode of mixing controls which phase predominates, and that the more continuous phase has the more pronounced transition. Most importantly, however, the appearance of two distinct transitions may be

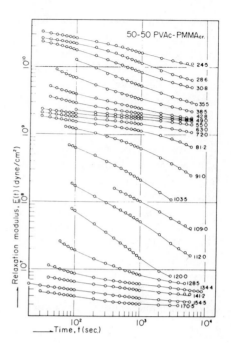

Figure 2.8. Stress-relaxation data for a PVAc/PMMA 50/50 polyblend. The PMMA portion was prepared from a lightly crosslinked latex to suppress flow at high temperatures. Numbers at right are temperatures in °C. (Takayanagi *et al.*, 1963.)

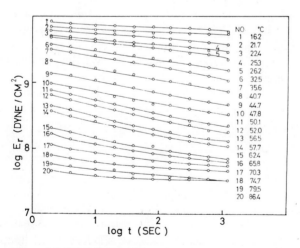

Figure 2.9. Relaxation modulus of a PVC/NBR 77/23 blend. Note nearly parallel relaxation lines. The numbers denote the temperature of measurement. (Takayanagi *et al.*, 1963.)

interpreted as meaning that each kind of polymer molecule is surrounded primarily with its own species—an interpretation which tends to confirm our impression of incompatibility.

The relaxation data for the poly(vinyl chloride)/poly(butadiene-co-acrylonitrile), PVC/NBR, system shown in Figure 2.9, on the other hand, appear essentially as a series of parallel lines (Manabe et al., 1969). The broad spectrum of relaxation times observed also suggests a lack of discrete phase domains.

As might be expected, master curves prepared from incompatible or semicompatible blends do not quantitatively obey the WLF equation, the experimental shift factors being considerably different. For example, the master curve prepared from the data in Figure 2.9 (shown in Figure 2.10) shows relaxation taking place over some 16 decades of time, whereas the WLF equation, Section 1.5.7, predicts a transition covering about nine decades of time. Further, although the Tobolsky–Aklonis–DuPre damped torsional oscillator theory (Tobolsky and Aklonis, 1964; Tobolsky and DuPre, 1968) predicts a slope of approximately -0.5 for a plot of log $E(t)$ vs. log t in the transition region, based on the following equation, the master curve in Figure 2.10 has a maximum slope of approximately -0.1:

$$E_r(t) = \frac{E_1 \tau_{\min}^{1/2}}{\tau_{\min}^{1/2} + t^{1/2}} + E_2 \qquad (2.14)$$

In this equation E_1 represents the glassy modulus (3×10^{10} dyn/cm^2), $\tau_{\min}$ is a characteristic relaxation time, E_2 represents the rubbery plateau modulus, and $E_r(t)$ is the relaxation modulus as a function of time.

It may be that in PVC/NBR materials the nearest neighbor chain composition is nearly (but not quite) random.

One further point should be made. The WLF equation can correctly predict the relaxation behavior of incompatible systems, but for temperature ranges limited to one transition. For incompatible polyblends that exhibit two transitions, the equation will yield satisfactory results if applied to each transition separately.

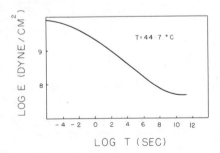

Figure 2.10. A master curve prepared by shifting each curve in Figure 2.9 along the abscissa until the curves formed a smooth line. The reduced temperature is 44.7°C. (Takayanagi et al., 1963.)

2.6.4. The Takayanagi Models

Takayanagi and co-workers transformed the spring and dashpot relaxation models (Section 1.5.6) to plastic and rubber elements in an effort to better explain the mechanical behavior of polyblends (Takayanagi *et al.*, 1963). Some simple combinations of the Takayanagi models are shown in Figure 2.11. The plastic phase is denoted by P and the rubber phase by R, while the quantities λ and φ are functions of the volume fractions of parallel and series elements, respectively.

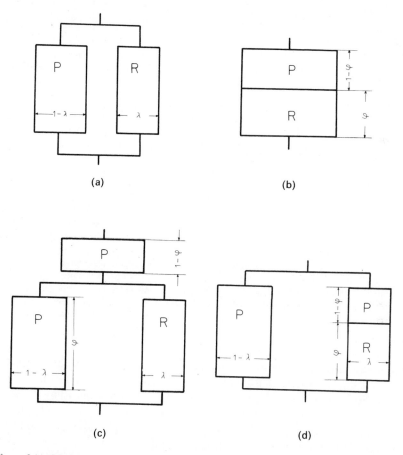

Figure 2.11. Models for two-phase polymer systems. Elements in (a) and (b) form the basic parallel and series models. Combinations shown in (c) and (d) represent two other possible models. Note the similarity to the spring and dashpot models often invoked in explaining homopolymer behavior. (After Takayanagi *et al.*, 1963.) Part (a) represents an isostrain model, (b) represents an isostress model, and (c) and (d) illustrate combinations of these limiting cases.

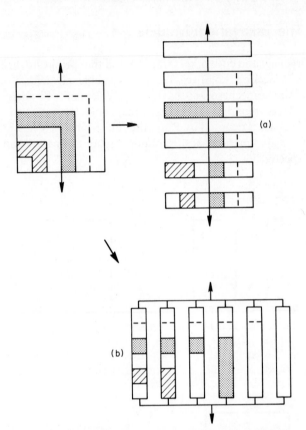

Figure 2.12. A model for the semicompatible polyblend: (a) series representation, equation (2.19); (b) parallel representation, equation (2.21). This semicompatible model represents a graded series of random copolymer compositions $1, 2, 3, \ldots, n$, each differing only very slightly from the other. (Takayanagi, 1972.)

In a manner similar to the application of springs and dashpots to the theory of linear viscoelasticity, we note that for units in parallel the total stress is $\sigma = \sigma_1 + \sigma_2 + \sigma_3 + \cdots$, and that for units in series the total strain is $\varepsilon = \varepsilon_1 + \varepsilon_2 + \varepsilon_3 + \cdots$. Finally, application of Hooke's law, $\sigma = \varepsilon E$, allows the complex modulus E^* of the Takayanagi models in Figures 2.11a–d to be represented by the following equations, respectively:

$$E^* = (1 - \lambda)E_P^* + \lambda E_R^* \tag{2.15}$$

$$E^* = \left(\frac{\varphi}{E_R^*} + \frac{1 - \varphi}{E_P^*} \right)^{-1} \tag{2.16}$$

$$E^* = \left[\frac{\varphi}{\lambda E_R^* + (1 - \lambda)E_P^*} + \frac{1 - \varphi}{E_P^*} \right]^{-1} \quad (2.17)$$

$$E^* = \lambda \left(\frac{\varphi}{E_R^*} + \frac{1 - \varphi}{E_P^*} \right)^{-1} + (1 - \lambda)E_P^* \quad (2.18)$$

Takayanagi *et al.* (1963) found that the model of Figure 2.11c, equation (2.17), most closely represented the behavior of incompatible polyblends.

One further model should be explored, since we are ever interested in semicompatibility. When the plastic and rubber phases show increasing miscibility, the model shown in Figure 2.12a (an extended form of Figure 2.11c) is found to most accurately express the relaxation behavior. The latter model can be expressed mathematically by the equations

$$\frac{1}{E} = \sum_{i=1}^{n} \lambda_i \left[E_i \left(1 - \sum_{k=i+1}^{n} \lambda_k \right) + \sum_{k=i+1}^{n} E_k \lambda_k \right]^{-1} \quad (2.19)$$

$$\sum_{i=1}^{n} \lambda_i = 1 \quad (2.20)$$

In the above, $\varphi_i = \lambda_i$ and the portion of the segmental environment i with the modulus E_i contributes to the modulus E with weight λ_i.

As depicted in Figure 2.13, equation (2.19) predicts a single, very broad transition. A real system exhibiting such behavior is shown in Figures 3.13, 8.13, and 8.14.

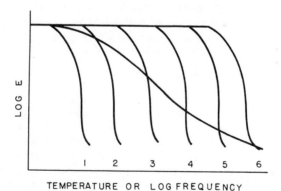

TEMPERATURE OR LOG FREQUENCY

Figure 2.13. The modulus–temperature or modulus–negative frequency plots of the several random copolymer compositions and blends. The blend is predicted to exhibit a very broad glass transition, in comparison with the corresponding random copolymers. (Takayanagi *et al.*, 1963.)

If we assume that the model in Figure 2.12 is modified into a parallel rather than a series model (see Figure 2.12b), then the modulus of the system is represented by the following equation (Takayanagi, 1972):

$$E = \sum_{i=1}^{n} \left\{ \lambda_i \left[\sum_{k=1}^{i} \frac{\lambda_k}{E_i} + \sum_{k=i+1}^{n} \frac{\lambda_k}{E_k} \right]^{-1} \right\}$$ (2.21)

with

$$\sum_{i=1}^{n} \lambda_i = 1$$ (2.22)

2.6.5. Free Volume Model

Takayanagi and co-workers (Takayanagi et al., 1963; Manabe et al., 1969) also studied glass transitions in polymer blends, basing their theory on communalization of free volume. They divided the blend system into cells having dimensions of several tens of angstroms, just large enough to contain the coordinated chain motions associated with the glass transition. Takayanagi points out that for perfectly miscible systems and random copolymers, the free volume is well communalized, and both components behave as one component with a common free volume. For partly miscible systems, on the other hand, communalization is incomplete and depends upon the method of measurement. The cell free volume f is assumed to be different from cell to cell in the blends.

Figure 2.14 gives a schematic representation of the distribution function of the free volume fraction $F(f)$ vs. f for blends of various degrees of compatibility. When the components are mixed to a heterogeneity limit of 10–20 Å, the curve of $F(f)$ will approach the curve for a molecularly mixed system.

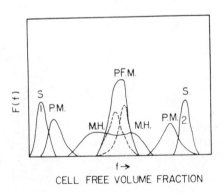

CELL FREE VOLUME FRACTION

Figure 2.14. Schematic representation of the distribution function of free volume fraction $F(f)$ for various blend systems. Curves 1 and 2 show the shapes of the functions of components 1 and 2, respectively. Curves S., P.M., M.H., and P.F.M. indicate two separated phases, partially miscible, microheterogeneous, and perfectly miscible systems, respectively. The volume fraction of each component is 50% for all these blends. The broken line shows the $F(f)$ of the components in the perfectly miscible system. (Manabe et al., 1971.)

If the free volume fraction of the ith component is f_i, then we have

$$f_i = f_{i,0} + \sum_{j=1}^{n} (a_i^j v_j) + \sum_{j=1}^{n} \sum_{k=1}^{n} (b_i^{jk} v_j v_k) \qquad (2.23)$$

$$\sum_{i=1}^{n} v_i = 1 \qquad (2.24)$$

where the subscripts i, j, and k indicate the corresponding components, the subscript 0 indicates the pure component, and a_i^j and b_i^{jk} are the coefficients of v_i and $v_j v_k$, respectively, and the v's represent the component volume fractions.

For a two-component model, Takayanagi derived the following relationships for the glass temperatures:

$$T_{g_1} = T_{g_{1,0}} - (a_1^2 v_2 / \Delta\alpha_1) - (b_1^{22} v_2^2 / \Delta\alpha_1) - \cdots \qquad (2.25)$$

$$T_{g_2} = T_{g_{2,0}} - (a_2^1 v_1 / \Delta\alpha_2) - (b_2^{11} v_1^2 / \Delta\alpha_2) - \cdots \qquad (2.26)$$

where $\Delta\alpha_i$ is the thermal expansion coefficient of f_i. With appropriate values of the constants, the T_g's can be predicted over wide ranges of composition.

One of the most important conclusions associated with the Takayanagi model is that the distribution function of the free volume fraction $F_f(f)$ is evaluated by the following equation based on the specific volume–temperature curve:

$$\left| \frac{d^2\bar{v}}{dT^2} \right|_{T=T_g} = \Delta\alpha \cdot F(T_g) \qquad (2.27)$$

and

$$F(T_g) \cong \Delta\alpha \cdot F_f(f) \qquad (2.28)$$

where $\Delta\alpha$ is the thermal expansion coefficient of fractional free volume f and is approximately equal to $\alpha_l - \alpha_g$. According to equations (2.27) and (2.28), it becomes possible to evaluate $F_f(f)$ by making the second-order differentiation of specific volume–temperature curves in a range of temperature close to T_g.

2.6.6. Other Models

While the Takayanagi models have proved useful because of their simplicity, the effects of changes in mechanical behavior with composition and phase structure may also be profitably explored using several analytical relations, which include equations derived by Kerner (1956b), Hashin and Shtrikman (1963), and Halpin and Tsai (Ashton et al., 1969, Chapter 5). The most widely applied of these is the Kerner equation, which presents the lower bound modulus change for spherical inclusions in a continuous matrix; with several modifications (Dickie, 1973) the Kerner equation has seen application to polymer blends and grafts. Such equations for composite systems are discussed in more detail in Section 12.1.1; for accounts of recent work and reviews in the areas of polymer blends and composites, see Bucknall and Clayton (1972), Bucknall *et al.* (1972a), Deanin (1973), Halpin and Kardos (1972), Kambour and Barker (1966), Kardos *et al.* (1972), Nielsen (1970a), Shen and Bever (1972), Szwarc (1973), and Work (1973).

2.6.7. Morphology–Modulus Interrelationships

We may now visualize more clearly the interrelationships between morphology (as revealed by electron microscopy) and the time- and temperature-dependent modulus. Both experiments yield information about phase domain formation and extent of molecular mixing, but in different ways.

The morphology reveals the size and shape of the domains; clearly this information cannot be deduced from the appearances of the glass–rubber transitions. Although some information of a qualitative nature about the extent of molecular mixing may be based on electron microscopy, the glass–rubber transition behavior tells much more. Inward shifts of the transitions, broadening and merging of the transitions, and increased damping between the two transitions are instrumental in deciding how much mixing takes place. Thus, while morphology controls the modulus–temperature behavior, knowledge of the morphological features that are easily detectable in the electron micrographs complements the information about mixing that is obtainable by means of modulus–temperature or –time experiments.

2.7. ANALOGY BETWEEN POLYMER BLENDS AND CRYSTALLINE HOMOPOLYMERS

In the earlier discussion of the properties of crystalline homopolymers in the bulk state (Chapter 1), it was shown that such polymers always contain some amorphous material, various segments of the same chain lying in crystalline or amorphous regimes (Krigbaum *et al.*, 1964). If the amorphous portion is above its T_g, i.e., is elastomeric, a direct analogy exists between such materials and polymer blends, blocks, and grafts, in which the formation of hard and soft domains is induced by the use of polymer components differing in chemical composition and inherent properties; thus, as pointed out by Haward (1970), polyolefins such as polyethylene or polypropylene may be regarded as a dispersion of crystals held together by rubber. On this basis there is an important correspondence in modulus and impact properties between rubber-reinforced plastics and polyolefins. The toughness of semicrystalline polymers such as polyethylene or nylon is common knowledge. An analogy also exists to particulate composites; see Chapter 12.

2.8. POLYMER BLEND CHRONOLOGY

To the surprise of many, synthetic polymeric materials of commerce date back to the 19th century and before. Because of the obvious utility of polymers, people used and studied these materials in many important ways even before Staudinger suggested the long-chain nature of polymers in 1920. Commercially important polymer blends in the modern sense date from after World War II. Kato's staining of blends in 1964 stands out as the most important research step in the elucidation of their two-phase morphologies. The following is a selective chronology for the development of polymer blends (Dunn and Melville, 1952; Platzer, 1971; Rodriguez, 1970, pp. 7, 8, 252, 391):

1912   First polymer blend prepared .
1933   First graft copolymer
1937   Styrene–butadiene rubber (Buna S)
1948   First commercial polymer blend (HiPS)
1948   ABS polymers
1952   First block copolymers
1960   Segmented polyurethanes (Spandex fibers)
1964 OsO_4 staining (Kato)
1965   Thermoplastic elastomers

APPENDIX A. COUNTERPART PHASE SEPARATION CHARACTERISTICS OF METALLIC ALLOYS AND INORGANIC GLASSES

An interesting analogy exists between metallic alloys, many inorganic glasses, and polymer blends. In fact, the latter have sometimes been referred to as polymer alloys (Rodriguez, 1970, pp. 252, 391). Both polymer blends and metallic alloys may exist as a single-phase system, a multiphase system, or in an intermediate state. The homogeneous alloys correspond in some important ways to the random copolymers, while the heterogeneous metallic alloys correspond more closely to the polymer blends. Both polymer blends and metallic alloys display improved properties with respect to their respective homopolymers or metals because of their multiphase nature, the combination of phases of differing properties resulting in important synergisms. Here we describe briefly a few of the important morphologies that exist in steel, one of the most important heterogeneous metallic alloys (Bramfitt and Marder, 1968; Keyser, 1968; van Vlack, 1964).

When carbon is dissolved in iron at temperatures around 2000°F at a level of about 0.8 % C and subsequently cooled to 1333°F, a eutectoid reaction takes place, two solid phases emerging. One phase is cementite or iron carbide, with a composition of Fe_3C. The other phase is ferrite, or α-iron, containing 0.025 % C. The resulting microstructure, known as pearlite, is shown in Figure 2.15. While ferrite is soft and ductile, cementite is very hard.

Another important two-phase structure in steel is observed in tempered martensite. The parent material, martensite, is a supersaturated single-phase

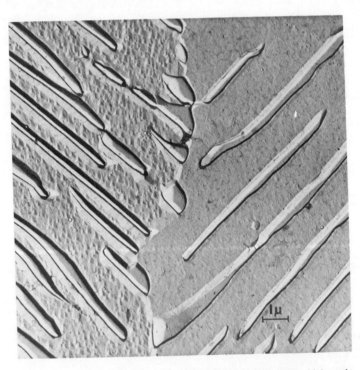

Figure 2.15. Surface replica of two adjacent pearlite colonies with a high-angle boundary. The microstructure is a lamellar mixture of ferrite (major phase) and carbide (minor phase). (Photo courtesy of B. L. Bramfitt and J. R. Gruver, Bethlehem Steel Corporation.)

solution of carbon in iron. With sufficient time at elevated temperatures (but below the eutectoid), the martensite transforms into ferrite and carbide,

$$\text{martensite} \rightarrow \text{ferrite} + \text{carbide}$$

In contrast to the pearlite structure, which is lamellar (Figure 2.15), tempered martensite contains the carbide particles as a spheroidal dispersed phase. While the tempered martensite is soft and tough, the parent martensite is hard and abrasion resistant.

In both martensite and pearlite two phases of contrasting mechanical properties are in close juxtaposition; such morphologies in steels and other alloys should be compared with those of the polymer blends discussed in the next several chapters. The major point is that toughening depends upon fine-structure detail for both metallic alloys and polymer blends. While the above analogy was developed for polymer blends, similar analogies can be developed for polymer composites, as in Chapters 11 and 12.

Many inorganic glasses also exhibit well-defined phase separation at the electron microscope level of magnification. The reader is referred to current research in Russia (Porai-Koshits, 1973) and the U.S. (Cahn and Charles, 1965; Halley et al., 1970; Rawson, 1967) for further details.

BIBLIOGRAPHY OF POLYMER BLEND SYMPOSIA

The following reports of symposia have appeared immediately before or during the preparation of this book. Although individual papers have been abstracted and included in the references at the end of this book, the following is set forth as a guide for readers wishing greater detail.

AGGARWAL, S. L., ed. (1970), *Block Polymers*, Plenum.

BRUINS, P. F., ed. (1970), *Polyblends and Composites* (J. Appl. Polym. Sci. Applied Polymer Symp. No. 15), Interscience.

BURKE, J. J., and WEISS, V., eds. (1973), *Block and Graft Copolymerization*, Syracuse.

CERESA, R. J., ed. (1973), *Block and Graft Copolymerization*, Vol. 1, Wiley–Interscience.

KESKKULA, H., ed. (1968), *Polymer Modifications of Rubbers and Plastics* (J. Appl. Polym. Sci. Applied Polymer Symp. No. 7), Interscience.

MOACANIN, J., HOLDEN, G., and TSCHOEGL, N. W., eds. (1969), *Block Copolymers* (*J. Polym. Sci.* **26C**), Interscience.

MOLAU, G. E., ed. (1971), *Colloidal and Morphological Behavior of Block and Graft Copolymers*, Plenum.

PLATZER, N. A. J., ed. (1971), *Multicomponent Polymer Systems* (Adv. Chem. Series No. 99), American Chemical Society.

PLATZER, N. A. J., ed. (1975), *Copolymers, Polyblends, and Composites* (Adv. Chem. Series No. 142), American Chemical Society.

SPERLING, L. H., ed. (1974), *Recent Advances in Polymer Blends, Grafts, and Blocks*, Plenum.

3

Rubber-Toughened Plastics

One of the oldest, and certainly one of the most important, methods of preparing polymer mixtures involves blending techniques. This chapter will discuss mechanical blending, graft-type blending, and related schemes for producing compositions that have a rubber phase dispersed within a plastic matrix. Some of these materials have become famous for their impact resistance, while others, although interesting morphologically, exhibit little if any toughening.

There are three principal ways of preparing blends with toughness or high impact resistance. The original method involved blending by mechanical techniques (method 1). Blends of poly(vinyl chloride) with poly(butadiene-co-acrylonitrile) are presently manufactured by a variation of this technique. While high-impact polystyrene is also produced by means of mechanical blending with linear polybutadiene, currently a solution-graft copolymer technique (method 2) is more often employed, in which the elastomer component is first dissolved in the styrene monomer and then the latter is polymerized with agitation. The technologically important ABS (acrylonitrile–butadiene–styrene) polymers are produced by emulsion polymerization (method 3), a variation of the solution graft copolymer technique in which the plastic component is polymerized onto a seed-latex particle of the elastomer component.

A most important concept emphasized throughout this chapter is that the mode of synthesis controls morphology, which in turn determines physical and mechanical behavior. Thus there is a direct one-to-one relationship between the chemistry of these materials and such properties as impact strength. The first part of this chapter considers synthesis and morphology, while several subsequent sections focus on the physical and mechanical behavior of the resulting blends and grafts. Finally, several current theories of toughening in blends and grafts are considered, and optical and degradation behavior is reviewed. Portions of this subject were recently reviewed by Amos (1974).

3.1. SYNTHESIS AND MORPHOLOGY

3.1.1. Impact-Resistant Polystyrene

The simplest method of polyblending involves equipment such as rolls or extruders, which can effect the mechanical blending of the two polymeric components in the molten state (Matsuo, 1968). High-impact polystyrene (HiPS) is an important example of a polyblend made by this technique. Such materials commonly contain 5–20 % of rubber, usually polybutadiene, dispersed in a polystyrene matrix. As shown in Figure 3.1, electron microscopy studies on specimens stained with osmium tetroxide reveal well-defined, irregular rubber particles (1–10 μm in diameter) dispersed in the polystyrene matrix. The elastomer domains appear dark because the osmium tetroxide stains the elastomer preferentially (see Section 2.4).

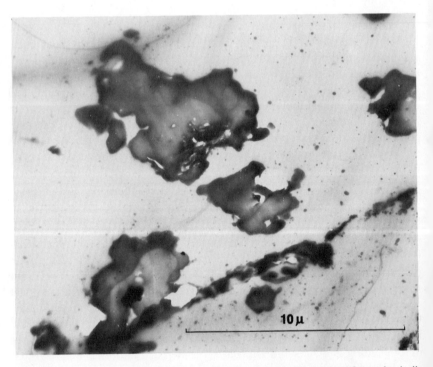

Figure 3.1. Electron photomicrograph (transmission) of an ultrathin section of a mechanically blended impact-resistant polystyrene, showing well-defined, but irregular, rubber particles. (Matsuo, 1968.)

While the impact resistance of the mechanical blends is clearly superior to that of the parent polystyrene, they have two important deficiencies that cause them to be inefficient. First, due to the high viscosity of the melts, the problem of attaining intimate mixing cannot be entirely overcome. As a result, the dispersed phase maintains a relatively large particle size, as shown in Figure 3.1. Second, the two phases are bonded together only by weak van der Waals forces, so that the material as a whole exhibits poor cohesion.

3.1.1.1. Solution-Type Graft Copolymers

Still higher levels of toughness in rubber-modified polystyrene can be obtained by the use of graft polymerization techniques, which give both more intimate mixing and better interfacial bonding than is possible by mechanical blending. In effect, the efficiency of the rubbery component in toughening the matrix is considerably increased. In general, the graft polymers are prepared by bulk polymerization processes, in which the rubber component is first dissolved in the styrene monomer to a level of 5–10% by weight. For such bulk processes, efficient mixing is required during polymerization of the styrene monomer until the mass has a fairly high viscosity, that is, until a conversion of about 50% is reached (Molau and Keskkula, 1968). During this time, the polybutadiene, while soluble in the pure styrene *monomer*, becomes insoluble as the polystyrene concentration increases and undergoes phase separation and later phase inversion. When this occurs, both polymer phases contain extensive quantities of styrene monomer.

After polymerization is complete, a complex cellular structure exists, as illustrated in Figure 3.2 (see also Figure 2.4). It is clear that polystyrene is the grossly continuous phase. However, the styrene monomer that remained within the polybutadiene phase after phase separation forms discontinuous droplets upon polymerization within the polybutadiene phase. It is believed that most of the grafted polymer formed is located at the interfacial regions of the cellular structures (see Section 3.1.1.3). If mixing is not carried out during the early stages of polymerization, a gross honeycomb-like or cellular structure may result, where the rubber remains as the continuous phase. This is true even though the rubber is present at a level of only 5–10% (Keskkula and Traylor, 1967). A product of this type is usually softer and lower in impact strength than the phase-inverted materials just discussed. The morphology of a typical nonagitated product is shown in Figure 3.3.

Upon comparing Figure 3.3 with Figure 3.2, the advantages of osmium tetroxide staining, coupled with transmission electron microscopy, become clear. It is easier to distinguish phases and to determine morphological

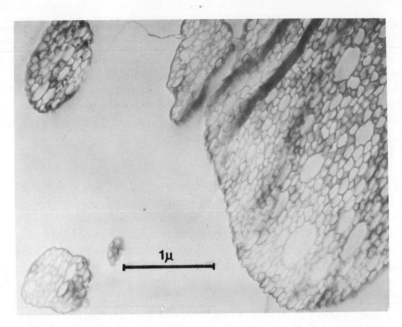

Figure 3.2. Electron photomicrograph (transmission) of an ultrathin section of a graft-type high-impact polystyrene made by a bulk polymerization process. Note the cellular structure within the polybutadiene phase, where polystyrene forms a discontinuous phase. Large, white areas correspond to the continuous matrix of polystyrene homopolymer. (From Huelck and Covitch, 1971.)

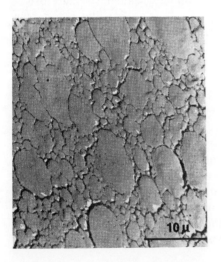

Figure 3.3. Electron micrograph (replica) of a graft-type high-impact polystyrene with polybutadiene as the rubbery component (Keskkula and Traylor, 1967). In the absence of agitation, phase inversion does not occur, and an interwoven cellular structure results, with polybutadiene remaining as the continuous phase. The specimen was prepared for electron microscopy by exposing a polished surface to isopropanol vapor, which preferentially swells the polystyrene phase; a double replication technique was then used. The reader should compare the results obtained by this technique to results obtained using thin-section transmission techniques (see, for example, Figure 3.2).

detail. Thus, for materials having residual double bonds, transmission electron microscopy after osmium tetroxide staining has become popular. In many two-polymer combinations, however, neither polymer can be directly stained with osmium tetroxide and other techniques must be employed, such as incorporation of a trace of a diene (Chapter 8). In general, there is a need for improved staining techniques suitable for use with saturated polymers.

3.1.1.2. Phase Inversion

The phenomenon of the phase inversion itself has been investigated by Molau (1965), who studied the graft-type polyblending of styrene with

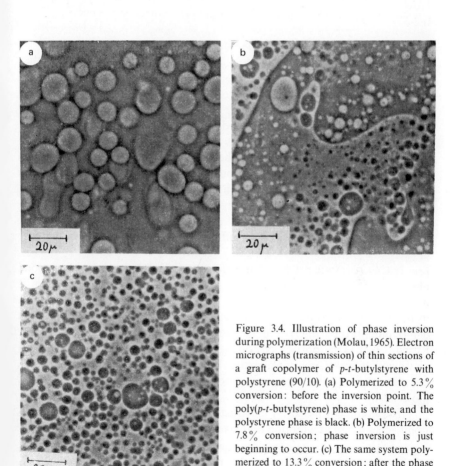

Figure 3.4. Illustration of phase inversion during polymerization (Molau, 1965). Electron micrographs (transmission) of thin sections of a graft copolymer of p-t-butylstyrene with polystyrene (90/10). (a) Polymerized to 5.3% conversion: before the inversion point. The poly(p-t-butylstyrene) phase is white, and the polystyrene phase is black. (b) Polymerized to 7.8% conversion; phase inversion is just beginning to occur. (c) The same system polymerized to 13.3% conversion; after the phase inversion point.

several polymers. To illustrate this phenomenon, he polymerized p-tertbutyl-styrene in the presence of 10% polystyrene. Figure 3.4 illustrates the morphology of the graft polymer before, at, and after the inversion point. It was suggested that the graft copolymer formed by chain transfer during the reaction behaves as an emulsifier; on this basis, the products were designated as polymeric oil-in-oil (POO) emulsions (Molau, 1965).

Of course, the exact point of phase separation depends upon the choice of the polymer pair, the molecular weight control, the temperature, etc. However, for many important systems, phase separation occurs in the region of 5–15% polymerization of polymer II. Phase inversion, brought about by continued mixing, should be made to occur shortly thereafter while the viscosity is still relatively low. A conversion of 30–35% is usually considered adequate for completion of the phase inversion process.

3.1.1.3. Grafting vs. Mechanical Entrapment

The composition of graft copolymers* is often quite different from that considered classically. Originally, it was thought that about 50% or more of the molecules were chemically grafted together in solution-type graft copolymers (Battaerd and Tregear, 1967; Burlant and Hoffman, 1960; Ceresa, 1962; Chapiro, 1962; Fels and Huang, 1970). It is now believed that the products of graft polymerizations contain many fewer graft points than originally believed to be the case. The older conclusion was based on two observations: (1) By the use of standard separation techniques based on extraction and precipitation, it was not possible to separate the two components, and (2) the chain transfer constants (Brandrup and Immergut, 1966, Section II, p. 77) for the transfer of growing chains of polymer II to the preexisting chain of polymer I were sufficiently large to suggest the occurrence of extensive grafting.

Let us reexamine the evidence (Rosen, 1973). Considering the second observation first, it is seen that the argument assumes that the system is homogeneous at all times. Clearly, however, this assumption is invalid from the onset of phase separation onward; in fact, only a modest fraction of monomer II is polymerized in the presence of polymer I (see Section 3.1.1.1; Rosen, 1973). (That portion of polymer II synthesized within the polymer I phase will, however, contain a greater than expected number of grafted sites, because of the close proximity and high concentration of polymer I molecules.)

The argument based on inextricability also assumes that the system is homogeneous, an assumption that was quite reasonable before the existence

---

* As usually used, the term "graft copolymer" refers to the product of a polymerization technique intended to produce a truly grafted molecule. The term does not necessarily imply that grafting did in fact occur to any significant degree.

of a phase-within-a-phase-within-a-phase structure (Figure 3.2) was demonstrated. However, we now know that the dispersed phase is largely mechanically occluded or entrapped (Wagner and Cotter, 1971). The polystyrene, being insoluble in the polybutadiene, cannot pass through the cellular structure of polybutadiene, even when swelled or dilated to a considerable extent. When ultrasonic vibration was employed to disrupt the cellular structure (Wagner and Cotter, 1971), however, the occluded polystyrene could then be extracted.

The polyblends in question may really be considered to comprise three components: a plastic phase, much of which is far removed from any elastomer, but partly contained within the rubber phase; the rubber phase itself; and the graft copolymer component. One might speculate that much of this third component, totalling no more than a few percent in many materials (Wagner and Cotter, 1971), is largely formed within the monomer-swollen elastomer phase after the initial phase separation has taken place. The importance of the graft copolymer component should be emphasized, however, because it is believed that the grafted molecules tend to migrate to the phase interface and bond the two principal phases together. The graft copolymer component serves to increase the compatibility between the two phases, thus increasing mutual wetting. In other sections of this monograph the formation of other types of grafted structures (including heavily grafted copolymers, Chapter 7) will be considered. Section 7.1 contains a further discussion of the problems associated with quantitatively estimating the degree of grafting.

3.1.2. ABS Resins

Like the rubber–polystyrene blends, ABS (acrylonitrile–butadiene–styrene) resins are two-phase systems in which an elastomeric phase is finely dispersed in a glassy matrix of a styrene–acrylonitrile (SAN) resin. The rubbery phase usually consists of polybutadiene, NBR (acrylonitrile–butadiene rubber), or a random copolymer of styrene and butadiene (SBR), which may itself be modified by grafting of, for example, acrylonitrile, in order to alter its compatibility with the matrix. A wide range of resins may be produced that combine good resistance to heat and chemicals with ease of processing, stiffness, and toughness. Indeed, the ABS resins are considered to be true engineering plastics, suitable for many applications requiring high levels of mechanical performance and durability (Baer, 1962a,b,c; Bovey et al., 1955; Duck, 1970; Grancio, 1971; Kato, 1968; Matsuo et al., 1969a).

The simplest type of ABS is made by mechanically blending the two polymers (Kato, 1968). In this case the two copolymers always have a certain fraction of one mer, styrene, in common to improve compatibility. ABS's produced by mechanical blending are sometimes known as type B resins (Kato, 1968) and are generally considered to be acrylonitrile-containing analogs of the HiPS materials prepared in a similar manner (see Section 3.1.1).

3.1.2.1. Emulsion Polymerization

The larger portion of ABS plastic now manufactured is prepared using an emulsion polymerization process, which yields an aqueous dispersion of colloid-sized polymer particles ranging from 1000 to 5000 Å in diameter (Grancio, 1971). In this case PB or SBR is polymerized first as a seed latex, and monomeric acrylonitrile (A) and styrene (S) subsequently added to continue polymerization within the seed particles. The key process lies in the grafting of a significant portion of the growing AS random copolymer radicals onto the double bonds of the existing elastomeric component. The grafting between the plastic and elastomer components lends compatibility to the system, resulting in a favorable state of dispersion, and also bonds the phases together. The graft-type ABS produced in this manner is sometimes called type G resin (see Figure 3.5).

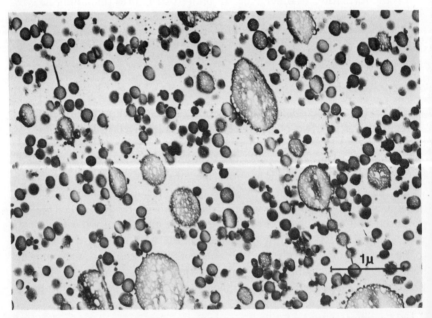

Figure 3.5. Electron micrograph of an ultrathin section of a type G ABS resin. Note the cellular structure within the rubbery particles. (Grancio, 1971.)

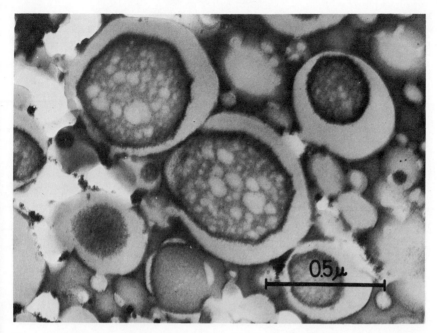

Figure 3.6. Electron micrograph of an ultrathin section of a type G ABS latex specimen, showing the latex particles completely surrounded with AS copolymer. (Matsuo *et al.*, 1969a.)

3.1.2.2. Structure of the Latex Grafts

Discussion of the detailed structure of the graft-type polyblend latex particle requires amplification. In the formation of the ABS type G resin, part of the AS copolymer forms a shell around the seed latex (Kato, 1968), as shown in Figure 3.6. As with other types of graft copolymers, some monomer dissolves within the seed latex. Upon polymerization, the second monomer mix phase-separates to yield the complex inner morphology observed.* After coagulation, the glassy AS polymer forms the matrix, while the portion occluded within the latex particles remains within the rubber phase (Figure 3.7).

First considerations led to the belief that the core–shell structure originated in the gross incompatibility of the AS copolymer with the PB. This, however, need not be the case. Grancio and Williams (1970) and Keusch and Williams (1973) showed that a core–shell structure can be formed with a styrene core and a styrene–butadiene shell containing only

* This morphology was appropriately nicknamed the "salami structure" by M. Matsuo (private communication).

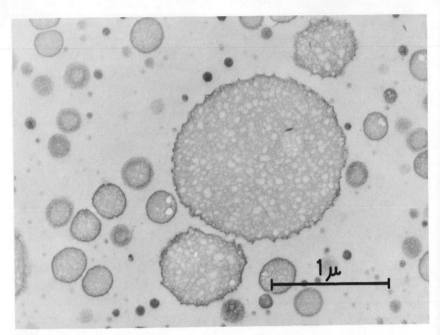

Figure 3.7. Electron micrograph of a thin section of ABS type G polyblend after coagulation (stained with osmium tetroxide). Note the distinct salami structure due to the occlusion of AS copolymer within the rubber phase. (Matsuo *et al.*, 1969a.)

enough butadiene (about 1 %) to permit staining. The implications of the core–shell structure obtained will be discussed further in Section 13.4.

3.1.3. Origin of the Cell Structure

Let us examine the origin of the cell structure that arises in solution-graft-copolymer blends such as HiPS, ABS, and similar materials. As discussed previously, polymer II separates from polymer I at an early stage of the reaction, causing the formation of two phases, both highly swollen with monomer II. Besides polymer–polymer incompatibility, cell formation depends upon the relative solvating power of polymers I and II for remaining monomer II. If monomer II prefers polymer II over polymer I (the usual case), the original phase-separated domains of polymer II will grow selectively, leading to cell formation. Otherwise a finer structure will develop (see Section 8.2).

To aid in comparing relative compatibility and solvating power of monomers and polymers, the solubility parameter δ, a measure of attractive forces between molecules (Tobolsky, 1960), may be used. This parameter

Table 3.1
Solubility Parameters of Monomers and Polymers[a]

Monomer	$\delta \, (cal/cm^3)^{1/2}$	
	Monomer	Polymer
1,3-Butadiene	7.1	8.1
Styrene	9.3	9.1
Ethyl acrylate	8.6	9.4
Methyl methacrylate	8.8	9.1
Acrylonitrile	10.5	12.5
Vinyl chloride	7.8	9.4

[a] Brandup and Immergut (1966).

is related to the molar energy of vaporization E_v and the molar volume $\tilde{v}_1$ of the liquid as follows:

$$\delta = (\Delta E_v / \tilde{v}_1)^{1/2} \tag{3.1}$$

In general, if two different polymers or monomers have similar values of δ, they will tend to be mutually soluble.

Using the case of HiPS formation as an example, it can be seen from Table 3.1 that the solubility parameter of styrene is significantly closer to that of polystyrene than to that of polybutadiene. Predictably, then, styrene will swell the nascent polystyrene in preference to polybutadiene, and the locus of polymerization will center increasingly in the polystyrene phase as polymerization progresses.

3.1.4. Poly(vinyl chloride) Blends

Poly(vinyl chloride) (PVC) homopolymer is a stiff, rather brittle plastic with a glass temperature of about 80°C. While somewhat more ductile than polystyrene homopolymer, it is still important to blend PVC with elastomer systems to improve toughness. For example, methyl methacrylate–butadiene–styrene (MBS) elastomers can impart impact resistance and also optical clarity (see Section 3.3). ABS resins (see Section 3.1.2) are also frequently employed for this purpose. Another of the more important mechanical blends of elastomeric with plastic resins is based on poly(vinyl chloride) as the plastic component, and random copolymers of butadiene and acrylonitrile (AN) as the elastomer (Matsuo, 1968). On incorporation of this elastomeric phase, PVC, which is ordinarily a stiff, brittle plastic, can be toughened greatly. A nonpolar homopolymer rubber such as polybutadiene (PB) is incompatible with the polar PVC. Indeed, electron microscopy shows

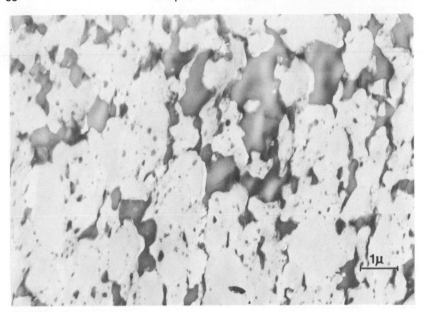

Figure 3.8. Electron micrograph of an ultrathin section of a PVC/PB blend (100/15) showing irregularly shaped, but well-defined, rubber particles (dark areas) dispersed in the PVC matrix. This system is incompatible. (Matsuo, 1968.)

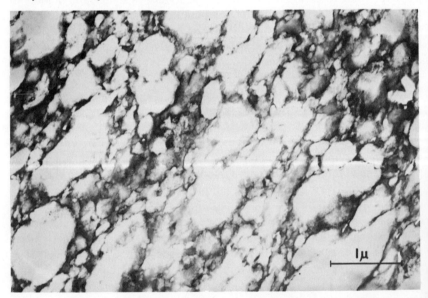

Figure 3.9. Electron micrograph of a thin section of a PVC/NBR-20 blend; the NBR contains 20% acrylonitrile. The rubber phase in this semicompatible mixture appears to form a network structure extending throughout the PVC matrix. (Matsuo, 1968.)

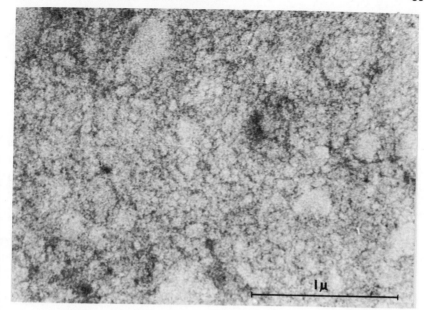

Figure 3.10. Electron micrograph of a thin section of PVC/NBR-40 blend; the NBR contains 40% acrylonitrile. While two clearly defined phases no longer exist, microheterogeneity still remains in the system, with phase domains of the order of 100 Å. Comparison of this figure with Figures 3.8 and 3.9 shows the progressive increase in molecular mixing. (Matsuo, 1968.)

well-defined two-phase systems for simple blends of PVC with PB, as shown in Figure 3.8. However, the introduction of a small amount of AN as a comonomer in the PB component results in the more polar, and more compatible, acrylonitrile–butadiene rubber (NBR). As shown dramatically by Matsuo (1968) (Figure 3.9), the addition of 20% AN to PB (NBR-20) causes the phase boundaries to become less distinct—an indication of enhanced compatibility.

Addition of AN to a level of 40% (NBR-40) destroys the phase boundaries entirely, resulting in the microheterogeneous system shown in Figure 3.10. The phase domains (~100 Å) shown in Figure 3.10 are clearly smaller than the polymer molecules themselves, yet the material is not totally compatible. Only a few cases are known in which the phase division in blends is so fine: such cases include the IPN's discussed in Chapter 8 and the poly(2,6-dimethyl phenylene oxide)/polystyrene blend described in Section 9.7.1.

3.1.5. Mixed Latex Blends

In addition to mechanical blending and graft copolymers, the blending together of two different types of latexes is sometimes useful, because of the low viscosity of the components (Merz *et al.*, 1956; Rosen, 1967). Simple methods of preparing such blends involve mixing the two dispersions, and film formation, coagulation, or spray-drying. The result is an intimate dispersion without the need for large amounts of mechanical energy for mixing. One drawback to this method, which is characteristic of most products of emulsion polymerization, is the difficulty of removing the emulsifier.

Although ideally one may imagine the resulting morphology to be a three-dimensional mosaic of the two polymer species with domain dimensions controlled by the latex particle diameter, Merz *et al.* (1956) showed that some particles were much larger than expected. As shown in Figure 3.11, many rubber particles appear to have coalesced sufficiently to form a large loose aggregate (Kato, 1968); the exact state of aggregation is, of course, controlled by the details of mixing.

If the latex components are crosslinked after the bulk product is formed, the material is known as an interpenetrating elastomer network (IEN), discussed in Section 8.5.

3.2. PHYSICAL AND MECHANICAL BEHAVIOR OF POLYBLENDS

The previous section considered some of the more important methods of synthesizing polyblends, together with their resulting morphology as revealed by electron microscopy. While electron microscopy most clearly shows the size and structure of the domains, dynamic mechanical spectroscopy better reveals the actual extent of molecular mixing in the blends. Shifts and broadenings of the glass transitions in both polymers are brought about by the presence of both molecular species as nearest neighbors.

3.2.1. The Effect of Compatibility on Transition Behavior

The temperature dependence of the dynamic storage modulus E' and the dynamic loss modulus E'' for polymer blends as a function of composition yields particular insight into the structure of semicompatible polymers (Matsuo *et al.*, 1969a). A good example is the more or less compatible PVC/NBR system studied by Matsuo (1968); in this system the effects of changing the concentration of acrylonitrile, and hence the degree of phase compatibility, were clearly seen in Figures 3.8–3.10. Figures 3.12–3.14 show the corresponding results of a modulus–temperature study by Matsuo *et al.* (1969a). The major features of such blends whose constituents vary in compatibility may be summarized as follows.

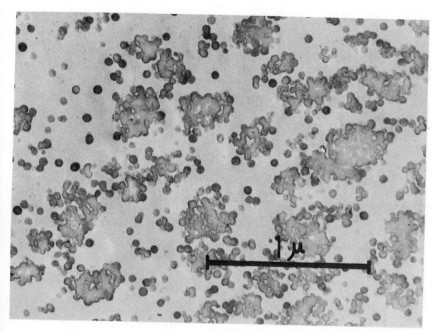

Figure 3.11. Electron micrograph of an ultrathin section of a type B ABS resin, stained with osmium tetroxide. The rubber particles are coalesced to some extent. (Kato 1968.)

For incompatible polymer pairs, each polymer has its own characteristic glass-transition temperature. Thus in Figure 3.12 the T_g for the minor component, PBD, in PVC/PBD is observed at temperatures in the range of $-100°C$, depending on PBD concentration. While only a slight decrease in E' is observed in this temperature range, values of E'' exhibit distinct peaks when PBD is present. At $+86°C$ the continuous phase, PVC, undergoes its glass transition, resulting in a much larger change in E' as the material softens from a plastic to viscoelastic mass, and E'' exhibits a separate loss peak.

Similarly, in the blends of PVC and NBR-20 (Figures 3.13), the presence of an E'' peak corresponding to the T_g of NBR-20 at $-40°C$ is evident; however, its upper tail overlaps with that of the PVC transition. This could indicate that although the rubber phase exists independently in this semi-compatible system, the interaction between the two phases is marked. The values of E' do not exhibit two transitions; instead the glass transition is broadened and shifted lower in temperature.

The PVC/NBR-40 system shows a third distinct type of behavior, somewhat reminiscent of the behavior of random copolymer compositions

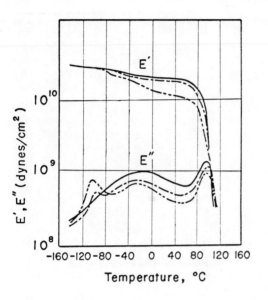

Figure 3.12. Temperature dependence of dynamic modulus E' and dynamic loss modulus E'' for PVC/PBD blends (an incompatible system): (—) 100/0; (– – –) 100/5; (– – – –) 100/15. Note the two distinct loss peaks near $-100°C$; these correspond to the PBD phase. This transition does not appear clearly in the E' curve because the PBD is the dispersed phase. (Matsuo *et al.*, 1969a.)

(Figure 3.14). Values of E' exhibit only one transition, which systematically moves down the temperature scale as the NBR-40 content is increased. Further, the glass transition for the NBR-40 blends assumes a sharpness characteristic of homopolymers or random copolymers. Clearly an increase in the AN content of NBR results in an increased compatibility with PVC. However, as shown earlier by electron microscopy, a definite degree of phase separation still exists for all of these compositions.

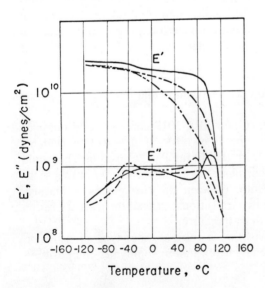

Figure 3.13. Temperature dependence of dynamic modulus E' and dynamic loss modulus E'' for PVC/NBR-20 blends (a semicompatible system): (—) 100/0; (– – –) 100/15; (– – – –) 100/25. Such behavior is preferred in order to achieve high impact strength. (Matsuo *et al.*, 1969a.)

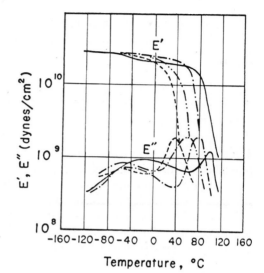

Figure 3.14. Temperature dependence of dynamic modulus E' and dynamic loss modulus E'' for PVC/NBR-40 blends (a nearly compatible system): (—) 100/0; (– – –) 100/10; (– – – –)100/25;(- - -)100/50. (Matsuo et al., 1969a.)

Idealizing slightly, we can imagine that a blend of two incompatible homopolymers results in two sharply defined phases, each with its own glass transition temperature. As the two polymers attract each other more so that extensive (but incomplete) mixing ensues, the two glass transitions shift inward toward each other and broaden. A point is reached where one very broad transition is noted, which may span the temperature range between the two homopolymer transitions. On the other hand, if the mixing is complete on the molecular level and a true solution is formed, the (now) single glass transition tends to revert to a narrow transition again, at a temperature controlled by the weight fraction of the components.

3.2.2. Impact Resistance and Deformation

As one might intuitively expect, the incorporation of rubber particles within the matrix of brittle plastics enormously improves their impact resistance. Indeed, the impact resistance imparted by the rubber is the principal reason for its incorporation (Rosen, 1967) in rubber–plastic blends and grafts. Toughening in such polymers is also observed under other loading conditions, such as simple low-rate stress–strain deformation and fatigue. It is believed that several deformation mechanisms are important in all such cases, though their relative importance may depend on the polymer and on the nature of the loading.

In this section, stress–strain behavior of typical rubber–plastic systems at low and high (impact) rates of static testing and under low-frequency cyclic loads (fatigue) are described; probable mechanisms of toughening, e.g., controlled crazing, are discussed in Section 3.2.3.1 (see also Sections 1.16 and 12.1.2.4 for related material).

3.2.2.1. Impact Behavior

As mentioned above, the degree of impact resistance obtained depends both on the quantity of rubber incorporated and the method of forming the polyblend. The effect of these two variables on the Izod impact strength of polybutadiene/polystyrene blends is shown in Figure 3.15. When the polymers are mechanically blended together, or when styrene is graft-polymerized in solution onto the dissolved rubber, an improvement in impact strength is observed, the greater improvement being noted for the graft polymer. In general, an optimum rubber concentration and phase domain size exists, the values depending on the polymer and rubber concerned. If, on the other hand, a polystyrene latex is blended with the relatively incompatible polybutadiene latex, only a modest degree of reinforcement is obtained. In such cases, the exact level of toughening attained with the latex blends depends upon the fineness of dispersion, with partly aggregated materials preferred (Section 3.1.5.). (It is interesting that although most mixed latex blends are relatively nonreinforcing, the corresponding dispersion of latex material within a concrete mix may result in some strengthening, as discussed in Section 11.3.2.) In commercial products, levels of rubber usually do not exceed 10–15%, since the addition of greater quantities results in an undesirable degree of softening.

The importance of graft-type bonding and the finer, more complex morphology developed in graft copolymers should be emphasized. Table 3.2

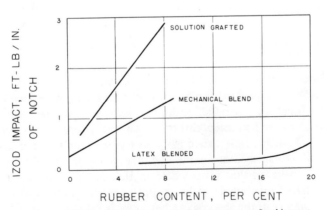

Figure 3.15. Effect of blend type and rubber content on Izod impact resistance of polybutadiene/polystyrene blends. (Haward and Mann, 1964; Lannon and Hoskins, 1965.)

Table 3.2
Izod Impact Strengths of Several Plastics[a,b]

Polymer	Impact strength
Polystyrene	0.25–0.4
High-impact polystyrene	0.5–4
ABS plastics	1–8
Epoxy resin (no filler)	0.2–1.0
Epoxy resin (glass-fiber-filled)	10–30
Cellulose acetate	0.4–5.2
Poly(methyl methacrylate)	0.3–0.5
Phenol–formaldehyde	0.20–0.36
Poly(vinyl chloride)	0.4–1
High-impact poly(vinyl chloride)	10–30

[a] *Modern Plastics Encyclopedia* (1973), Lannon (1967).
[b] ASTM test D256 was followed, and the values reported have the units ft-lb/in. of notch.

lists impact strength data for several homopolymers and blends (Lannon, 1967; Modern Plastics Encyclopedia, 1973). Materials having values of impact strength about 1.0 ft-lb/in. of notch or greater are generally classified as impact resistant. A major factor in impact strength appears to be the ability to yield and undergo cold-drawing at the high rates of loading concerned. Vincent (1967) has shown a correlation between tensile yield stress and impact strength for a wide variety of polymers, and Petrich (1972) has found a similar correlation for rubber-modified PVC, with a limiting maximum yield stress of 7000–8000 psi, above which high impact strength is not obtained. Such a correlation is also supported by measurements of stress–strain behavior during actual impact tests (using an impact tester equipped with a suitable transducer to measure force as a function of time). Thus, for example, Bucknall et al. (Bucknall, 1967a,b; Bucknall and Smith, 1965; Bucknall and Street, 1967) noted three types of fracture behavior in HiPS tested at temperatures between −100 and +70°C (Figure 3.16):

Type I.   No yielding and no stress-whitening; typical brittle fracture (−100 to −68°C); impact strength <0.5 ft-lb/in. of notch.

Type II.   Some yielding and stress-whitening during crack initiation; brittle fracture during crack propagation stage (−68 to +10°C); impact strength rises with temperature, from 0.5 to 1.5 ft-lb/in. of notch.

Type III.   Considerable yielding and stress-whitening throughout the fracture process; rather ductile failure (+10 to +70°C); strength rises with temperature, from 1.5 to 4.5 ft-lb/in. of notch.

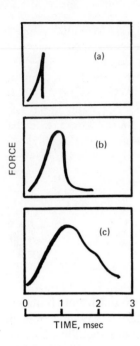

Figure 3.16. Schematic indicating three types of fracture during impact testing of HiPS at different temperatures: (a) Type I ($-70°C$), brittle throughout, (b) type II ($+10°C$), some yielding and ductility during crack initiation, and (c) type III ($+50°C$) general yielding and ductility. (After Bucknall, 1967a,b.)

It was proposed that the particular type of fracture behavior observed depends on the ability of the rubber to relax at the temperature and time scale of the experiment. Thus, in type I fracture, the rubber cannot relax at any stage; in type II, the rubber can relax during crack initiation but not during the fast crack propagation stage; and in type III, the rubber can relax throughout the entire fracture process. Clearly, the effectiveness of a rubber in toughening a brittle matrix must depend on its T_g; the lower the T_g, the greater the rate of loading that can be accommodated without vitrification of the rubber (see also Section 3.2.3.2). Crazing (Section 3.2.3.1) is also associated with the presence of the rubber, and may be expected to occur to an extent paralleling the "rubberiness" of the rubber, as measured by its ability to relax under the conditions of loading used.

The three types of force–time behavior noted for HiPS fractured at different temperatures also apply to other polymer blends and grafts, where values of the impact strength (or fracture energy) were measured as a function of temperature. Such behavior has been observed by Bucknall and Street (1967) not only for ABS (Figure 3.17), but also for rubber-modified PVC, HiPS, and a methacrylate–butadiene–styrene (MBS) copolymer. Not surprisingly, the concentration of rubber is important with respect to both the absolute value of impact strength (Figures 3.16 and 3.17) and the type of fracture observed. As shown in Figure 3.16, type II fracture occurs at lower temperatures when the rubber content is higher; at 20°C, such fracture is observed only at rubber concentrations higher than about 10%.

Although impact behavior is of great technological importance, fundamental studies of fracture at high rates of loading are difficult to conduct. Most studies of yielding and toughness in rubber-modified plastics have therefore been concerned with much lower rates of loading, with the assumption that the basic mechanisms involved will be similar in all cases, though perhaps present to different extents. The question of low-strain-rate tensile and creep behavior is discussed in more detail below, along with possible mechanisms of toughening.

3.2.2.2. Tensile and Creep Behavior

The total energy required to break brittle plastics such as polystyrene in a nonimpact tensile mode is also greatly increased by polyblending (Amos, 1974; Bergen, 1968; Bucknall, 1967a,b; Bucknall and Smith, 1965; Bucknall and Street, 1967; Kambour, 1973; Newman and Strella, 1965). A study of stress–strain behavior reveals that, as shown in Figure 3.18, the rubber-containing materials undergo not only yielding, but also a high degree of drawing prior to ultimate failure. The area under the curve, of course, is a direct measure of the energy to break, and associates the ability to yield and cold-draw with toughness in polyblends (Bergen, 1968). Such an increase in toughness is also found in measurements of apparent fracture energy γ; Broutman and Kobayashi (1971) have reported a 100-fold increase

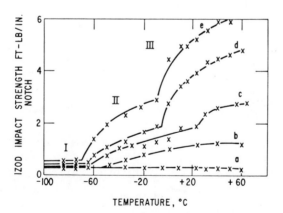

Figure 3.17. Izod impact strength as function of temperature for ABS polymers of different rubber contents: (a) 0%, (b) 6%, (c) 10%, (d) 14%, (e) 20%. (Bucknall and Street, 1967.)

in γ for poly(methyl methacrylate) on incorporation of a rubbery phase. (For the correlation between yielding and impact behavior see Section 3.2.2.1.)

Several other observations can be made. First, both HiPS and ABS resins exhibit a significant degree of stress-whitening when strained to even small extents. The onset of stress-whitening appears to be related to the occurrence of yielding and drawing, and has been shown in ABS and HiPS to be a direct consequence of the initiation of small crazes (Section 3.2.3.1) in the matrix adjacent to the rubber particles (Bucknall, 1967a,b; Bucknall and Smith, 1965; Bucknall and Street, 1967; Haward, 1970; Kambour, 1973). Stress-whitening is also usually seen in rubber-modified PVC (Petrich, 1972), though not necessarily in specimens of PVC that contain no rubbery phase, but which exhibit a high impact strength due to elevation of the temperature. In the latter case, in which crazing appears to be uncommon, Petrich (1972) has suggested that differences in refractive index between the rubber and drawn (but not crazed) matrix may be responsible for enough scattering of light to develop an apparent whitening. Cavitation in the matrix or interfacial failure may play a role, as proposed by Schmitt (1968); the former seems unlikely since PVC does not dilate readily on straining, but the latter would be possible for less than well-bonded systems.

Second, the ductility normally evident in HiPS was shown by Biglione *et al.* (1969) to be essentially completely suppressed by the application of

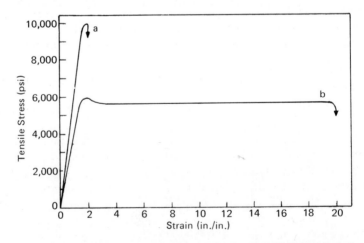

Figure 3.18. Tensile stress–strain curves at 23°C for (a) 70/30 SAN resin and (b) a mechanical blend of 70/30 SAN with 25 wt % NBR. Although the final breaking stress is lower, the polyblend requires much more work to break than the simple SAN copolymer. (Bergen, 1968.)

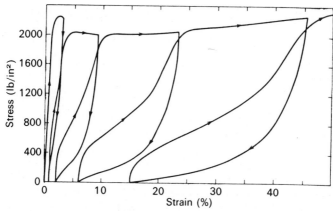

Figure 3.19. Cyclic stress–strain behavior of HiPS (Bucknall, 1967a,b) Note the development of a second yield stress after the first deformation cycle, and the large increase in hysteresis.

a 2-kbar hydrostatic pressure, while the ductility of ABS and rubber-modified poly(methyl methacrylate) was only partially reduced. Since hydrostatic pressure will tend to suppress a dilatational phenomenon such as crazing, but not yielding in shear, it may be concluded that both shear and crazing phenomena may be involved in the deformation of rubber-toughened plastics, though not to the same extent in all cases and probably not to the same extent when under impact loading (Bucknall, 1967a,b, 1972; Bucknall and Drinkwater, 1973; Bucknall et al., 1972b). (For a more complete discussion of crazing and shear effects, see Section 3.2.2.4.)

Third, in HiPS, the density is known to decrease by about 8 % on straining (Merz et al., 1956) and Poisson's ratio to decrease from 0.33 to 0.15 (Nielsen, 1965). On the other hand, the density of PVC and rubber-modified PVC is a complex function of strain (Petrich, 1972); in the former case, density first increases and then levels off, while in the latter density tends to decrease after an initial rise. Again, the diversity of density effects presumably reflects the various balances achieved between crazing and shear effects in different polymers under different loading conditions.

The importance of crazing in the development of yielding in HiPS can be illustrated dramatically by cyclic stress–strain measurements. As shown in Figure 3.19, Bucknall (1967a,b) conducted a sequence of stress–strain cycles on a single specimen, the stress in the first cycle being high enough to induce yielding and stress-whitening. On successive cycles, the polymer becomes softened, progressively larger hysteresis loops are observed, and a second, lower yield stress appears, with a magnitude which decreases as cycling proceeds. The development of crazing is associated with the upper yield point, and yielding of already crazed material with the lower (Kambour, 1972); in fact, the initial portion of the hysteresis curve resembles a typical curve obtained by Kambour and Kopp (1969) for deformation of a single craze. By analysis of the hysteresis curves, Bucknall (1969) estimated the elastic, viscous, and plastic components of the work per cycle. The work of plastic deformation in crazing appeared to be the major component (65 %), while viscous energy (as indicated by the hysteresis loop) and recoverable elastic energy amounted to 23 % and 15 %, respectively. [These estimates vary quantitatively from estimates for crazes in PMMA; it is not known if the variance reflects a real difference or uncertainty in the approximations used in calculation (Kambour, 1972).]

Creep measurements can be helpful in determining the relative importance of shear yielding vs. crazing* in a polymer (Bucknall, 1972). An extensive and elegant series of studies by Bucknall and others (Bucknall, 1972; Bucknall and Clayton, 1972; Bucknall and Drinkwater, 1973; Bucknall *et al.*, 1972a, 1973) has demonstrated the usefulness of this technique with a variety of rubber-modified plastics, including HiPS, ABS, and ABS-modified poly(2,6-dimethyl-1,4-phenylene oxide) (PPO). Typical curves are shown in Figures 3.20 and 3.21. The determination of the extent of crazing or shear effects is based on the assumption that shear yielding occurs without a volume change (other than predicted from the value of Poisson's ratio) but that crazing does involve an additional expansion. By plotting volume strain ($\Delta V/V$) at a given time against longitudinal strain ($\Delta L/L$), the two effects can be distinguished and their magnitudes assessed. The slope of the curve is a measure of the extent to which crazing contributes to the deformation; thus a slope of unity implies 100 % crazing, and a slope of zero, no crazing, only shear yielding. Thus, in HiPS (Figure 3.20), virtually all the deformation at all strains may be attributed to crazing, while in blends of HiPS with PPO, shear deformation becomes increasingly important as the concentration of PPO is increased. In comparison (Figure 3.21), a typical high-impact ABS resin exhibits a significant degree of crazing only at strains above about 2.5 %, and then only at stresses above about 27 MN/m^2. Thus at low stresses and strains, ABS tends to deform predominantly by shear yielding, while HiPS tends to craze under all conditions; the balance, however, changes as stresses and strains are increased, with the contribution of crazing in ABS rising from zero to about 85 %. The matrix also plays a role; in general, the more

---

* In general, the development of crazes is associated with dilatational stresses (Kambour, 1973). In one case, crystalline poly(ethylene terephthalate), so-called shear crazes have been reported to lie along shear bands induced by yielding. While such crazes are not yet understood, it is reasonable to assume that a dilatational stress component must somehow be involved. If such crazes exist in other systems, however, the argument in this section should not be affected.

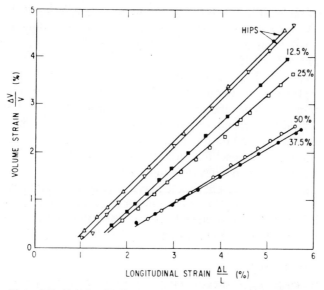

Figure 3.20. Relationship between volume strain $\Delta V/V$ and longitudinal strain $\Delta L/L$ for creep of HiPS and blends of HiPS with PPO (from 12.5 to 50% ABS) (Bucknall *et al.*, 1972a). Note that the fraction of deformation due to crazing as opposed to shear drops from about 100% in the case of HiPS alone to about 60% for a 50–50 blend with PPO. It is interesting that the HiPS, itself a craze-prone polymer, promotes shear yielding in the somewhat ductile PPO.

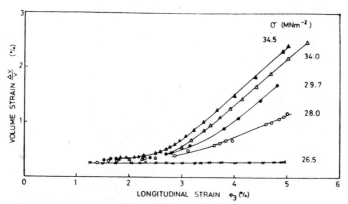

Figure 3.21. Relationship between volume strain $\Delta V/V$ and longitudinal strain e_3 for creep of a high-impact ABS resin, showing mechanism of creep as a function of strain at five different stresses. (Bucknall and Drinkwater, 1973.)

ductile the matrix, the less crazing is seen. It is interesting that in ABS-modified PVC, crazing contributes only about 10 % of the creep deformation (Bucknall, 1972), an observation rather consistent with results of other deformation studies discussed above.

In any case, the response of a rubber-modified plastic to low-strain-rate tests is complex, depending on the polymer and rubber concerned, as well as the test conditions. Since the underlying deformation mechanisms, crazing and shear yielding, depend on factors such as stress and strain rate to different extents, one cannot quantitatively apply results of tensile or creep studies to the case of impact fracture (Bucknall, 1972; Bucknall and Drinkwater, 1973). It does seem likely that under impact loading (especially with a notched specimen) crazing will tend to play a greater role than shear yielding, and to play a greater role than in creep studies. Elucidation of the precise role of the two mechanisms must, however, await the development of tests suitable for use at high rate of loading.

3.2.2.3. Fatigue Behavior

Since engineering plastics are often subject to cyclic or repeated stresses, the question of fatigue behavior is important. In general, relatively little attention has been given this phenomenon in plastics, with the exception of measurements of lifetime as a function of stress, frequency, and number of

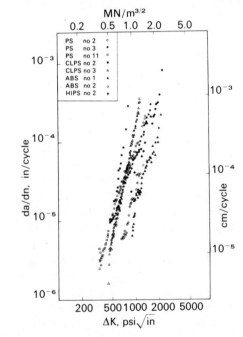

Figure 3.22. Relationship between fatigue crack propagation rate per cycle da/dn and the range of applied stress intensity factor ΔK for various modifications of polystyrene (Hertzberg et al., 1973). Note the toughening effect of incorporating a rubbery phase, as in HiPS or ABS, compared to the weakening effect of crosslinking.

cycles (Bucknall and Clayton, 1972). However, in studies of fatigue crack propagation in rubber-modified plastics, Manson *et al.* (Manson and Hertzberg, 1973*a,b*; Hertzberg *et al.*, 1970) found that the toughening effect (as seen in the stress–strain and impact behavior of HiPS and ABS resins) was paralleled by a similar toughening effect in fatigue. As shown in Figure 3.22, the rate of fatigue crack propagation at a given range of stress intensity factor ΔK (Section 1.5) was found to be reduced by the incorporation of a rubbery phase into both HiPS and ABS resins (Hertzberg *et al.*, 1970). At high values of ΔK, a reduction of almost an order of magnitude was noted for ABS in comparison to homopolymer polystyrene; HiPS was somewhat less effective, as might be expected based on its generally lower value of impact strength. In recent fatigue studies, Manson and Hertzberg (1973*b*) found extensive equatorial crazing around the rubber particles in HiPS. (For further details, see Section 3.2.3.1.)

As with tensile and impact behavior of rubber-toughened plastics, a major energy-absorbing mechanism appears to be crazing. Thus, at least qualitatively, low-frequency fatigue behavior of rubber-modified plastics appears to involve the same phenomena as are seen in tensile and impact loadings.

3.2.3. Toughening Mechanisms

It is now generally accepted that two-phase polymer systems such as the rubber-modified plastics form tough engineering plastics because they possess several mechanisms for deformation, which can dissipate strain energy that would otherwise be available to extend an existing flaw or crack. The principal toughening mechanisms are believed to involve shear yielding and crazing, interaction between shear yielding and crazing, and the diversion and multiplication of the growing crack itself. Some of these mechanisms can also be effective in resins filled with high-modulus materials (Chapter 12). However, the incorporation of an incompressible phase such as a rubber (Poisson's ratio $\cong 0.5$) tends to greatly encourage the development of a dilatational stress field at the plastic–rubber interface, so that crazing tends to be much more common in rubber-modified plastics than in, for example, glass-filled plastics; shear flow may also be encouraged in more ductile matrices (Section 3.2.2.1).

In this section, the role of crazing and shear yielding in conferring toughness in polymers will be discussed, as well as some aspects of the crack propagation itself. For further details the reader is referred to a recent comprehensive and critical review of crazing and fracture in polymers by Kambour (1973) and a discussion by Kambour and Robertson (1972), as well as to the specific references cited. The virtue of heterogeneity in the control of fracture is also discussed in a review by Eirich (1965).

3.2.3.1. Crazing and Shear Phenomena

When a stress is applied to a polymer specimen, the first deformation typically involves shear flow of the polymer molecules past one another—a response familiar to anyone who has ever conducted a tensile or creep test. Eventually, as the strain (or stress) increases, a crack may begin to form, presumably at a flaw of some kind, and then propagate at an ever-increasing rate up to a very high value, at which catastrophic rupture occurs. Intuitively, one would expect that the greater the propensity for a shear-type deformation, the greater in general should be the toughness, and, indeed, this is the case. A second major deformation mechanism, as pointed out by Bridgman (1952), occurs in the presence of a triaxial stress field, and is dilatational in nature (Bridgman, 1952; Sternstein et al., 1968). In the case of metals and ceramics, such a response involves first cavitation, and then coalescence of cavities to form cracks at right angles to the principal stress. In the case of polymers, what would otherwise be a void is spanned by segments of long molecules, the whole entity being termed a craze (Kambour, 1966, 1968, 1973; Kambour and Russell,1971).* Thus crazes in polymers differ from crazes in ceramics in that the former are small cracks held together with fibrils while the latter are simply small, open cracks (Figures 3.23 and 3.24). That crazing in polymers involved more than simple cracks was first shown by Sauer et al. (1949); later, the current view, as shown in Figure 3.23, was demonstrated by Spurr and Niegisch (1962). Many others have studied and discussed the formation, characteristics, and behavior of crazes, for example, Newman and Wolock (1957), Murray and Hull (1969), Hull (1970), Hull and Owen (1973), Berry (1964), Kambour (1966, 1968, 1973), Kambour and Russell (1971), Gent (1970). Indeed, it seems that crazing is a very general precursor to actual fracture in many cases; see reviews by Kambour (1973) and Rabinowitz and Beardmore (1973).

Realizing that crazing is a common response to an applied stress in polymers makes it easier to gain some insight into the possible role of rubber inclusions in toughening otherwise brittle plastics. One of the first mechanisms for toughening was proposed by Merz et al. (1956), who noted that elongation

---

* The phenomenon of crazing in pottery has been known since ancient times; the term is derived from the Middle English "crasen," meaning "to shatter or render insane," which in turn i probably derived from the Old Norse krasa, meaning "to shatter" (Oxford English Dictionary 1933). Thus, use of the term in the context of fracture has a long and honorable history, ap parently predating the implication of mental difficulties.

(a)   (b)

Figure 3.23. Schematic difference between a crack (a) and a craze (b). A craze may be visualized as a cave containing joined stalactites and stalagmites. These structures are strands of highly oriented polymeric material.

in HiPS was accompanied by stress-whitening. These observations led to the suggestion that many small cracks were formed on straining but that the rubber particles spanned the cracks and hindered further crack growth. The higher energies required to fracture were attributed to the ability to absorb a great deal of energy by stretching the rubber particles. Although we now know that the total impact fracture energy cannot be accounted for by such stretching, the ideas provoked much further attention. Schmitt and Keskkula (1960) later suggested an alternate energy-absorption mechanism—that rubber particles serve as stress concentrators and thus initiate

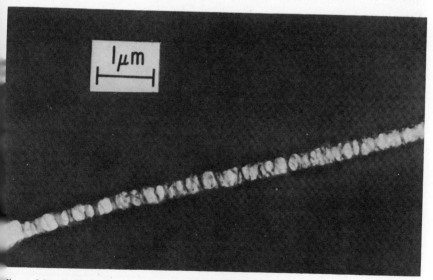

Figure 3.24. Transmission electron micrograph of a craze in polystyrene. (Courtesy of R. P. Kambour, General Electric Co.)

the development of many microcracks around the rubber particles. In addition, it was suggested that the rubber particles help stop cracks from growing catastrophically. Later, Newman and Strella (1965) proposed that a rubbery inclusion serves to induce sufficient triaxial tension to permit shear yielding in the matrix; under the influence of the induced tension, the T_g of the matrix is expected to be lowered sufficiently to permit facile cold-drawing. A similar reduction in T_g ("devitrification") due to the hydrostatic component of applied stress has been suggested by Gent (1970) to explain the development of crazing in glassy matrices.

In view of what is now known about crazing, it is now generally accepted that the microcracks proposed above (Merz *et al.*, 1956; Schmitt and Keskkula, 1960) are actually crazes, and that crazing (as well as shear yielding) provides a mechanism for the absorption and dissipation of impact (or other) energy, at least in many cases (Bucknall, 1967a,b; Howard, 1970; Kambour, 1973; Matsuo, 1969; Newman and Strella, 1965; Turley, 1968). As pointed out by Kambour (1972), four possible contributions to fracture energy can be identified with crazing (in approximate order of importance): (1) plastic work of craze formation, (2) viscoelastic work of craze extension under stress, (3) surface energy of the holes created, and (4) bond breakage. Approximate calculations for the case of poly(methyl methacrylate) suggest that the first two components account for most of the energy absorption during crazing, and presumably for a major portion of the observed fracture energy. Thus, even though crazes are generally weaker than the uncrazed polymer, considerable energy can be dissipated in their development.

Considerable morphological evidence confirms the development of crazes around rubbery inclusions in glassy matrices, thus supporting inferences based on observations of stress-whitening and density decreases on straining (Section 3.2.2). For example, Bucknall and Smith (1965) observed the development of crazes in HiPS, originating at or near the plastic–rubber interface (Figure 3.25). It was suggested that the rubber particles served as stress concentrators to initiate a multiplicity of small crazes, mainly equatorially about the particles [cf. the earlier proposals by Merz *et al.* (1956)]. As was the case with the elucidation of phase structure in polyblends, the development of techniques for the electron and scanning electron microscopy of polymers greatly facilitated the study of crazing. Matsuo (1968) demonstrated that stress-whitening in ABS was associated with the development of crazes that begin at one rubber particle and terminate at another (Figure 3.26). Matsuo (1966, 1969) also observed crazes in a methacrylate–butadiene–styrene terpolymer and rubber-modified PVC. Matsuo employed the Kato technique (Kato, 1968) of staining the rubber particles with osmium tetroxide, which also stains the crazes themselves (Bucknall, 1967a,b; Kambour, 1972). Another technique has been used by Kambour and Russell

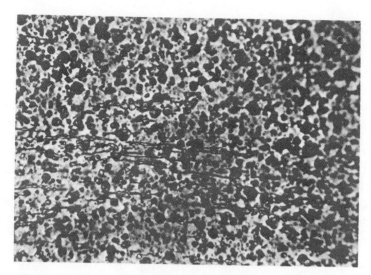

Figure 3.25. Crazes in stress-whitened high-impact polystyrene. (Bucknall and Smith, 1965.)

(1971), who used an iodine–sulfur eutectic to impregnate and preserve the craze structure in HiPS during sectioning for microscopy; the eutectic was then sublimed off, leaving the fibril–void craze structure intact. This study also illustrated the growth of crazes at other than equatorial positions, presumably due to the complex interaction of stress fields around the rubber particles; such effects have been observed in other studies as well (Bucknall and Smith, 1965; Manson and Hertzberg, 1973b; Matsuo et al., 1972). In addition to the use of transmission or replica electron microscopy, scanning microscopy of actual fracture surfaces has been useful in the case of HiPS. As shown in Figure 3.27, Manson and Hertzberg (1973b) found direct evidence for equatorial crazing around the rubber particles, as well as evidence for some crazes in other planes, extensive deformation of the rubber particles, and some interfacial cavitation.

Thus crazing has been directly identified in typical rubber-modified plastics in a plane perpendicular to the direction of the applied stress and initiated at the rubber–matrix interface. These findings are quite consistent with macroscopic model studies performed by Matsuo et al. (1972), who studied the behavior of polystyrene-containing rubber balls (as an analog of ABS morphology). It was found that equatorial crazes developed in tension, as expected, and also that stress field interaction occurred when the balls were close together, resulting in a heavier craze density between the balls. This shows that a principal role of rubber particles is to induce many

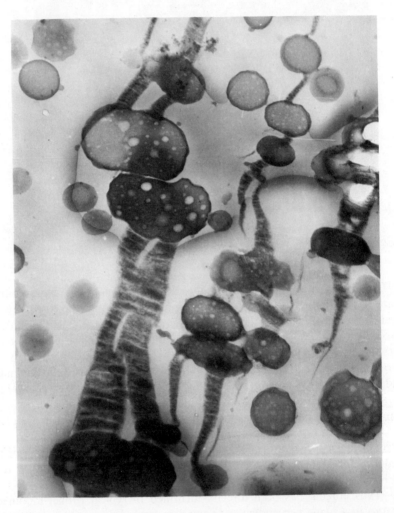

Figure 3.26. Electron micrograph of an ultrathin section cut parallel to the stress-whitened surface of a commercial ABS plastic. Dark portions represent both rubber particles and craze lines stained by osmium tetroxide. Note that the rubber particles appear to encourage a high density of crazing between them. (Matsuo, 1968.)

small crazes, thus dissipating energy and diverting the ultimate catastrophic crack. If the rubber particles are sufficiently close together, the interaction of the stress fields may increase the craze density, and hence increase the ultimate fracture energy.

The rubber must also do more than initiate crazing; otherwise voids or other stress concentrations should also be effective in toughening the matrix. Since good impact strength requires good interfacial bonding, it seems likely that the rubber serves to share the load with the matrix after crazing has occurred (Bucknall, 1967a,b). The breaking of rubber–plastic

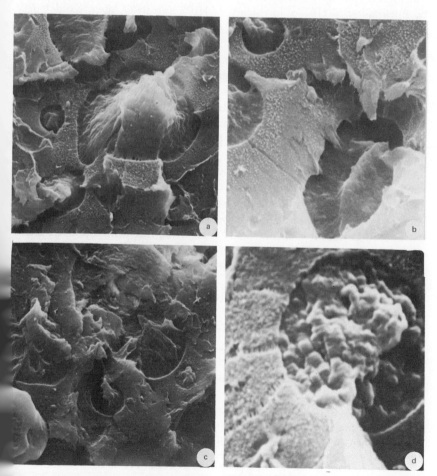

Figure 3.27. Scanning electron micrographs of fracture surfaces of high-impact polystyrene subjected to cyclic loading (10 Hz). (a,b) Crack propagation rate $\sim 2.5 \times 10^{-4}$ cm/cycle; $\times 2300$. (c) Crack propagation rate $\sim 2.5 \times 10^{-5}$ cm/cycle; $\times 1830$. (d) Crack propagation rate $\sim 2.5 \times 10^{-5}$ cm/cycle; $\times 8000$. (Manson and Hertzberg, 1973b.)

graft bonds also serves to degrade the mechanical energy. Thus a controlled number of grafting sites results in optimum toughness.

As shown by Bucknall *et al.* (Bucknall and Clayton, 1972; Bucknall and Drinkwater, 1973; Bucknall *et al.*, 1972a, 1973) crazing is not always the dominant mechanism for deformation. For example, shear yielding is more important in ABS than in HiPS, and rubber-modified PVC does not necessarily exhibit crazing on impact (Petrich, 1972). Shear yielding is important for two reasons. First, shearing provides an effective means of energy dissipation (probably more effective than crazing, which involves considerable shearing but leads to a weakened structure). Second, as shown by Bucknall *et al.* (1972a) for blends of HiPS and PPO, shear bands* may serve to terminate crazes (and presumably hinder growth of the ultimate cracks through the crazes). Such an effect has been noted in glassy polymers themselves (Kambour, 1973; Newman and Wolock, 1957). Thus shear banding may take over to some extent the crack-stopping function of the rubber particles. These observations are consistent with the fact that in HiPS, which deforms essentially only by crazing (Bucknall *et al.*, 1972a), a much larger rubber particle is required (10–20 μm) for optimum toughening, in comparison to a typical ABS (∼1 μm), which exhibits a much greater extent of shear yielding.†

Based on the evidence so far, it seems probable that impact strength in rubber-modified plastics depends on the relative importance of crazing and shear yielding induced by the rubber particles,‡ as well as on other parameters such as the rubber characteristics (below). As mentioned above, however, extrapolation from low-strain-rate to high-strain-rate tests cannot be made quantitatively. Research into the deformation mechanisms at high strain rates should be fruitful, though difficult to accomplish.

3.2.3.2. Characteristics of the Rubber

In the preceding section, it was shown that rubber inclusions in a glassy matrix can induce some combination of shear yielding and multiple crazing, at least at low strain rates and presumably at high rates as well. Clearly, however, characteristics of the rubber itself and its interaction with the matrix are important in determining the quantitative effectiveness of

---

* Shear bands may be seen as birefringent entities at 45° to the direction of applied stress; their occurrence indicates that molecular slippage has taken place in a highly localized region, with concomitant high orientation of the polymer.

† It should be pointed out that ABS polymers themselves vary considerably with respect to the relative roles of crazing and yielding (Bucknall and Drinkwater, 1973).

‡ According to Bucknall, the preferred mechanism of toughening is shear yielding, since less damage to the specimen results. In the ideal case, crazing serves as an additional mechanism to prevent failure when shearing alone cannot stop crack formation and growth.

the rubber. Several important factors are (1) the extent of mixing of the two components, (2) the concentration and state of dispersion of the rubber, (3) the phase structure within the rubber particles, and (4) the glass transition temperature of the rubber.

Let us examine the effects of compatibility first, using the PVC/NBR blends as an example. The Charpy impact strength of PVC/NBR blends as a function of acrylonitrile content (Matsuo, 1968) is shown in Figure 3.28. The point at 0% acrylonitrile (AN) corresponds, of course, to the PVC/PB blends, which do not mix well, i.e., are incompatible. (See Section 3.1.4 for a discussion of morphology.) Impact resistance increases as the AN content is increased, the phase boundaries become indistinct, and some mixing takes place. At still higher concentrations of AN, mixing becomes too extensive, and impact resistance again declines. The compositions that are semi-compatible (AN content in the range of 10–20% in the NBR) exhibit the greatest toughening. This evidence supports the notion that limited molecular mixing in blends improves impact resistance. It may well be that enough mixing is required to develop a strong interface, which can first transmit the applied stress to the matrix and enhance crazing and then help share the load with the matrix after crazing (Bucknall, 1967a,b); a weak interface would tend to permit easy cavitation at the interface rather than crazing in the matrix, and would be expected to lower the fracture energy. In addition, an ill-defined boundary may serve to blunt or slow down growing cracks more effectively; in general, the fracture path tends to proceed through the interfacial region (Cross and Haward, 1973).

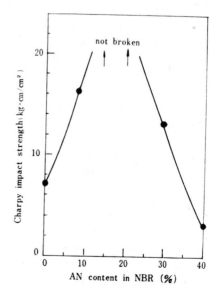

Figure 3.28. Effect of AN content in NBR on Charpy impact strength of PVC/NBR blends. Impact resistance reaches a maximum in the semicompatible region. Room temperature, 25°C; PVC/NBR = 100/15. (Matsuo, 1968.)

The general effect of rubber concentration has already been mentioned in Section 3.2.2.2; the effect of particle size is also important, since it will determine the number and spacing of particles at constant composition. In general, for a particular polymer pair there will be some size of the rubber domains that yields optimum toughness. Larger or smaller domains result in reduced toughness.

Several arguments may be developed to rationalize this behavior. First, if the particle size is reduced, the interparticle distance is correspondingly reduced, so that a given craze (and the crack associated with it) will tend to strike a rubber particle or interact with another craze before it has a chance to accelerate and develop stress concentration at the crack tip large enough to circumvent or fracture the rubber particle (Kambour, 1973). Second, as pointed out above, interaction of the stress fields of particles that are close together tends to enhance the development of crazes between the particles. Third, the greater the surface area of the rubber particles, the greater will be the actual surface area of the ultimate crack, and hence the apparent fracture energy (see also Section 12.1.2.4). Fourth, if the rubber particles are sufficiently close together that a critical extent of crazing is exceeded, failure may be caused by adiabatic heating due to the rapid conversion of strain to thermal energy (Cross and Haward, 1973).

In any case, the existence of highly dispersed short crazes in association with rubber particles can divert, branch, or slow down a crack that could otherwise become catastrophic. The minimum effective particle size, to be sure, depends on the polymer system; the greater the tendency to undergo shear deformation rather than crazing, the smaller the minimum particle size required, at least for ABS and HiPS (Bucknall and Drinkwater, 1973). Presumably effective termination of crazes and cracks requires a larger rubber particle if shear bands are not available to serve as crack stoppers.

An alternate quantitative approach to the question of the effect of rubbery inclusions on crack propagation has been based upon the prediction that rubber particles initiate branches in very fast cracks. As shown by Yoffe (1951) and later by Sih (1970) and Sih and Irwin (1970), a single crack will be unable to dissipate enough energy to remain stable at speeds greater than about one-half the speed of sound in the material concerned (the speed of sound in plastics is about 1240 m/sec), and will tend to split itself into several branches. In fact, crack branching has been found to occur in polymers at high speeds of crack propagation (Rosenfeld and Kanninen, 1973), though not always.

Based on this combination of theory and experiment, Bragaw (1970, 1971) has proposed that rubber particles can serve as nuclei for crack (or craze) branches because of their lower sound velocity; the proposal has been supported by Grancio (1971). Now the modulus of typical plastics is ordinarily three orders of magnitude higher than for an elastomeric inclusion, and the speed of sound in materials varies as the square root of the modulus. Bragaw and Grancio point out that if the speed of a crack or craze in a polyblend is relatively constant, it may be stable in the plastic phase. However, upon propagating across the phase boundary into the elastomeric phase, the crack or craze may suddenly become unstable and branch, because of the suddenly lower maximum crack velocity required. They suggest that 1–5 μm is required to accelerate a craze in the plastic phase to a velocity sufficient to permit the craze to branch efficiently upon entering the rubber phase. To achieve such branching effectively, a rubber particle diameter of at least 0.5 μm would be predicted for a typical rubber-reinforced system. While such a prediction is in accord with experience, it is also true that multiple crazing is observed with low-velocity cracks as well. In any case, it has been shown that the presence of rubber particles reduces the terminal crack velocity in a glassy polymer below the value predicted (Peretz and DiBenedetto, 1972).

It is also interesting to consider the effect of particle size at constant content of rubber per se rather than at constant content of the rubbery phase, as was the case in the discussion just completed. Since the rubbery phase of a graft-type polyblend contains both rubber and occluded plastic (see Figure 3.2), the two measures of concentration are not equivalent. In this case, the impact strength of HiPS tends to *increase* as the rubber particle volume is increased by the inclusion of more polystyrene within the rubbery phase (Bucknall and Hall, 1971). This effect may also be explained in terms of a reduced interparticle distance and a consequent retardation of crack growth rate. In addition, the greater the fraction of polystyrene in the rubbery phase in HiPS, within limits, the greater is the energy dissipation associated with the T_g of the polybutadiene—a fact which may be related to the impact strength.

Given these observations, it would seem that to ensure high toughness in a rubber-modified plastic one would want many rubbery particles, each enlarged as much as possible by inclusion of the glassy polymer in the rubber. As pointed out by Kambour (1973), there is, however, a limit to the improvement of toughness in HiPS in this way, for several possible reasons: (1) a tendency of rubber particles to break up on processing if they have a high polystyrene content (Baer, 1972); (2) lower crazing efficiency due to the apparent increase in rubber-phase modulus (Bucknall, 1969); and (3) the occurrence of crazes within the rubber particle inclusions, thus facilitating crack growth through rather than around the particles [K. Lawrence, quoted by Kambour (1973), p. 53].

One might speculate that the crack growth through the rubber portion could be used to help explain the greater toughness of graft-type polyblends in comparison to mechanical blends. With respect to graft-type polyblends, it can easily be visualized that a crack passing through a composite rubber particle with its many internal interfaces must follow a much more tortuous path than in a pure rubber particle in a mechanical blend. Such diversion has been shown for HiPS by Seward (1970), and may also be seen in Figure 3.27. Moreover, the strength per se of the rubber particle will be greater in the graft copolymer case due to reinforcement by the stiffer polystyrene, which will tend to hinder passage of a craze or crack through the rubber particle.

At constant content of rubber, the greater toughness of graft-type ABS materials over their HiPS analogs may also be partially attributed to the smaller rubber phase domains in the former, which limit the size of the crazes formed.

Finally, the T_g of the elastomer phase must be considered. It is well known that the T_g must be significantly lower than the test temperature if the composite is to have a significant value of impact strength (Bucknall and Smith, 1965; Matsuo et al., 1970a; Imasawa and Matsuo, 1970). As shown in Table 3.3, the lower the T_g of the elastomer phase, the greater the toughening. Bucknall and Smith (1965) have proposed that the rubber must be able to relax at the very high rate of loading in an impact test for an increase in toughness to occur. Since, however, the deformation of the rubber per se contributes relatively little to the fracture energy, the basic problem must be the inability of the rubber to induce crazing or shear yielding if the rubber behaves in a glassy manner. In any case, it is interesting to estimate the effect of crack velocity on T_g. Qualitatively, the glass temperature observed depends on the rate of testing. If the "test" is an onrushing crack or craze, it must remain elastomeric under that condition to be effective. We may consider a simple model in which the growing crack tip causes a deformation which may be likened to the first 90° segment of a cyclic deformation (Sperling, 1975) (Figure 3.29). As shown in Figure 3.29a, a point at the leading edge of the crack (or craze) must move some 1000 Å in a direction at right angles to the motion of the crack. This motion must occur in the time span that the crack moves forward the 1000 Å indicated. (The crack head diameter of 2000 Å is estimated from electron microscopy, as shown above.) If one assumes that the point in question is executing the first 90° segment of a cyclical deformation, the point may be modeled as moving in the vertical direction the indicated distance in Figure 3.29b. In a plastic matrix having Young's modulus equal to 3×10^{10} dyn/cm^2, Bragaw (1970, 1971) estimated the maximum crack velocity attainable as about 620 m/sec. With a crack radius of about 1000 Å (see Figure 3.29) the equivalent cyclic deformation frequency corresponds roughly to about 10^9 Hz. If the glass temperature increases at about 6 or 7°C per decade of frequency (Nielsen, 1962, Chapter 7)

Table 3.3
Impact Resistance of ABS Polymers at 20°C[a]

Sample No.	Composition of rubber component		T_g of rubber component, °C	Charpy impact strength, kg-cm/cm^2
	BD	ST		
GT-1	35	65	40	0.75
GT-2	55	45	−20	18
GT-3	65	35	−35	30
GT-4	100	0	−85	40[b]

[a] Matsuo (1969).
[b] Estimated from Matsuo *et al.* (1970a), Fig. 1.

the effective glass temperature of the rubbery phase at 10^9 Hz is calculated to increase about 60°C above values typical of measurements at low frequencies (10^{-1} Hz). This calculation suggests that the T_g of the rubber phase must be about 60°C below ambient, which correlates well with the evidence in Table 3.3 that the T_g of the elastomeric domains must be below about

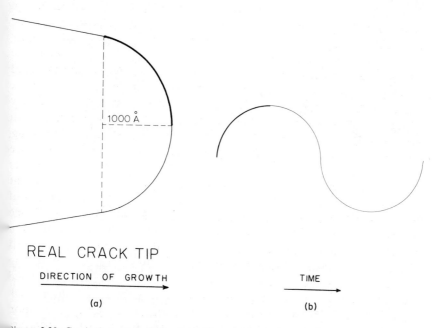

1000 Å

REAL CRACK TIP

DIRECTION OF GROWTH

TIME

(a)

(b)

Figure 3.29. Crack tip growth deformation likened to a sinusoidal cyclic motion. (a) Dark portion resembles first 90° of deformation; (b) hypothetical crack tip. (Seward, 1970.)

– 20 to – 40°C in order to attain significant impact resistance (Bucknall and Smith, 1965) (see also Figure 3.17). The calculation and empirical data, however, are difficult to reconcile with the fact that at very high crack speeds, localized temperature rises of several hundred degrees are commonly observed near the crack tip (Cross and Haward, 1973), unless it is assumed that crazing takes place prior to cracking rather than just ahead of the moving crack. If this is so, then the principal effect of a high rate of loading is to vitrify the rubbery phase and inhibit the development of crazes prior to actual crack initiation.

In summary, although much is still unknown about the mechanism of toughening of plastics by rubber and only a few systems have been studied in detail (usually at low strain rates), several features do appear to be important. The elastomer must have a low enough T_g to remain rubbery at the loading rate concerned and to initiate a combination of crazing and shear yielding in the matrix. The concentration, particle size, and phase composition should be such that many small crazes can form and interact, or branch in the matrix, not in the rubber particle itself, and that a growing crack or craze can be diverted or branched by interaction with the rubber particles. As for the matrix, clearly even a modest degree of ductility would appear to greatly enhance the role of shear yielding rather than the less efficient crazing.

In the final analysis, we may compare the properties of the three most common types of impact-resistant polymer blends and grafts. Referring to Table 3.2, the impact strengths of many HiPS materials are usually near 1.5–2 ft-lb/in. of notch, ABS values are usually near 4 ft-lb/in. of notch, and high-impact PVC has values over 10 ft-lb/in. of notch. These values may be related to the principal methods of toughening: HiPS mainly crazes on stressing, ABS exhibits a combination of crazing and shearing, and rubber-toughened PVC exhibits primarily shearing. Thus the shearing mechanism of deformation is preferred over crazing.

3.3. OPTICAL PROPERTIES OF POLYBLENDS

Since most polyblends are opaque or cloudy, because of their two-phase nature, they are unsuitable for certain applications. The clarity of polyblends can be improved by lowering the size of the dispersed particles below that of the wavelengths of visible light, but there is a limit if toughness is to be retained.

A superior method of obtaining high clarity involves proper selection of refractive indices of the two components. If both phases have identical refractive indices, a clear polyblend will be obtained regardless of the details of the phase separation morphology. In fact, a clear ABS-type resin can be made from a mixture of styrene–butadiene rubbery copolymer with a methyl methacrylate–styrene–butadiene copolymer (Gesner, 1967). Alternately, a clear, impact-resistant acrylic can be prepared from a methyl methacrylate–styrene–acrylonitrile copolymer containing a rubbery methyl methacrylate–butadiene copolymer as a dispersed phase (Gesner, 1967).

The development of clear poly(vinyl chloride) plastics has been given considerable attention. For example, Petrich (1972) showed that methacrylate–butadiene–styrene (MBS) impact modifiers for PVC can impart toughness as well as clarity. Acrylic modifiers are also described by Ryan and Crochowski (1969), Ryan (1972), and Souder and Larson (1966).

However, unless the two refractive indices exhibit the same temperature dependence, the clarity will depend on temperature (Conaghan and Rosen, 1972; Rosen, 1967), as illustrated in Figure 3.30. At the point where the two refractive indices are identical, a maximum in clarity is obtained. Usually this last aspect is not very important unless the plastic will be used over a very broad temperature range.

3.4. OXIDATION AND WEATHERING OF POLYBLENDS

All types of polymers tend to degrade on weathering, at least to some extent (Gesner, 1965; Hirai, 1970; Shimada and Kabuki, 1968). The analysis

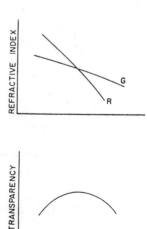

Figure 3.30. The effect of temperature on refractive indexes and transparency in a rubber–glass two-phase system. (Rosen, 1967.)

of weathering behavior becomes quite complicated in the case of polyblends, because not only do the two phases age at different rates, but continuing interactions between the two phases remain of prime importance.

The most important target of oxidative attack involves carbon–carbon double bonds. In fact, at the initial stage of oxidation it is possible to consider that only the rubber phase oxidizes, with no attack in the plastic portions. This is illustrated by Figure 3.31, where materials high in double bond concentration show more oxygen uptake than those having a low or zero concentration of double bonds (Hirai, 1970).

According to Shimada and Kabuki (1968), the mechanism of attack includes the formation of UV light-induced peroxide radicals (Imp = impurity):

$$RH + h\nu \xrightarrow{k_i} R\cdot + H\cdot \tag{3.2}$$

$$R\cdot + O_2 \xrightarrow{k_a} ROO\cdot \tag{3.3}$$

$$ROO\cdot + RH \xrightarrow{k_p} ROOH + R\cdot \tag{3.4}$$

$$ROO\cdot + Imp \xrightarrow{k_e} \text{inert product containing OH or CO} \tag{3.5}$$

$$ROO\cdot \xrightarrow{k_d} \text{inert product containing OH or CO} \tag{3.6}$$

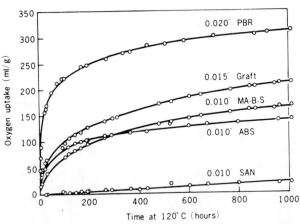

Figure 3.31. Oxygen uptake for polybutadiene rubber (PBR), grafted polybutadiene rubber (Graft), methacrylonitrile–butadiene–styrene resin (MA–B–S), acrylonitrile–butadiene–styrene (ABS), and styrene–acrylonitrile copolymer (SAN). Only the last polymer contains no double bonds. (Hirai, 1970.)

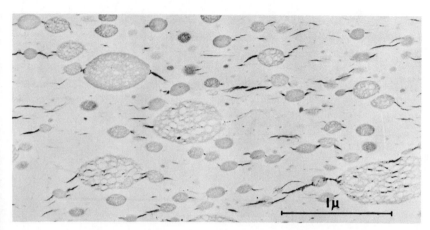

Figure 3.32. Electron micrograph of ultrathin section of irradiated ABS plastic (osmium tetroxide staining). Note cracking, which occurred after 400 h exposure to Fade-o-Meter irradiation. (Hirai, 1970.)

Hirai (1970) found that the rubber particles in ABS materials harden on oxidation and develop microcracks. These structures propagate in the continuous plastic phase parallel to the irradiated surface (Figure 3.32).

Hirai also showed that except for regions very close to the surface, the oxygen diffusion rate was the controlling step in the aging of ABS plastics. Since ultraviolet absorbers are effective only for the interior of the specimens, this method of prevention leaves something to be desired for surface protection (Hirai, 1970). Hirai suggests that oxidative deterioration in polyblends could be effectively prevented by coating with low-permeability materials, such as Saran. According to Vollmert (1962), the oxidative deterioration of the rubber phase through attack on double bonds can be avoided by the use of saturated elastomers such as poly(butyl acrylate). When a saturated rubber is employed, oxidative degradation is much reduced (see Section 9.1). Coefficients of thermal expansion of polymer blends and grafts are treated in Section 12.1.3.3.

4

Diblock and Triblock Copolymers

Block copolymers, among the newer creations of modern chemistry, provide one of the more exciting classes of materials in terms of both intellectual stimulation and practical utility. This chapter will be primarily concerned with AB and ABA types of block copolymers, where the A's and B's represent fairly long subchains. The **ABABABAB**... types with short repeating blocks, exemplified by the segmented polyurethanes, will be treated in Chapter 5.

4.1. SYNTHESIS

The most elegant method of preparing block copolymers utilizes anionic polymerization (Henderson and Szwarc, 1968; Lenz, 1967, Chapters 13, 17; Szwarc, 1973). This scheme may be represented as follows:

$$R-M + CH_2=C\overset{\displaystyle H}{\underset{\displaystyle X}{\big<}} \;\rightarrow\; R-CH_2-C\overset{\displaystyle H}{\underset{\displaystyle X}{\big<}}{}^{\ominus} \;\cdots M^{\oplus} \qquad (4.1)$$

where $R-M$ represents a basic compound such as butyl lithium. In many reactions of this type, all the chains may be initiated simultaneously, and provided that the rate of initiation is high with respect to propagation and that chain transfer is negligible, a very narrow (Poisson) molecular weight distribution can be obtained. In the absence of polar impurities the growing chains cannot terminate, leading to their common nickname, "living polymers." If, after all of the first monomer has been consumed, a second, different monomer is added, chain growth may be resumed, thus leading to block copolymer formation.* Of course, after consumption of the second monomer, more of the first monomer may be added, or a third monomer, etc. Finally, the chain may be terminated conveniently by exposure to air, or quenching with water or alcohol.

4.1.1. Dilithium Initiators

An especially simple way to prepare triblock polymers having identical end blocks involves the employment of difunctional initiators, such as 1,4-dilithio-1,1,4,4-tetraphenylbutane. The growing chain has the structure

$$
\text{Li}^{\oplus} \cdots {}^{\ominus}\!\overset{\displaystyle \overset{H}{|}}{\underset{\displaystyle \underset{X}{|}}{C}} - R - \overset{\displaystyle \overset{H}{|}}{\underset{\displaystyle \underset{X}{|}}{C}}{}^{\ominus} \cdots \text{Li}^{\oplus} \tag{4.3}
$$

where —R— represents the central, completed portion of the molecule. This procedure also was employed to synthesize the otherwise difficult to prepare poly(α-methylstyrene-b-isoprene-b-α-methylstyrene) (Fetters and Morton, 1969).

4.1.2. Mechanochemical Methods

A few other, older methods also deserve mention. One involves the mechanochemical degradation of polymer I in the presence of monomer II (Casale *et al.*, 1971; Ceresa, 1962, Chapter 5). In such systems the extensive shearing action tears the polymer molecules in two, creating free radicals and other active species, which can initiate polymerization to form irregular block copolymers. Such systems require much power, and produce materials containing extensive homopolymer in addition to blocks. As mentioned earlier (Section 2.1.1), prolonged mastication of two previously synthesized polymers also leads to block copolymer formation by recombination of the free radicals formed. Active groups on the ends of the polymer molecules (Tobolsky and Rembaum, 1964) and bond interchange reactions between

---

* Addition of a small portion of a second monomer just before exhaustion of the first, or copolymerization of two monomers of very different addition rates, yields graded block copolymers (Aggarwal, 1969) having such junction structures as

$$
\text{=A—A—A—A—A—B—A—B—B—A—B—B—B—B=} \tag{4.2}
$$

The graded block copolymers have indistinct phase boundaries, while maintaining two incompatible phases. Because of the graded phase boundaries, a certain toughness, not otherwise available, can be built in.

two polymer molecules (Burlant and Hoffman, 1960, Chapter 5) also lead to block copolymer formation.

4.2. SOLUTION BEHAVIOR OF BLOCK COPOLYMERS

While block copolymers are essentially linear molecules, chemical differences between the blocks complicate their solution behavior. For example, one block may be soluble in a given solvent and the other insoluble. Alternatively, the mixing thermodynamics may give rise to different chain conformations for the two blocks in solution. Important variables include temperature and concentration. Sadron and Gallot (1973) have identified three concentration ranges of importance : (1) a range limited to low concentration in which monomolecular micelles exist in true solution, (2) a range in which the less soluble block component exists in the form of multimolecular micelles, and (3) a range in which periodic regular structures exist.

Since ranges 2 and 3 will be considered in more detail beginning in Section 6.1.2, at this point we shall only indicate that well-defined and

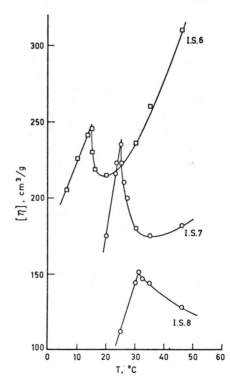

Figure 4.1. Intrinsic viscosity as a function of temperature for poly(isoprene-b-styrene) block copolymers I.S. 6, I.S. 7, and I.S. 8 dissolved in cyclohexane. (Girolamo and Urwin, 1971.)

characteristic phase separations usually take place, and that the exact morphology will depend on the relative solubility of the two blocks in the solvent.

Because the solubility and conformation of the chain portions vary independently in dilute solution, the temperature dependence of such systems can be used to examine the various interactions among the components. For example, Girolamo and Urwin (1971) found a sharp maximum in the intrinsic viscosity–temperature relationship of poly(isoprene-b-styrene) dissolved in cyclohexane, a θ-solvent for the polystyrene component at 34°C. (See Figure 4.1.) The authors attribute the anomalous viscosity behavior to a transition that marks the change from the phase-separated form for the polystyrene component to the dissolved, random conformation.

4.3. PLASTIC COMPOSITIONS

It is convenient to divide the block copolymers into two major categories: those that are plastic at room temperature, and those that are elastomeric. Midrange-composition block copolymers exhibit leathery behavior. Elastomeric compositions will be discussed in Section 4.4.

In a now classic paper on polymer blends and blocks, Matsuo (1968) described a broad series of styrene–butadiene block copolymers; compositions of one series are given in Table 4.1. The modulus–temperature behavior of these compositions (Matsuo et al., 1968) is shown in Figure 4.2. Two sharp glass transitions are observed in each case, suggesting phase separation. As the percentage of polybutadiene in the blocks increases, the materials become progressively softer at temperatures between the two glass transitions. This results, of course, from the drop in the modulus at $-80°C$ becoming more pronounced.

Table 4.1

Characterization of Poly(styrene-b-butadiene-b-styrene) Plastic Compositions Studied by Matsuo (1968)

Group sample	S/B Mole ratio	$\overline{M}_n \times 10^{-5}$ (osmotic pressure)	DP of each sequence		
			S	B	S
SBS-1	80/20	1.69	720	350	720
SBS-2	70/30	2.02	790	685	790
SBS-3	60/40	5.14	1840	2450	1840
SBS-4	50/50	4.08	1300	2580	1300
SBS-5	40/60	2.75	740	2240	740

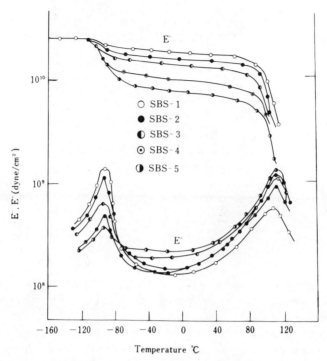

Figure 4.2. Dynamic mechanical behavior of SBS triblock polymers as a function of the styrene–butadiene mole ratio. (Matsuo, 1968; Matsuo *et al.*, 1968.)

Electron microscopy studies confirm phase separation in these materials, and give details as to the size and fine structure of the domains, as shown in Figure 4.3. With SBS-1 (20% butadiene) similar patterns were observed in both normal and parallel sections, indicating spherical particles (Figure 4.3a). The normal and parallel sections shown in Figure 4.3b and 4.3c for SBS-3 (40% butadiene) indicate the existence of a rodlike structure. With SBS-5 (60% butadiene), the normal and parallel sections show the formation of alternating butadiene and styrene-block lamellae (Matsuo, 1968, 1970; Matsuo *et al.*, 1969b). It should be pointed out that due to the difficulties in cutting soft elastomeric materials with an ultramicrotome, electron micrographs for the compositions of Figure 4.3 are much easier to obtain (Matsuo, 1969b) than for the corresponding compositions having polybutadiene as the continuous phase (next section).

To generalize, when the elastomer segment is small with respect to the glassy segment, uniform spheres a few hundred angstroms in diameter pervade the polystyrene matrix. As the length of the elastomer segment is

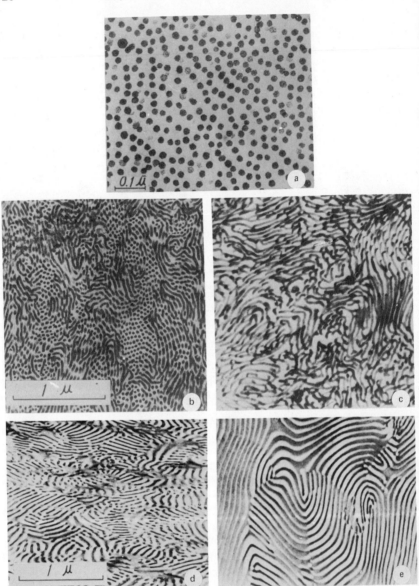

Figure 4.3. Variation of block copolymer morphology with composition: (a) SBS-1 (20 % butadiene); (b, c) SBS-3 (40 % butadiene); (d, e) SBS-5 (60 % butadiene): Morphology changes from spheres (a) to cylinders (b, c) to alternating lamellae (d, e) as the butadiene content increases from 20 to 60 % from (a) through (e). Polymers cast from toluene solution and stained with osmium tetroxide, so that the polybutadiene phase is black and the polystyrene phase is white. (Matsuo, 1968.)

increased, the spheres do not grow in diameter beyond a certain point, but instead are transformed into uniform cylinders. At a characteristic, still higher fraction of elastomer, the cylinders become platelets, or lamellae, while at midrange compositions (40–60% of each component), the material consists of alternate layers of styrene and butadiene. As the percentage of elastomer is increased still further (not shown in Figure 4.3), the phase structure goes through the same changes in reverse—with the elastomer constituting the continuous phase. At the point where the plastic polystyrene phase domains are reduced to short cylinders or spheres, a new class of elastomeric material appears (Section 4.4). The phase structures observed in Figure 4.3 might profitably be compared to structures illustrated for metallic alloys (see Appendix A of Chapter 2).

Spheres, cylinders, and alternating lamellae were also observed in the crystalline block copolymers, which will be considered in Chapter 6.

4.4. THERMOPLASTIC ELASTOMERS

The thermoplastic elastomers are triblock polymers, the central portion being elastomeric, with short, glassy blocks on either side (Holden *et al.*, 1969; Robinson and White, 1970). While many combinations of monomers are possible, most important commercial systems comprise styrene–butadiene–styrene (SBS) and styrene–isoprene–styrene (SIS) compositions. Below the glass transition of the plastic component, tough, highly elastic

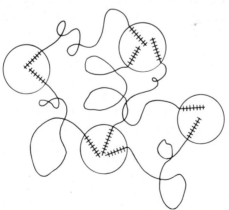

Figure 4.4. Phase arrangement in thermoplastic elastomers. —, elastomeric component; ╫, plastic component. (Holden *et al.*, 1969*a*).

Table 4.2
Comparison of SBS and BSB Polymers[a]

A. Composition		
Segmental mol. wt., g/mol $\times 10^{-3}$	10S–52B–10S	28B–20.5S–28B
Styrene content, %	27.5	27
Total mol. wt., g/mol $\times 10^{-3}$	73	76
B. Mechanical properties		
Stress at 100% elongation, kg/cm^2	16.9	4.9
Extension at break, %	860	120
Ultimate tensile strength, kg/cm^2	277	4.9
Shore A hardness	65	17

[a] Matsuo (1968).

materials are formed, while at higher temperatures, the flow behavior of linear polymers is approached.* This curious phenomenon can be analyzed by observing that although no chemical crosslinks exist in the system, the plastic phase domains behave as multifunctional crosslink sites. Each chain has either end embedded in these domains, as illustrated schematically in Figure 4.4.

In addition to serving as crosslink sites, the plastic domains in triblock polymers behave as reinforcing fillers, much as finely divided carbon blacks or silicas do. These experimental facts can be illustrated by examining other arrangements of blocks containing nominally the same percentages of elastomer and plastic. For example, the plastic portion of diblock polymers (A–B) can be embedded in only one sphere. The molecules in such polymers are incompletely anchored and give tensile properties characteristic of unvulcanized conventional synthetic elastomers, exhibiting flow. That two plastic blocks are required and that two elastomeric blocks will not do is illustrated by Table 4.2. The BSB, while still containing spheres of polystyrene embedded in a continuous polybutadiene matrix, is quite inferior to the SBS, and indeed has quite different properties.

Let us examine the modulus–temperature behavior of polymers having different block arrangements, which lead to large differences between the properties of these materials. Figure 4.5 (Robinson and White, 1970) illustrates the glass transition behavior and rubbery plateau modulus of styrene–isoprene–styrene blocks (tough elastomer) with isoprene–styrene–isoprene blocks (weak elastomer). For the materials of high styrene content little difference exists between the modulus–temperature spectra of the two sets of samples. This is because the polystyrene phase is continuous (see preceding

* However, viscosity remains higher, as discussed in Sections 4.11 and 9.6.

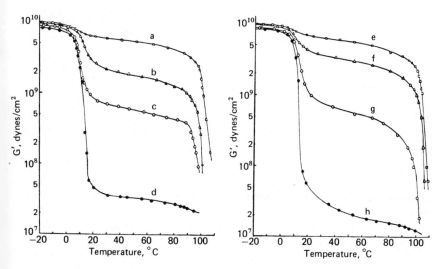

Figure 4.5. Variation of shear modulus G' of block copolymers of styrene and isoprene with composition and temperature at 1 Hz. (a) SIS/20; (b) SIS/40; (c) SIS/60; (d) SIS/80; (e) ISI/80; (f) ISI/60; (g) ISI/40; (h) ISI/20. The numerical value corresponds to the weight percent of the middle block. (Robinson and White, 1970.)

section) with spherical polyisoprene inclusions in each case. These compositions are, of course, in the realm of impact-resistant plastics. However, at lower styrene contents, the SIS arrangement is stiffer, especially at higher temperatures, because of the superior degrees of crosslinking and entanglement effected by the styrene end blocks. A further study of block copolymer and model systems is presented in Section 10.13. By incorporating polystyrene latexes into an elastomer homopolymer component, model morphologies similar to those presented here are attained; however, the covalent bonds between the blocks are effectively removed, forming a model system.

4.5. LONG-RANGE DOMAIN ORDER

Electron microscopy of the SBS blocks discussed in the previous section revealed the presence of a rather high degree of long-range order. In Figure 4.3a, for instance, spheres with an average spacing of about 400–500 Å are shown to be embedded in a continuous phase of polystyrene (Matsuo, 1968). When the composition ratio is reversed and polybutadiene forms the continuous phase, there might be two reasons to expect the development of greater long-range order: (1) The butadiene segments, being elastomeric, can achieve a relaxed state more rapidly than the counterpart polystyrene plastic component, and (2) when the butadiene phase is continuous, each chain has two ends embedded in two (presumably different) polystyrene phase domains so that the disturbances in the statistical chain end-to-end distance can play a direct role in ordering, assuming a sharp distribution of chain lengths. (See Sections 1.2.1 and 1.5.4.) Forces due to rubber elasticity would not be expected to play as important a role when long (plastic) ends are embedded in a continuous plastic phase.

The development of long-range order in thermoplastic elastomers has been explored simultaneously by means of small-angle x-ray scattering by different research teams in England and the U.S. (Campos-Lopez *et al.*, 1973; Holden *et al.*, 1969; Keller, 1971; Keller *et al.*, 1970; McIntyre and Campos-Lopez, 1970; Robinson and White, 1970). Keller *et al.* (1970) investigated an extruded rod of an SBS block polymer containing 25% polystyrene and having segmental molecular weights of 1×10^4, 5.5×10^4, and 1×10^4 g/mol, respectively. The x-ray diffraction pattern obtained is shown in Figure 4.6. Although some diffraction was obtained from the rod as received, lengthy annealing increased the intensity and improved the distinctness of the patterns. Keller (1971) and Keller *et al.* (1970) concluded that the polystyrene discontinuous phase was distributed as cylinders with hexagonal symmetry along the extrusion direction. In another study of a triblock copolymer, McIntyre and Campos-Lopez (1970) examined the low-angle x-ray diffraction of solution-cast films of an SBS polymer (segment molecular weights,

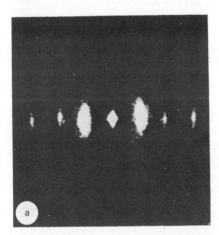 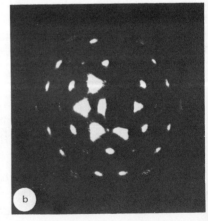

Figure 4.6. Low-angle x-ray diffraction patterns from heat-treated cylinders of an SBS block copolymer: (a) beam perpendicular to cylinder axis (vertical); (b) beam parallel to cylinder axis. (Keller *et al.*, 1970.)

in the sequence SBS, were 2×10^4, 6×10^4, 2×10^4 g/mol; 38 wt %
polystyrene). Based on results such as shown in Figure 4.7, McIntyre and
Campos-Lopez concluded that the polystyrene phase formed a face-centered
orthorhombic lattice of dispersed spheres (McIntyre and Campos-Lopez,
1970; Campos-Lopez *et al.*, 1973), as shown in Figure 4.8. It was noted that
the distance between nearest-neighbor surfaces, only 566 Å, is comparable

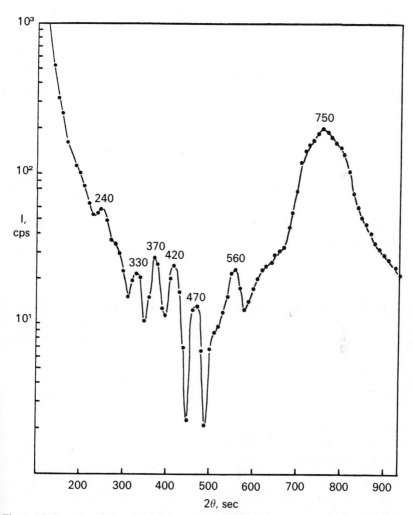

Figure 4.7. Intensity of scattering, corrected for infinite slits, against scattering angle (2θ) for a
solvent-cast film of triblock polymer, styrene–butadiene–styrene, having segment molecular
weights of 21,100, 63,400, and 21,100 g/mol. (McIntyre and Campos-Lopez, 1970.)

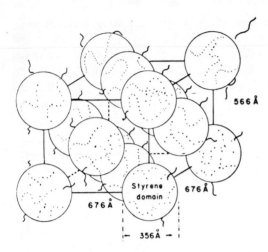

Figure 4.8. Scale model of proposed macrolattice of SBS block copolymer containing 38.5% styrene; segmental molecular weights are 21,100, 63,400, and 21,100 g/mol in sequence SBS. (McIntyre and Campos-Lopez, 1970.)

to the expected random coil end-to-end distance of the polybutadiene portion of the chain (see Table 1.2). However, the ends are not required to occupy domains at only this shortest distance; indeed, a distribution of distances between ends seems quite likely.

An as yet unresolved discrepancy between these two studies should be pointed out. Keller *et al.* (1970) observed cylinders for a material containing 25% polystyrene even after annealing; on the other hand, McIntyre and Campos-Lopez observed spheres for a material containing a higher concentration of polystyrene, 38.5%. In view of the work of Meier (1969) and others, spheres should have been obtained at the lower concentration, and cylinders at the higher. (See Section 4.7.) It is possible that the inversion of phase shape may be related to the fact that the specimens studied by Keller *et al.* (1970) were extruded rather than cast; certainly the shear forces present in extrusion can themselves affect particle shape (VanOene, 1972) (see Section 9.6).

4.6. THERMODYNAMICS OF DOMAIN CHARACTERISTICS

In the polyblends considered previously in Section 3.1, domain dimensions and fine structure were controlled by preparation techniques, such as by applying shearing forces during mixing, grafting reactions, etc. Most of the polymer molecules remained ungrafted, and were not constrained to appear at any particular location such as domain interfaces.

The picture in the case of block copolymers is quite different, for the dissimilar segments in each molecule are chemically linked together. If phase separation is reasonably complete, each block must be allowed residence in its appropriate domain. This requires that domain interfaces must be present at distances absolutely not greater than twice the contour length of the individual blocks, because of space-filling requirements. Simple knowledge that polymer molecules prefer coiled conformations suggests that domain interfaces are constrained to appear at intervals measured in hundreds, or at the most a few thousands, of angstroms, and that these distances will be controlled primarily by the molecular weights of the blocks. The phase domain surfaces in block copolymers must also have some peculiarities. Since all of the molecules on one side of the phase boundary are joined to molecules on the other side, the interface may be considered the locus of A–B bonds. However, if a two-dimensional quasi-lattice is constructed at the interface, it is obvious that not all the sites in the lattice can contain A–B bonds, because of the crowded conditions that would result. More realistically, most of the sites on this lattice must contain portions of chain segments that lie flat on it for a distance, as postulated in Figure 4.9. In fact, simple calculation using the data of McIntyre and Campos-Lopez (Section 4.5) shows that only about 5 % of the surface lattice sites are occupied by A–B junctures (Sperling, unpublished).

In the previous section on block copolymer morphology, the major domain shapes were identified as spheres, rods, and alternating lamellar structures. The next section will consider such problems as the thermodynamic criteria for phase separation, the factors controlling domain shape and size, and how the molecules are arranged within their respective domains.

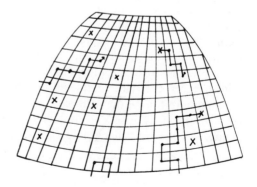

Figure 4.9. Quasilattice structure erected at phase interface. The crosses indicate —A–B—junctures. Molecules lying in the plane are indicated by –•–•–, with arrows showing sites of departure.

4.7. THERMODYNAMIC CRITERIA FOR PHASE SEPARATION

4.7.1. Zeroth Approximation

The basic thermodynamic equation giving the change in the Gibbs free energy of any process can be written

$$\Delta G = \Delta H - T\Delta S \qquad (4.4)$$

and the condition for equilibrium is of course that $\Delta G = 0$. In any problem involving the critical conditions for phase separation, appropriate values must be found for ΔH and $T\Delta S$; the general problem of mixing polymer molecules was first considered by Scott (1949). (See Section 2.5.)

Krause (1969, 1970, 1971) recently refined the Scott treatment by the addition of several terms specific to blocks. The final equation can be expressed as

$$\frac{\Delta G}{kT} = -\frac{V}{V_r} v_A^c v_B^c \chi_{AB}\left(1 - \frac{2}{Z}\right) - N_c \ln[(v_A^c)^{v^c A}(v_B^c)^{v^c B}]$$

$$+ 2N_c(m - 1)\ln\left(\frac{Z - 1}{e}\right) - N_c \ln(m - 1) \qquad (4.5)$$

where V is the total volume of the system; kT is Boltzmann's constant times the absolute temperature; V_r is the volume of a lattice site; Z is the lattice coordination number (number of first neighbors); v_A^c and v_B^c are the volume fractions of mers A and B in blocks; χ_{AB} is the Flory interaction parameter; m is the number of blocks in the block copolymer; e is the base of natural logarithms; and N_c is the total number of copolymer molecules in the system.

Let us consider the meanings of the terms in equation (4.5). The first term on the right is derived from the Van Laar heat of mixing,* and the second term on the right expresses the entropic decrease in available volume for each block during microphase separation, the volume available to each block being restricted to the volume fraction of the phase in question. The third term on the right represents the additional entropy decrease caused by immobilization of the A–B pair juncture at the phase interface. The last term on the right arises for copolymers containing three or more blocks, because of the interchangeable status of identical blocks in multiblock systems.

---

* The Van Laar heat of mixing term involves the reaction between 1, 1 contacts and 2, 2 contacts to form 1, 2 contacts: $[1, 1] + [2, 2] = 2[1, 2]$.

Table 4.3

$(\chi_{AB})_{cr}$ from Equation (4.5) for a Total Degree of Polymerization Equal to 400^a

Z	m	$(\chi_{AB})_{cr}$ if $v_A^c = 0.25$	$(\chi_{AB})_{cr}$ if $v_A^c = 0.50$
6	2	0.052	0.040
	3	0.080	0.060
	4	0.112	0.084
	6	0.181	0.137
	10	0.328	0.246
8	2	0.047	0.036
	3	0.071	0.053
	4	0.100	0.075
	6	0.161	0.121
	10	0.292	0.218

[a] Krause (1970).

At the critical condition for phase separation, χ_{AB} may be obtained in terms of DP, Z, m, v_A^c, and v_B^c; critical values of χ_{AB} for phase separation are tabulated in Table 4.3. Since the experimental value of χ_{AB} of the polystyrene–polybutadiene system is ~ 0.5, this polymer pair is quite incompatible. These values should further be compared to those obtained from the Scott (1949) treatment for blends having similar molecular parameters. For instance, calculation for a total DP of 400, $v_A^c = 0.25$, Z very large, and $m = 2$ yields $(\chi_{AB})_{cr} = 0.012$.

It may be concluded that blocks are more compatible than the corresponding blends, primarily because restricting the A–B link to the interface between the two phase domains decreases the free energy of phase separation, thus increasing $(\chi_{AB})_{cr}$. It may also be predicted from Table 4.3 that at equal composition and molecular weight, an increase in the number of blocks m increases the compatibility. The limit of large m corresponds to an alternating copolymer, in which case one stable phase is usually obtained. While the Krause approach has been successful in yielding valuable criteria for phase separation, neither the types of domain (spheres, cylinders, or alternating lamellae) nor their fine structure are suggested by the above thermodynamic treatment. These features will be considered below.

4.7.2. Dilute Solution Approach

Inoue *et al.* (1970*a,b*) considered phase separation of solutions of a block copolymer in hypothetical solvents as a function of polymer concentration in order to ascertain phase domain structure. As the concentration of polymer was increased, a critical concentration C^* was reached at which each block underwent phase separation and aggregated into characteristic molecular micelles (spheres, cylinders, or alternating lamellae) (Sadron and Gallot, 1973). The micelle structures thus formed, although perhaps not thermodynamically stable at higher polymer concentrations, would tend to be maintained because of the low diffusion rate of one block type through the other phase. The Gibbs free energy ΔG per unit volume for formation of the micelle is given by (Inoue et al., 1970a,b)

$$\Delta G = H \, \Delta W - T(\Delta S)N \tag{4.6}$$

where H is the interfacial contact area between the two components per unit volume, ΔW is the interfacial contact energy per unit area, ΔS is the entropy change per chain in micelle formation, and N is the number of copolymer chains per unit volume.

For AB-type polymers, the A–B junction point is taken at the micelle interface (Inoue et al., 1970a,b), as before. The ends of the molecules are, admittedly, unrealistically placed; for example, for spheres, the free end of the A block is at the center of the micelle and the bound end of the B block is constrained to a position on a sharply defined micelle spherical surface. Corresponding restrictions are placed on cylinder and lamellae formation. These assumptions, of course, lead to anomalous values for calculated densities, being high in the middle and low near the surface of the micelle.

The characteristic dimensions of the blocks were calculated assuming random flight statistics. Some of the interesting conclusions reached by Inoue et al. (1970a,b) are as follows.

1. The minimum free energy of micelle formation is positive when compared with the corresponding mechanical mixtures, i.e., the block copolymers are more compatible than the corresponding blends. This is a direct consequence of the existence of A–B junction points. (See Section 4.7.1.)

2. The equilibrium size of the micelles increases as the cube root of the degree of polymerization n ($n = n_A + n_B$) and with increasing ΔW.

3. The block segments are preferentially oriented along the direction perpendicular to the interface between the two phases.

4. The second phase will be in the form of spheres, cylinders, or lamellae, depending on the relative weight fractions of the A and B blocks, as illustrated in Figure 4.10. Spheres are predicted up to a weight fraction of about 0.16 for the second component, and cylinder formation between weight fractions of about 0.16 and about 0.39 of the minor component. Midrange compositions (40–60 % of each component) are predicted to have lamellar structures (Note the results in Figure 4.3.)

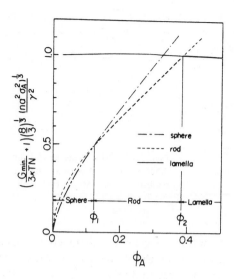

Figure 4.10. Relationship between the relative minimum free energy of micelle formation (A dispersed in B) and the weight fraction ϕ_A of A blocks, with the following assumed values: $\sigma_B/\sigma_A = 1$, $a_A = a_B = a$, and $M_A/M_B = 104/68$ (Inoue et al., 1970b). Values given for spheres, rods (cylinders), and alternating lamellae.

4.7.3. Diffusion Equation Approach

A somewhat more sophisticated approach to domain formation and fine structure was taken by Meier (1969, 1970). As with Inoue et al. (1970a,b) and Krause (1969, 1970, 1971), the A–B junction was restricted to a location somewhere in the interfacial region.* Meier's model (Figure 4.11) assumes that random flight statistics and regular solution theory hold, that statistical chain segments (not block lengths) are of equal size, and that chain perturbation is characterized by the usual parameter α:

$$\alpha = (\sigma_l^2)^{1/2}/(\sigma_l^2)_0^{1/2} \tag{4.8}$$

where σ_l is the end-to-end distance in the block in question (see Section 1.2). In contrast to previous workers, however, Meier employed a diffusion

---

* The problem of the distribution of A–B junctions in the interface has been treated recently by Meier (1971). The essence of the problem reduces to finding the junction at location X in an interface of thickness λ. The probability $P(x)$ is given by

$$P(x) = \frac{16}{\pi^2} \sum_{m,n \text{ odd}} \sin\frac{m\pi x}{T_A} \sin\frac{n\pi(\lambda - x)}{T_B} \exp -\frac{\pi^2}{6}\left(\frac{\sigma_A l^2 m^2}{T_A^2} + \frac{\sigma_B l^2 n^2}{T_B^2}\right) \tag{4.7}$$

where the T's are the lamella thicknesses for the A and B phases. As one might expect, the most probable position for the junction is the middle of the interface. Other workers have so far assumed a negligible distribution of A–B junctions at the interface.

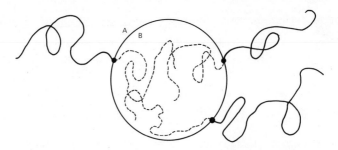

Figure 4.11. The Meier model for a spherical domain. Use of the diffusion equation permits chain ends to be randomly placed in a realistic manner. (Meier, 1969, 1970.)

equation to generate chain conformations that allow chain ends to fall throughout the domain in a realistic manner. Meier also recognized the physical requirement of constant density throughout a domain. In this light he devised density distribution functions that were solved by computer for various choices of domain size and chain dimensions. The results are presented schematically in Figure 4.12. From such calculations, Meier

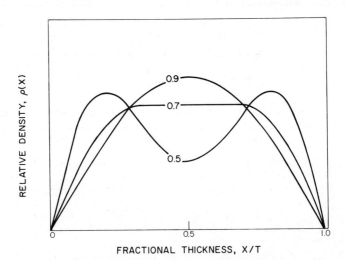

Figure 4.12. Domain density $\rho(x)$ as a function of domain position for three calculated values of chain end-to-end distance. Values on the curves are of $(\sigma_I^2)^{1/2}/T$. The value of $(\sigma_I^2)^{1/2}/T = 0.7$, yielding the most even (and hence most realistic) density, represents the best choice. (Meier, 1971.)

obtained the characteristic domain dimensions in terms of molecular parameters:

Sphere $\quad R = 1.33(Nl^2)^{1/2} = 1.33\alpha(Nl^2)_0^{1/2} = 1.33\alpha\bar{K}M^{1/2}$

Cylinder $\quad R = 1.0(Nl^2)^{1/2} = 1.0\alpha(Nl^2)_0^{1/2} = 1.0\alpha\bar{K}M^{1/2}$ $\qquad$ (4.9)

Lamella $\quad R = 1.4(Nl^2)^{1/2} = 1.4\alpha(Nl^2)_0^{1/2} = 1.4\alpha\bar{K}M^{1/2}$

where $\bar{K}$ is the experimental constant relating the unperturbed root-mean-square end-to-end distance to molecular weight, R is a characteristic dimension—the radius for spheres and cylinders and the half-thickness for lamellae—and n is the number of bonds of length l of the block within the specified phase. For some common polymers, values of $\bar{K}$ for R in micrometers are as follows:

Polystyrene	$7.5–8.6 \times 10^{-4}$
Polyisobutylene	$9.1–10.7 \times 10^{-4}$
Poly(dimethyl siloxane)	$7.9–8.1 \times 10^{-4}$
Poly(methyl methacrylate)	6.5×10^{-4}

Values of α vary between 1.0 and 1.5 for most cases of interest.

Meier also concluded that there is a mutual perturbation of chain dimensions, which is dependent on the ratio of the block molecular weights:

Sphere $\qquad \alpha_B \cong \alpha_A$

Cylinder $\qquad \alpha_B/\alpha_A = (M_B/M_A)^{1/6}$ $\qquad$ (4.10)

Lamella $\qquad \alpha_B/\alpha_A = (M_B/M_A)^{1/2}$

These chain perturbations are uniaxial, and not the isotropic expansion factors employed in solution theory. It should be noted that while Inoue (Section 4.7.2) found dimensions increasing as the cube root of the molecular weight, values estimated by Meier increase in a manner closer to the square root of the molecular weight, and depend upon the relative lengths of the blocks. It should be emphasized that the mutual perturbations arise from the necessity that space be filled by chains connected together such that the numbers of A and B chains per unit area of interface must be the same (Meier, 1972).

Meier calculated the contributions to the free energy, from the uniaxial expansion factors (α's) [see equation (4.10)], the surface free energy, and the constraint free energy. The latter term arises from the constraints that keep the A and B segments confined to restricted volumes. The relative free energies for the appearance of spheres, cylinders, and lamellar structures are shown

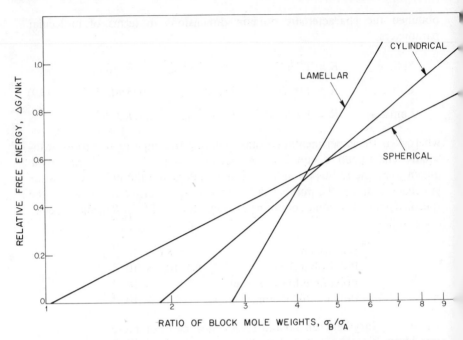

Figure 4.13. Relative free energy as a function of block molecular weight ratio. (Meier, 1971.)

in Figure 4.13.* Since the fine structure that has the lowest free energy will be the one to appear, these results predict spheres at molecular weight ratios M_B/M_A from infinity to four, a short range of compositions giving rise to cylinders, and the existence of alternating lamellar structures at midrange compositions. Figure 4.13 also shows that the molecular weight ratio is in fact the dominant parameter. Both Figures 4.10 and 4.13 predict spheres, rods, and lamellar structures for different relative block lengths, but differ in detail with respect to the composition ranges corresponding to each morphological variant.

In their calculations, Scott (1949), Krause (1969, 1970, 1971), and Inoue *et al.* (1970*a,b*) all assumed complete phase separation immediately following the critical condition. No allowance was made for partial phase separation, where polymer 1 enriches phase 2 to some degree, and vice versa. Although the question of partial phase separation in polymer blends generally has remained unresolved, Meier (1971) has treated the question as part of the general problem of the block copolymer interface; he found that if molecular weights are high and the interaction parameter χ_{AB} is large, then the fraction

---

* Meier later modified his surface energy term since he found little relationship between the interfacial properties of mixed homopolymers and block copolymers (Meier, 1971); value given in Figure 4.13 have therefore been recalculated accordingly.

Table 4.4
Fraction of Polymer in Interface[a]

$M\Omega$	Interfacial fraction
10^4	>0.5
10^5	0.25
10^6	0.05

[a] Meier (1971).

of material partially separated remains negligible. Conversely, if the values of the molecular weight, M, and χ_{AB} are low, a large fraction of the system will become mixed. The crucial parameter in Meier's calculations is $M\Omega$ where $\Omega = 2(\delta_1 - \delta_2)^2$ and δ represents the solubility parameter (Tobolsky, 1960, pp. 64–66). Table 4.4 gives values of the parameter $M\Omega$ vs. fraction of the material in the interfacial region. The values of δ for polybutadiene and polystyrene are 8.1 and 8.56, respectively (Tobolsky, 1960, pp. 64–66) showing that the fraction of material in the A–B interface of typical block copolymers can be appreciable.

4.8. EFFECT OF SOLVENT CASTING ON MORPHOLOGY

While the equilibrium thermodynamic approaches of Meier (1969, 1970, 1971) and Inoue et al. (1970a,b) predict that particular compositions will have particular fine structures, several investigators have shown that materials cast from different solvents and subsequently dried differ from each other and from materials prepared from the melt. As an example, let us examine the effects of the following solvents on a typical styrene–butadiene–styrene block copolymer: benzene/heptane 90/10 tetrahydro-furan/methyl ethyl ketone 90/10, and carbon tetrachloride (Beecher et al., 1969). The particular compositions were chosen to give selective solvating behavior. While benzene dissolves both blocks, the heptane component, which evaporates last, swells only the butadiene block. Tetrahydrofuran is also a mutual solvent; it evaporates first, leaving methyl ethyl ketone, which swells only the polystyrene block. Pure carbon tetrachloride is a mutual solvent. (Examples of swelling crystalline block copolymers are considered in Chapter 6.)

The damping vs. temperature characteristics of films dried from these three solvent systems indicate distinct differences in morphology, particularly in the degree of phase separation and phase continuity obtained, as shown in Figure 4.14 (Beecher et al., 1969). The results can be explained qualitatively

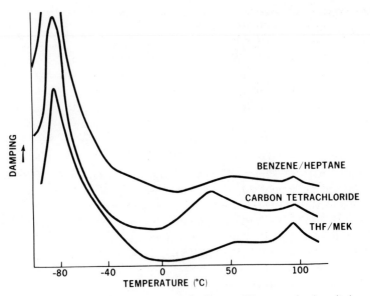

Figure 4.14. Damping vs. temperature for Kraton 101 on samples deposited
from different solvent systems. (Beecher *et al.*, 1969.)

by noting that the heptane-dried material has the largest low-temperature
maximum, suggesting that the polybutadiene component is more continuous. The use of methyl ethyl ketone gives a continuous polystyrene phase,
which results in a sharp maximum at 100°C. Carbon tetrachloride, on the
other hand, promotes phase mixing, which results in a maximum at an
intermediate temperature, near 30°C. This intermediate damping peak is
apparently due to a glass transition of polystyrene segments that are not
completely separated from the polybutadiene phase, but rather are in a
molecular solution. (Compare Figure 4.14 with Figures 4.2 and 4.5.)

A serious doubt arises about the phase domain stability of solvent-cast
materials. Certainly not all of them (perhaps none!) are in true thermodynamic equilibrium. Yet the difficulty of migration of one block type by
diffusion through the A phase of the other, especially at room temperature
and below, probably results in a series of pseudoequilibrium states that can
maintain their identity for long periods of time.

4.9. EFFECT OF DEFORMATION ON MORPHOLOGY

In the previous sections, the morphology of block copolymers was
shown to be complex, with the existence of two distinct phases or domains
being typical. The high strength of the thermoplastic elastomers was ascribed

to the presence of spheres or cylinders of plastic embedded within a continuous matrix of rubber. How exactly does this reinforcement occur? To answer this question, Beecher *et al.* (1969) took some remarkable photographs of an SBS block copolymer as it was deformed by stretching; the polymer was solvent-cast as a thin film so that a fibrous glassy polystyrene network evolved. Three distinct stages of deformation were observed, as illustrated in Figures 4.15–4.17. At first, strings of polystyrene spheres are seen to become aligned perpendicularly to the direction of the applied force. Subsequently, these strings of beads tend to orient in the force field and form V-shaped patterns. At higher force fields, the connections between particles break, and the particles become ellipsoidal in shape.

It is interesting to compare the effects of these morphological changes on mechanical properties, as illustrated in Figure 4.18, which compares the stresses developed on the first vs. the second deformation. The high strength of these materials is thought to originate from the inelastic deformation (creep or flow) of these spheres, a process that dissipates large amounts of strain energy and allows the remaining strain to be distributed more evenly throughout the mass.

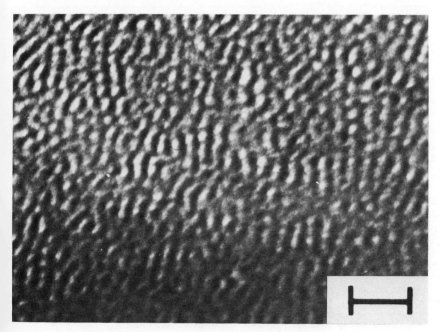

Figure 4.15. Transmission electron micrograph illustrating the first stage in the deformation process of SBS block copolymers (Beecher *et al.*, 1969): the formation of rows of spherical domains. Films were solvent-cast to a thickness of 500–800 Å, subjected to tensile stretch, and stained with osmium tetroxide ($\vdash\!\dashv$ = 1000 Å).

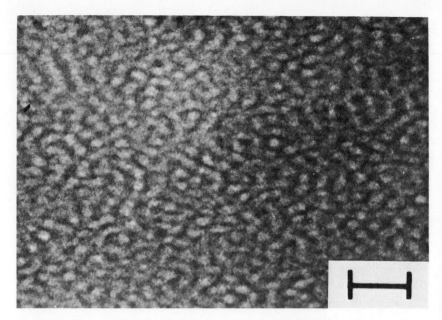

Figure 4.16. The second stage in the deformation process of SBS block copolymers: the distortion of rows into V's ($\vdash\dashv$ = 1000 Å). (Beecher *et al.*, 1969.)

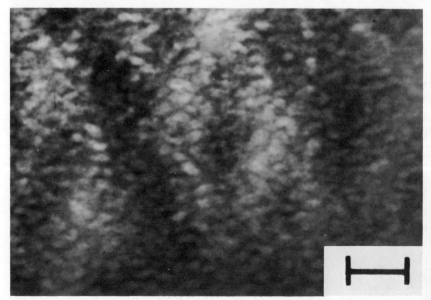

Figure 4.17. The third stage in the deformation process of SBS block copolymers: the breakdown of ordered structures into isolated ellipsoidal domains ($\vdash\dashv$ = 1000 Å). (Beecher *et al.*, 1969.)

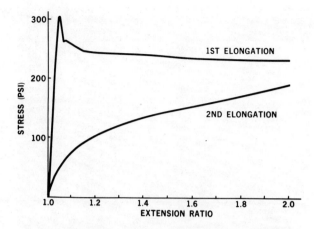

Figure 4.18. Stress–strain curve for SBS block copolymer (Kraton 101) deposited from solution (90/10: THF/MEK). Note behavioral changes brought about by altered morphology. (Beecher *et al.*, 1969.)

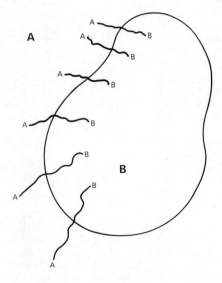

Figure 4.19. A–B block copolymers at the AB phase boundary. The effect is that of oil-in-oil surfactant emulsification.

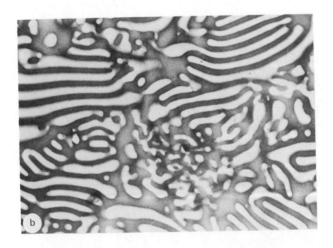

Figure 4.20. Electron micrographs of 40/60 poly(styrene-*b*-isoprene) systematically diluted with polystyrene and polyisoprene homopolymers. The two homopolymers have the same molecular weights as their respective blocks, and are added in a 40/60 ratio in order to maintain the same overall composition. The major effect is a progressive increase in phase domain dimensions, which eventually approach those of the cast blend structure (Kawai and Inoue, 1970). Compositions are shown in Figure 4.21 ; films are cast from 5 % solutions in toluene and are stained with osmium tetroxide.

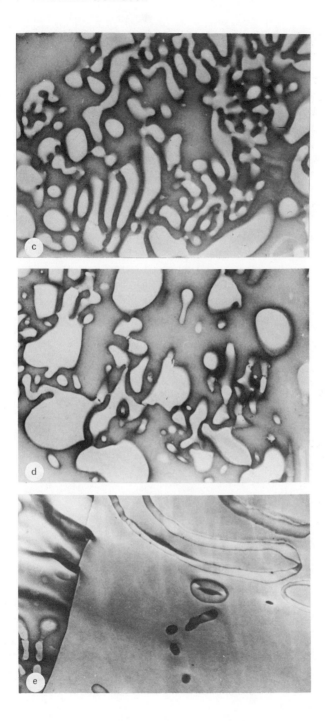

4.10. MIXTURES OF A–B BLOCKS WITH A AND B MECHANICAL BLENDS

The effects of adding various quantities of block copolymers to mechanical polyblends will be considered in this section. To a significant extent, the substance of this section will tie together portions of Chapters 3 and 4. Although the morphologies of the complete range of compositions will be discussed, the limiting compositions of low block content and low blend polymer content appear to be the more important ranges (Kawai and Inoue, 1970).

At low block copolymer concentrations, the blocks act as a substitute for graft-type molecules, serving to convert a mechanical blend into a pseudo-graft-type blend. The blocks tend to be concentrated at the phase boundaries, improving interfacial bonding and wetting. (See Figure 4.19.) As the block concentration is increased from zero, the phase domains become progressively smaller. In a broader sense, the blocks behave as emulsifiers, allowing further control over these tailor-made structures, which are of considerable commercial importance (Kohler *et al.*, 1968; Molau and Wittbrodt, 1968; Riess *et al.*, 1967).

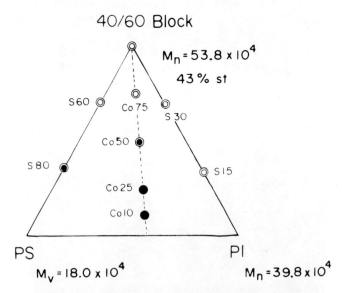

Figure 4.21. Ternary composition diagram showing positions of compositions in Figure 4.20. (Kawai and Inoue, 1970.)

At high block copolymer levels, the two homopolymers tend to dissolve within the block copolymer phase domains. Block copolymer-type morphology predominates, the main effect of the homopolymer additions being to increase the dimensions of the corresponding phase domains. In a certain sense the homopolymers behave as a "filler" for the blocks. Figure 4.20 shows the several morphologies existing in solution-cast films prepared by the systematic dilution of a block copolymer, poly(styrene-b-isoprene) by polystyrene and polyisoprene homopolymers; the compositions concerned are shown in the ternary composition diagram in Figure 4.21. (The morphologies shown in Figure 4.20 were obtained from ultrathin sections of films cast from 5% toluene solutions followed by staining with osmium tetroxide.)

4.11. RHEOLOGICAL BEHAVIOR OF BLOCK COPOLYMERS

The flow behavior of block copolymers differs from that of the parent homopolymers. Let us first examine the temperature dependence of the viscosity η for the thermoplastic elastomers. Below the glass transition temperature of polystyrene (about 110°C) the triblock material has a viscosity intermediate between that of the parent homopolymers, as shown in Figure 4.22. This is normal and expected. However, at a temperature where flow is well developed in the polystyrene, 140°C, an inversion occurs, the block copolymer assuming the higher viscosity (Holden et al., 1969b). The reason for this inversion lies in the difficulty of pulling styrene blocks out of their normal phase and into and through the polybutadiene phase, and vice versa. Motions of this type are required for viscous flow, and

Table 4.5

Viscosity of Block Copolymers vs. Homopolymers[a,b]

Polymer[b]	Viscosity, kP
80B	3.2
10S–53B–10S	29
16S–42B–16S	118
33S–18B–33S	28
83S	5.5

[a] Shear stress of 2×10^5 deg/cm^2 and a temperature of 175°C. Holden et al. (1969a).
[b] Molecular weights of blocks in thousands.

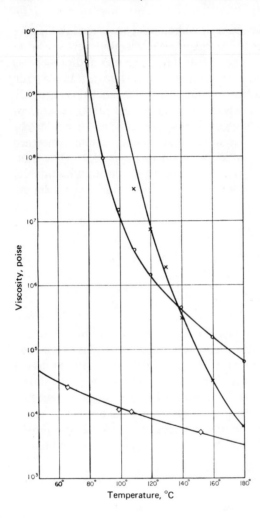

Figure 4.22. Viscosity of a triblock polymer compared to homopolymers of similar total block length. Constant shear stress was employed. (×) Polystyrene; (○) 10S-52B-10S; (◇) Polybutadiene. (Holden *et al.*, 1969*b*.)

hence flow remains impeded. Application of the WLF equation, which assumes a single phase, results in serious error.

The viscosities of these materials depend on the polystyrene/polybutadiene ratio. Table 4.5 compares the viscosities of polybutadiene, polystyrene, and three SBS triblock materials (Holden *et al.*, 1969*a*), all compositions having similar total molecular weights. It is clear that the highest viscosity is developed in materials having a midrange composition. One might imagine qualitatively that the destruction of the phase domains, necessary for normal flow, would be most difficult when the block lengths are similar.

The melt viscosities of block copolymers are also very significantly different from those of mechanical blends (Doppert and Overdiep, 1971; Meissner *et al.*, 1968; Work, 1972a). In the latter case the two phases are not joined chemically, and flow does not depend on destruction of the phases. Doppert and Overdiep (1971) point out that at high shear stresses the dispersed phase droplets are deformed into long streaks, and that the viscosity of the blend is *below* the weighted mean expected viscosity. This is because the lower viscosity polymer undergoes most of the deformation. A general treatment of the viscosity of two-component polymer systems is given in Section 9.6.

5

Multiblock Copolymers, Including Ionomers

As one of the blocks becomes polar relative to the other, the Krause theory (Section 4.7.1) predicts that phase separation will occur with shorter block lengths. Given the shorter block lengths, we are required to consider multiblock polymers of the form ABABAB ··· in order to maintain high molecular weight and concomitant polymer properties. This chapter will consider two important classes of multiblock copolymers, the polyurethanes and the ionomers.

Although polyurethane plastics and elastomers have seen extensive application for many years, the concept of phase separation in these materials and the establishment of its importance awaited the unraveling of the two phase morphologies in the diblock and triblock copolymers, the subject of the previous chapter. In retrospect, the elasticity of the polyurethane fibers and the toughness of polyurethane plastic foams are easy to explain with the realization that the highly polar urethane units undergo phase separation from the softer ether or ester components. The problem is complicated (and made more interesting) by the appearance of crystallinity in some of these materials, especially on stretching.

Differences in polarity are carried to the extreme in the ionomers, which feature several percent of salt-containing mers distributed randomly along an otherwise nonpolar, hydrocarbon chain. Although ionomers are actually a subclass of the random copolymers, phase separation and toughening resulting from domains of high salt concentration allow classification of these materials with the block copolymers. In this case, the "hard" blocks are reduced to but a single mer.

5.1. SEGMENTED POLYURETHANE ELASTOMERS

The unit that gives the polyurethanes their name stems from the basic urethane structure:

$$-N \overset{H}{\underset{}{\underset{\textstyle|}{}}} - \overset{O}{\underset{}{\overset{\textstyle\|}{C}}} - O-$$

However, the overall structure of the polymers as normally prepared is much more complicated. Two basic types of polyurethanes exist: the polyether-urethanes (Harrell, 1969, 1970),

$$G = +OCH_2CH_2CH_2CH_2 \,\overline{)_x} O-$$

$$B = -OCH_2CH_2CH_2CH_2O-$$

and the polyester-urethanes (Cooper and Tobolsky, 1966a,b; Estes et al., 1970):

(II)

Both of these materials consist of alternating "hard" and "soft" segments,* the lengths of which can be varied to produce desired properties. In most preparations, the hard and soft segments are both amorphous, differing in their glass transition temperatures. However, given sufficient regularity and molecular weights of 1000 or more in each segment (Huh and Cooper, 1971), the hard segment can crystallize (Flory, 1947), as shown by Harrell (1969). He prepared polyurethanes with a monodisperse hard segment distribution, systematically varied the hard portion of structure I from $n = 1$ to $n = 4$, and, as shown in Figure 5.1, measured the melting temperature T_m by means of differential scanning calorimetry (DSC). The hard segments exhibit sharp endothermic fusion peaks, the melting point increasing with the number of repeat units in the segments. Surprisingly, because of the short chain lengths involved, these melting points obey the Flory melting point depression equation (Flory, 1947)

$$\frac{1}{T_m} - \frac{1}{T_m^0} = \frac{2R}{\overline{H}_n n} \tag{5.1}$$

where the average heat of fusion per repeat unit $\overline{H}_n$, ~ 9.5 cal/g, is nearly the same as the value for the homopolymer composed only of the hard units, 11.4 cal/g. (When $n = \infty$, the homopolymer melting temperature will be reached.) The structures employed by Harrell were exceedingly

---

* The nomenclature prevalent in the literature refers to "segmented" polyurethanes, rather tha "block copolymer" urethanes. Not all types of polyurethanes form segmented, and henc block-copolymer-type structures, however.

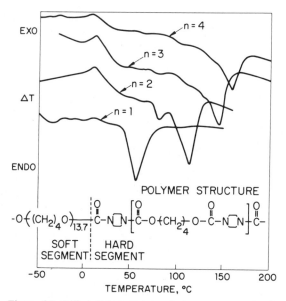

Figure 5.1. Differential scanning calorimetry scans of poly-urethanes with monodisperse hard segments of different sizes. (Harrell, 1969.)

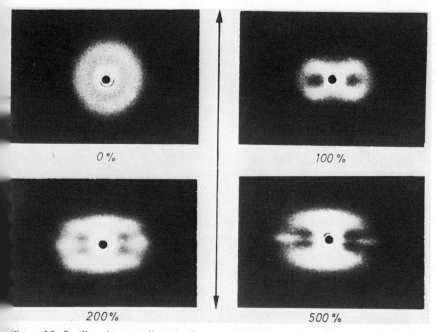

Figure 5.2. Small-angle x-ray diagrams for a polyether-based polyurethane as a function of elongation. (Bonart, 1968.)

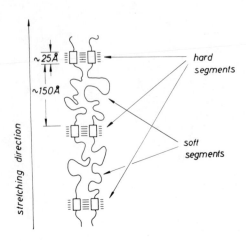

Figure 5.3. Principle of the buildup of a segmented urethane elastomer of hard and soft segments. (Oertel, 1965.)

regular; apparently most commercial hard polyurethanes are less crystalline. However, Bonart (1968) and Heidemann *et al.* (1967) (see also Oertel, 1965) showed that the polyether "soft" portion crystallizes on extension to yield a clear fiber diagram (Figure 5.2). This behavior is characteristic of other crystallizing rubbers, such as natural rubber or *cis*-polybutadiene (see Section 10.12).

The dimensions of the two phases can be estimated from small-angle x-ray diagrams of the type shown in Figure 5.2. As shown in Figures 5.3 and 5.4, Bonart (1968) estimated that material stretched approximately 500% had crystalline domains 25 Å thick separated by amorphous regions of 100–200 Å dimensions. Koutsky *et al.* (1970) verified the domain sizes obtained by Bonart (see Figure 5.5), using transmission electron microscopy with polyester- and polyether-urethane multiblock copolymers. Relatively thick films of the polyurethanes were placed on grids and solvent-etched with vapor of refluxing tetrahydrofuran to achieve workable thicknesses; optical contrast was improved by staining in an atmosphere of sublimed iodine. The dark domains are presumed by Koutsky *et al.* (1970) to be the hard aromatic-urethane microphase. The major conclusion drawn from these several investigations is that the phase structure of the polyurethanes is surprisingly like that of the diblock and triblock polymers, though the domains are smaller.

5.1.1. Modulus and Swelling Behavior

Cooper and Tobolsky (1966a) showed (Figure 5.6) that the modulus–temperature behavior of polyurethanes is characteristic of block copolymers

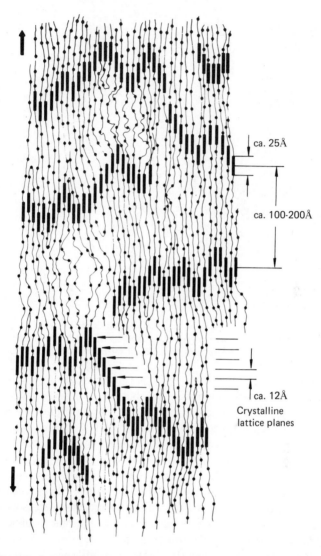

ca. 25Å

ca. 100-200Å

ca. 12Å
Crystalline
lattice planes

Figure 5.4. Schematic drawing of the structure of segmented polyurethanes stretched approximately 500%. The hard segments have turned into the direction of elongation and form paracrystalline layer lattices. The elongation crystallization of the soft segments has been reduced or has disappeared. On relaxation, the orientation of the hard segments is largely maintained. (Bonart, 1968.)

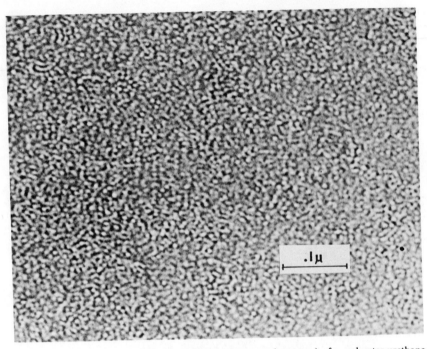

Figure 5.5. A bright-field image (transmission electron microscopy) of a polyester-urethane block copolymer film that was solvent-etched and stained by iodine. The dark domains are about 30–100 Å in width. Some samples exhibited variations in domain size. (Koutsky *et al.*, 1970.)

in general. Figure 5.6 compares the behavior of two typical commercial polyurethanes with schematic curves for linear amorphous, semicrystalline, and crosslinked polymers. Cooper and Tobolsky reasoned that if two phases really existed in these materials, each phase ought to swell independently. To test this assumption, they employed three plasticizer reagents: Carbowax, which solvates the polyester chain portion; dimethyl sulfoxide (DMSO), which, being highly polar, solvates the aromatic polyurethane (see structure II, p. 154); and Aroclor, which exhibits an intermediate solvating power. The modulus–temperature behavior of these swelled materials is summarized in Figure 5.7. Carbowax is seen to primarily affect the lower (polyester) transition, reducing T_g. DMSO has little effect on the polyester T_g, but lowers the modulus in the pseudorubbery plateau region. This results from a softening of the "hard" regions or perhaps from a reduction in crystallinity. The Aroclor plasticizer has an intermediate effect. These results support the hypothesis that polyurethanes do exhibit two distinct phases.*

* Stretched elastomeric polyurethanes obviously do crystallize; the degree of crystallinity developed in unstretched materials is uncertain, but certainly less.

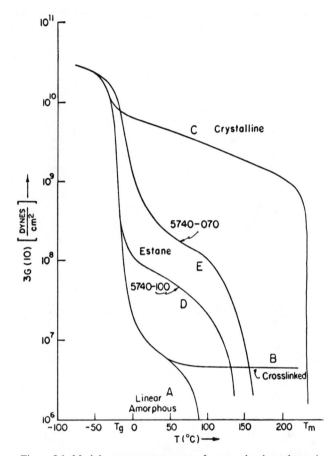

Figure 5.6. Modulus–temperature curves for several polyurethanes in comparison with various classes of polymer: (A) Schematic curve for a linear amorphous polymer; (B) schematic curve for a crosslinked polymer; (C) schematic curve for a semicrystalline polymer; (D) poly(ester urethane), Estane 5740-100; (E) poly(ester urethane), Estane 5740-070. Note that the polyurethanes exhibit two distinct glass transitions. (Cooper and Tobolsky, 1966a.)

5.1.2. Stress–Strain Behavior

Many types of graft and block copolymers, including the polyurethanes, stress-soften on repeated extension, as shown in Figure 5.8 (Estes *et al.*, 1970). The mechanism of stress-softening is believed to arise due to the deformation (and perhaps disruption) of the hard domains during the initial application of stress. As a result, lower moduli and stress levels

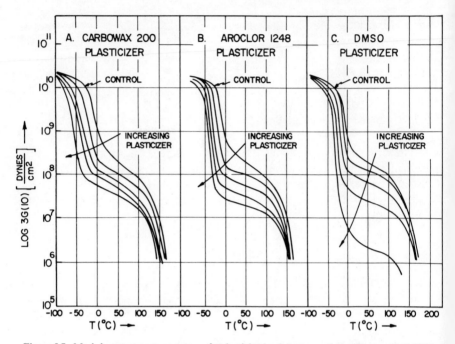

Figure 5.7. Modulus–temperature curves of a plasticized poly(ester urethane) (Estane 5740-070):
(A) Carbowax 200; (B) Aroclor 1248; (C) DMSO. (Cooper and Tobolsky, 1966a.)

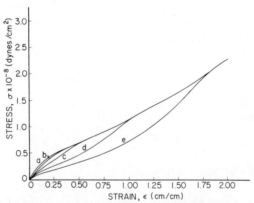

Figure 5.8. Stress–strain response of a polyether-urethane
(37.7% aromatic urethane) as a function of strain history
at 25°C. Prestrain: (a) 0%; (b) 25%; (c) 50%; (d) 100%;
(e) 200%. (Estes et al., 1970.)

are observed on subsequent testing; see for example, the discussion in Section 4.9. An interesting feature of Figure 5.8 is that once the prestrain value is exceeded, the stress–strain curve assumes values representative of virgin material.

5.1.3. Stress-Optical Behavior

Because these materials stress-soften on extension, their stress-optical behavior depends upon strain level and degree of prestrain (Estes *et al.*, 1969, 1970; Koutsky *et al.*, 1970). The classical equation for the stress-optical coefficient (SOC) derived from the theory of photoelasticity (Treloar, 1958) is not obeyed:

$$\text{SOC} = \frac{\Delta n}{\sigma} = \frac{2\pi}{45kTn}(n^2 + 2)^2(b_1 - b_2) \tag{5.2}$$

where σ is the true stress, Δn is the birefringence, kT is Boltzmann's constant times temperature, n is the refractive index of the unstrained sample, and $(b_1 - b_2)$ is the difference in polarizability of a polymer segment parallel and perpendicular to the direction of stretching. Instead, a semiempirical relationship concerning the *strain*-optical coefficient (STOC) $\Delta n/\varepsilon$ is obeyed,

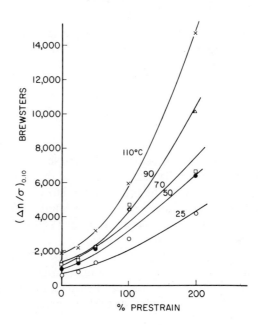

Figure 5.9. Effect of strain history on the stress-optical coefficient (measured at 10% strain) for a segmented polyester-urethane at various temperatures. (Estes *et al.*, 1969.)

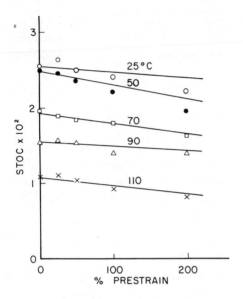

Figure 5.10. Temperature dependence of the strain-optical coefficient for a segmented polyester-urethane. Note that the STOC values are independent of percent prestrain. (Estes *et al.*, 1969.)

where ε is the strain. The STOC is obtained from the initial slope of the birefringence–strain plot (Estes *et al.*, 1969). This is because stress and birefringence do not reflect the same physical mechanisms, whereas strain and birefringence do. The dependences of SOC and STOC on temperature and prestrain are presented in Figures 5.9 and 5.10 for segmented polyester-urethanes.

5.1.4. Tensile Strength and Abrasion Resistance

The segmented polyurethane elastomers display very high tensile strength and abrasion resistance even though they are neither crosslinked nor filled in the classical sense of the term. That their tensile strength is associated with the presence of a hard phase is shown in Figure 5.11 (Aver, 1964; Havlik and Smith, 1964; Smith, 1970). The tensile strength is seen to level off at a concentration of 30–40 wt % hard groups, a level at which the material is fully reinforced.

The two-phase structure of polyurethane elastomers bestows unusual abrasion resistance (Baumann, 1969a,b). In fact, were it not for their high cost, polyurethanes would be excellent candidates for tire tread materials. For instance, under conditions described in ASTM test C-501 (Taber

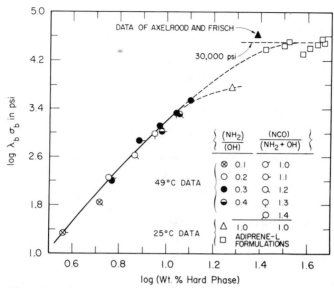

Figure 5.11. True tensile strength data for several types of segmented diisocyanate-linked polyurethane elastomers as function of urethane content. (Havlik and Smith, 1964.)

Abrasion wheel CS-17, 1000-g load, 5000 revolutions), polyurethane elastomer compares favorably with other materials (Table 5.1).

5.1.5. Some Generalizations

Although the segmented polyurethanes represent a particular class of the urethane polymers, we may speculate that many commercial polyurethanes are in fact block copolymers, and do exhibit phase separation. The controlling features would be the presence or absence of chain extension,

Table 5.1
Abrasion Resistance of Various Polymers[a]

Polymer	Weight loss, 10^{-3} g
Polyurethane elastomer	0.5–3.5
Natural rubber (tread formulation)	146
Styrene–butadiene rubber (tread formulation)	181
Ionomer	12

[a] Baumann (1969a, b).

or prepolymer formation of the ester or ether components. While nearly all of the scientific work has been concerned with polyurethane elastomers that have the rubber phase continuous, it is worthwhile to speculate further that at least some of the polyurethane plastics may also contain two phases, with the plastic phase continuous. In this case, however, chain-extended elastomer segments are seldom employed.

One may draw analogies with epoxy materials profitably at this point. Epoxy resins are cured with two types of compounds. The first comprises simple amines or acid anhydrides, which are low-molecular-weight compounds. The second comprises several kinds of polymers, which have high molecular weights. Experimentally, the latter types are much tougher than the former. Although good experimental evidence is lacking, the latter probably are phase-separated materials. The amide-cured epoxies have supermolecular structures of the AB crosslinked, or joined IPN type, as illustrated Sections 2.3 and 8.7.

5.2. CARBOXYLIC RUBBERS AND IONOMERS

In the preceding portions of this chapter and the previous chapter, two important classes of block copolymers have been considered:

1. Hydrocarbon or low-polarity materials. These polymers ordinarily contain a few relatively long blocks. An example given was the ABA triblock thermoplastic elastomers.

2. Blocks differing in polarity. The segmented polyurethanes containing the highly polar

$$-\text{N}\overset{\text{H}}{\diagup}\text{C}\overset{\text{O}}{\diagup}\text{O}-$$

units alternating with less polar groups exemplify this class of material. The block lengths are much shorter than in class 1, and these materials usually have many blocks of the form ABABABAB···.

As the polarity difference between the two blocks increases, $(\chi_{AB})_{cr}$ in the Krause (1969, 1970) theory (see Section 4.7.1) will become smaller, or alternatively, phase separation will occur for shorter block lengths. A limiting case in this argument involves the carboxylic rubbers and ionomers, in which one block consists of isolated single salt (ionic) mers, incorporated in an otherwise hydrocarbon-type backbone. It should be emphasized that the ionic block length is one monomer unit long. Phase separation and formation of ionic clusters occurs because of the extreme difference in polarity of the two blocks. The thermodynamic reason for the development of ionic clusters is elegantly summarized in the words of Tobolsky *et al.* (1968): "The tendency to segregate and form clusters is motivated by the highly unfavorable thermodynamic situation of ionic salts essentially dissolved in a hydrocarbon medium. The aggregation of the ionic groups from different chains relieves this energetically unfavorable condition. The long-range coulombic interactions between ions undoubtedly assists in setting up the clusters."

5.2.1. Carboxylic Rubbers

The carboxylic rubbers, which have been known since 1954 (Goodrich, 1954) to form a distinctive class of material (Tobolsky *et al.*, 1968), may be prepared by copolymerizing butadiene or other elastomer-forming monomers with small amounts of a carboxylic acid such as methacrylic acid. Crosslinking is accomplished by incorporation of a base or metallic oxide. High strengths are developed without the addition of reinforcing fillers, as illustrated in Table 5.2 (Tobolsky *et al.*, 1968). The dibasic zinc ions yielded higher tensile strengths than the equivalent quantity of monovalent sodium ions, probably because they behave more like a "crosslink."

Although no classical type of covalent crosslinking is introduced in the carboxylic rubbers, a broad and useful rubbery plateau region results. As shown in Figure 5.12, this plateau does not develop before zinc oxide neutralization, presumably because salt formation is required for phase separation. As also illustrated in Figure 5.12, hard plastic behavior results if the salt component predominates (Nielsen, 1969a; Nielsen and Fitzgerald, 1964). In this case, a small percentage of hydrocarbon units are incorporated

Table 5.2
Influence of Ions on the Bulk Properties of a Carboxylic Copolymer[a]

Polymer and treatment[b]	Tensile strength, psi	Elongation, %
Butadiene–methacrylic acid copolymer (0.12 ephr COOH)	100	1600
Treated with NaOH (0.12 ephr NaOH)	1700	900
Treated with ZnO (0.12 ephr ZnO)	6000	400
Sulfur vulcanizate (gum)	<500	—

[a] Tobolsky *et al.* (1968).
[b] ephr: equivalent parts per hundred.

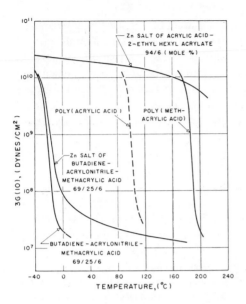

Figure 5.12. Modulus–temperature curves of poly(acrylic acid), a butadiene–acrylonitrile–methacrylic terpolymer (69:25:6), and their zinc salts. The polyacids have normal modulus – temperature behavior, while their salts are either soft elastomer or hard plastics, depending on salt content. (Tobolsky *et al.*, 1968.)

within an otherwise polyionic material. Although few data are available, it may be that the soft mers separate to form clusters, which cause a toughening action, reminiscent of the type encountered in high-impact plastics.

5.2.2. Ionomers

The ionomers (Bonotto and Bonner, 1968; Davis *et al.*, 1968; Delft and McKnight, 1969; Eisenberg, 1970; Kinsey, 1969; Longworth and Vaughan, 1968; Otocka and Kwei, 1968; Ward and Tobolsky, 1967) are usually formed by copolymerizing ethylene with about 5 mol % methacrylic acid, followed by neutralization. As shown by Kinsey (1969) (Figure 5.13),

Figure 5.13. Schematic structure of a plastic ionomer containing three phases. (Kinsey, 1969.)

these polymers are thought to contain three distinct phases: (1) amorphous polymer, (2) crystalline polymer, and (3) ionic clusters or domains.

Fairly direct evidence for phase separation in ionomers has been obtained by means of low-angle x-ray scattering (Ward and Tobolsky, 1967), as shown in Figure 5.14. Figure 5.14a depicts a common situation, in which inhomogeneities of all types (crystals, phase separation, impurities) become more and more frequent as their size decreases, with no maximum. The maximum in Figure 5.14b, on the other hand, suggests the presence of uniformly sized particles. If spherical domains are assumed, a diameter of 83 Å can be calculated (Guinier and Fournet, 1955).

On the basis of recent studies with both sodium and zinc-type ionomers, Cooper and co-workers (Marx et al., 1973) have developed a new "aggregate" model for the morphology of ionomers. The model emphasizes the presence of small aggregates homogeneously distributed throughout the amorphous phase. While the increase in aggregation observed with increasing ionic level may be deduced from general principles, the finding that plasticization with water increases the degree of aggregation is somewhat unexpected. In addition to presenting the new model, this paper (Marx et al., 1973) also effectively reviews the state of the art.

Ionomers are very tough materials, usually requiring several times more energy to break than polyethylene (Kinsey, 1969). Other properties, however, appear of paramount importance. Creep is considerably suppressed by the action of the ionic crosslinks, yet these materials retain thermoplastic characteristics. In fact, because of their thermoplastic nature, their maximum

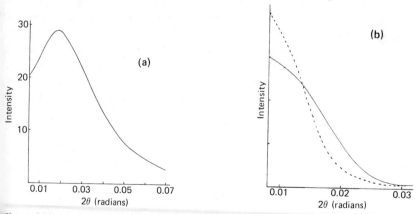

Figure 5.14. (a) Angular dependence of low-angle x-ray scattering pattern from low-density polyethylene (- - -) and an un-ionized acid polymer (—). (b) Angular dependence of low-angle x-ray scattering pattern from 59% ionized cesium salt. (Delft and McKnight, 1969.)

use temperature is only 160–180°F. The presence of ionic groups on the surface of the material improves adhesion, particularly for metal laminate applications. Their very small domain size lends transparency, and hence furthers their use as packaging materials.

5.2.3. Phase Structure of Ionomers

The crystallinity present in commercial, polyethylene-type ionomers complicates their morphology and mechanical behavior, such as creep. In an effort to obtain more easily interpreted data, Eisenberg and co-workers (Eisenberg, 1973) recently investigated the behavior of polystyrene-based ionomers, with sodium methacrylate serving as the ionic constituent. As the ion concentration was increased to 6 mol %, the glass temperature was noted to increase normally, as would be expected of a random copolymer; however, at approximately 6 mol % ionic concentration, the glass temperature suddenly jumped. Also, time–temperature superposition of relaxation data failed to hold above this composition. At low ion concentration, the water uptake was approximately one molecule per ion pair. Above 6 mol % ions, however, the water uptake jumped to the level of 3–5 molecules per ion pair (Eisenberg and Navratil, 1972). Such information, of course, strongly suggests a phase transformation at the crucial concentration. Indeed, a small-angle x-ray diagram of cesium ionomers indicates regular spacings about 70 Å apart (Eisenberg, unpublished). Eisenberg and co-workers (Eisenberg, 1973) concluded that below 6% ionic concentration, two, three, or four ion pairs cluster together, while above 6%, aggregation on a larger scale develops. Almost certainly, however, some organic segments are present among the ion pairs, giving rise to the broadened mechanical spectra observed.

In summary, although the ion-containing polymers actually consist of random mer sequences, the individual ionic mer components differ so greatly chemically from their hydrocarbon surroundings that phase separation ensues. Thus the ionomers behave in a manner similar to the multiblock copolymers.

6

Crystalline
Block
Copolymers

Most of the polymer blends, grafts, and blocks examined in the preceding chapters formed two amorphous phases. When one phase was plastic and stiff, and the other rubbery and soft, we observed that toughened materials resulted. In this chapter we examine several types of block copolymers in which one or *both* components crystallize; in particular we consider three possible combinations of such blocks:

1. A glassy/crystalline combination in the form of polystyrene/poly-(ethylene oxide), PS/PEO. This block copolymer, which has been examined in some detail in the literature, will also serve as a model system.

2. A rubbery/crystalline combination consisting of poly(dimethyl siloxane)/poly(diphenyl siloxane). The crystalline, high-melting diphenyl siloxane blocks alternate with rubbery dimethyl siloxane blocks, and the resulting copolymer behaves like a thermoplastic elastomer.

3. A crystalline/crystalline combination, polyethylene/polypropylene. This type of crystalline/crystalline block copolymer is sometimes referred to as a "polyallomer."

As in previous chapters, synthesis, morphology, and physical and mechanical behavior will be considered. In addition, dilute solution behavior and kinetics of single crystal formation will be treated, also using PS/PEO block copolymers as models, since these block copolymers have been investigated the most thoroughly.

6.1. CRYSTALLIZABLE BLOCK COPOLYMERS: STYRENE–ETHYLENE OXIDE

In the 1950's the synthesis of block copolymers was a major topic of interest in polymer science. One of the most general techniques was the "living polymer" anionic polymerization developed by Szwarc and his group (Szwarc, 1956; Szwarc et al., 1956). By this technique Richards and Szwarc (1959) prepared A–B block copolymers of styrene (A) and ethylene oxide (B). More detailed studies of the preparation and properties of block copolymers A–B, B–A, A–B–A, and (A–B)$_n$ were undertaken by a group of scientists at the Centre de Recherches Sur les Macromolecules, Strasbourg (Finaz et al., 1961, 1962; Franta et al., 1965; Gervais and Gallot, 1970; Grosius et al., 1970; Kovacs, 1967; Kovacs et al., 1966; Lotz, 1963; Lotz and Kovacs, 1966, 1969; Lotz et al., 1966; Manson and Kovacs, 1965; Sadron, 1962a,b, 1963, 1966; Skoulios and Finaz, 1962; Skoulios et al., 1963). These investigators studied block polymers in which one component could crystallize, and were also among the first to examine phase separation and transition (T_g and T_m) relationships (usually, though not always, in solvated systems) in block copolymers generally. Although other block copolymers have been studied by this group, e.g., styrene with isoprene or 2-vinyl pyridine (Franta et al., 1965), major emphasis has been placed on the ethylene oxide types.

In addition to phase and transition phenomena, the Strasbourg studies have included the following topics: the morphology of gels and of the crystals themselves (Finaz et al., 1961; Kovacs, 1967; Kovacs et al., 1966; Lotz, 1963; Lotz and Kovacs, 1966; Lotz et al., 1966; Manson and Kovacs, 1965); criteria for crystal stability and fractionation (Lotz and Kovacs, 1966); and kinetics of crystallization (Kovacs et al., 1966; Lotz, 1963; Lotz and Kovacs, 1966; Lotz et al., 1966; Manson and Kovacs, 1965). Behavior in the solid state has received some attention (Crystal et al., 1969; Erhardt et al., 1969; Litt and Herz, 1969; O'Malley et al., 1969; Pochan, 1971); transition and flow behavior of melts and solids has also been examined in some detail (Crystal, 1971; Crystal et al., 1969, 1970; Erhardt et al., 1969, 1970; O'Malley et al., 1969, 1970; Pochan, 1971).

Since observations and conclusions should be relevant not only to the case of block copolymers but to graft copolymers as well, and also to ordering and crystallization phenomena in general, considerable attention will be given to this interesting type of block copolymer in the following pages.

6.1.1. Synthesis and General Properties

As mentioned above, synthesis of A–B and B–A–B types* may be accomplished by the living polymer technique, as follows (see Section 4.1 (Finaz et al., 1962; O'Malley et al., 1969; Szwarc, 1956; Szwarc et al., 1956)

---

* In this discussion, the nomenclature of Finaz et al. (1962) is retained, that is, A for styrene and B for ethylene oxide sequences.

A–B type:

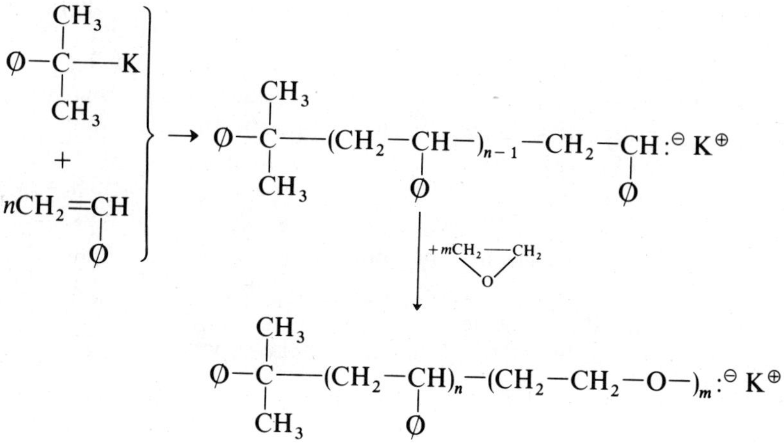

B–A–B type:

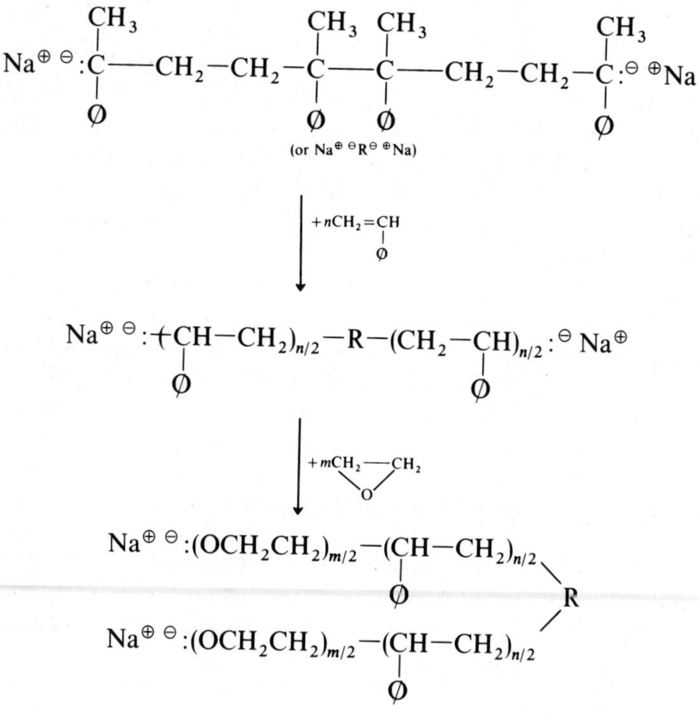

$$(6.2)$$

In equation (6.1), a simple anion is employed, while in equation (6.2), a dianion is used to generate both B blocks simultaneously. Other block structures may be synthesized by different techniques. For example, the A–B–A type may be prepared by the reaction of PS containing terminal acyl chloride groups with poly(ethylene glycol) (Finaz et al., 1962).

A wide range of compositions may be achieved by systematically varying the molecular weights of each component. Typically, values of w, defined as $M_n(PS)/M_n(PS–PEO)$, and equal to the weight fraction of polystyrene, may span the range 0.1–0.9 and number-average molecular weights in the ranges $(5–50) \times 10^3$ (PS sequence) and $(8–90) \times 10^3$ (PEO sequence) are common. These ranges may be obtained in PS–PEO combinations by varying the quantities of monomer added; in the case of PEO–PS–PEO, the lengths of the PEO sequences must, of course, be equal. In any case, the polymerization can be readily adopted to make copolymers rich in either PS or PEO.

6.1.2. Mesomorphic Gels

The state of organization of the PS–PEO block polymers in the swollen, dispersed, or dissolved states has been shown by Sadron, Skoulis, and others (Finaz et al., 1961; Gervais and Gallot, 1970; Kovacs et al., 1966; Lotz, 1963; Lotz and Kovacs, 1966; Lotz et al., 1966; Manson and Kovacs, 1965; Sadron, 1962b, 1963; Skoulios and Finaz, 1962) to depend primarily on three factors: the relative affinity of the solvent for both polymer sequences, the polymer concentration, and the temperature. One may obtain a wide variety of forms, ranging from mesomorphic gels* to spherulites and single crystals, depending on the above conditions.

Let us examine the dilute solution state first (Battaerd and Tregear, 1967, pp. 138, 139). Each block may be expected to adopt a conformation and end-to-end distance more or less similar to that expected for the homopolymer in the same solvent. Thus, as shown in Figure 6.1, one block of the chain may be highly extended while the other is relatively tightly coiled (Finaz et al., 1961; Kotaka et al., 1972; Saam et al., 1970; Skoulios and Finaz, 1962). Clearly the more highly extended portion contributes the major share to such properties as intrinsic viscosity. In the limiting case of poor solubility for one component, that component may ball up as a colloidall insoluble phase. The above discussion applies to block copolymers generall (see Section 4.2); however, if the insoluble block is capable of crystallization

---

* The term "mesomorphic gel" refers to a swollen material exhibiting a type of supermolecula organization, such as crystallinity (Battaerd and Tregear, 1967, pp. 138, 139; Mabis, 1962).

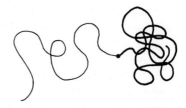

Figure 6.1. Schematic of a block copolymer in dilute solution, with one portion tightly coiled and the other highly extended.

such as in the case of PEO, a crystalline colloidal phase may form. It should be emphasized that in either event, with sufficient solvating power, the other block remains in true solution, and hence precipitation in the classical sense does not occur (Richards and Szwarc, 1969). Thus at low concentrations, block copolymers with one soluble and one insoluble block may be considered as monomolecular micelles.

As the concentration is increased in a solvent preferential for one component, the molecules tend to become aggregated in an organized manner, with local phase separation of the more insoluble component in the form of mesomorphic gels. Depending on the temperature, solvent, and concentration, spheres, cylinders, or alternating lamellar units (Figure 6.2) may be detected by methods such as x-ray diffraction; other things being equal, the spheres and cylinders tend to occur at lower concentrations than the alternating lamellar units (Finaz *et al.*, 1961, 1962; Gervais and Gallot, 1970; Skoulios and Finaz, 1962).

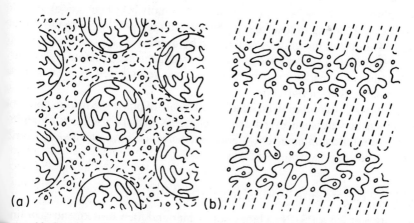

Figure 6.2. Structure of a copolymer of type A B made from polystyrene and poly(ethylene oxide): (a) in nitromethane (cylindrical structure) and (b) n-butyl phthalate (layer structure). Nitromethane dissolves preferentially the poly(ethylene oxide) part, butyl phthalate the polystyrene part (Sadron, 1962b). (—) polystyrene part, (- - -) poly(ethylene oxide) part, (○) solvent.

Because of inherent polar characteristics, PS tends to be soluble in nonpolar solvents, while PEO is soluble in more polar ones (in fact, PEO homopolymer is soluble in water). Thus at room temperature a concentrated solution of a PS–PEO block copolymer [$w = 0.41$; $M(PS + PEO) = 1.4 \times 10^4$] in nitromethane—a *poor* solvent for polystyrene—was shown to give a mesomorphic gel containing parallel cylinders of polystyrene units in a matrix of swollen and disordered poly(ethylene oxide). The cylinders were approximately 100 Å in diameter, and of varying lengths, with an axial separation of approximately 200 Å (Figure 6.2a). On the other hand, when the same copolymer was dissolved in butyl phthalate (a *good* solvent for polystyrene), the mesomorphic gel then obtained consisted of alternating lamellae, with layers of crystalline poly(ethylene oxide) approximately 100 Å thick, interspersed with layers of swollen polystyrene, also approximately 100 Å thick (Figure 6.2b). In certain other cases, spheres may be obtained (Franta *et al.*, 1965; Skoulios *et al.*, 1963) in either random or cubic arrays. Thus under appropriate conditions a wide variety of organizational types may be obtained, with one component or the other, or both, being continuous. These findings are roughly comparable to the behavior of amorphous diblock copolymers (see Chapter 4), the special condition of crystallinity being imposed for PEO in solvents poor for it.

It is interesting to note that the mesormorphic arrangements can be preserved in a rigid matrix by replacing the solvent with a polymerizable monomer such as acrylic acid, and polymerizing the new matrix. In one such example, Sadron (1962b) and Skoulios and Finaz (1962) were able to embed cylinders of polystyrene in polyacrylic acid by first forming a mesomorphic gel of polystyrene cylinders in a poly(ethylene oxide) matrix, acrylic acid serving as a solvent for PEO, and then polymerizing the acrylic acid. The technique could presumably be applied to ionized polymers as well. Indeed, in prior studies of micellar arrangements, Husson *et al.* (1961) successfully embedded micelles of a potassium soap (based on a substituted undecanoic acid) in polystyrene.

One may expect that, especially if the scale of phase separation could be extended at will, such preparative techniques may be useful in obtaining types of multicomponent polymeric systems which have unusual dielectric, optical, and mechanical properties.

6.1.3. Crystalline Forms from the Melt

By melting a PS–PEO block copolymer and then subsequently cooling (Figures 6.3 and 6.4), Kovacs (1967) showed that spherulites could be obtained with copolymers containing at least up to about 50% PS. Even though such block polymers contain a relatively large fraction of an uncrysta

Figure 6.3. Spherulitic texture of a thin film of a styrene–ethylene oxide block copolymer ($w = 0.40$) obtained on quenching to 20°C (Kovacs, 1967). Photomicrograph taken with film between crossed Nicols ($\sim 100 \times$).

lizable group, the driving force for crystallization may overcome the hindrance otherwise expected. If the hindrance becomes large, as when $w > 0.50$, the addition of a solvent preferential for PS, e.g., ethylbenzene, may still make spherulitic growth possible (Kovacs, 1967).

6.1.4. Crystalline Forms Cast from Solution: Spherulitic Morphology

When films of PS–PEO block copolymers were cast from solution, wide variations in texture were reported by Crystal et al. (1969, 1970). In these

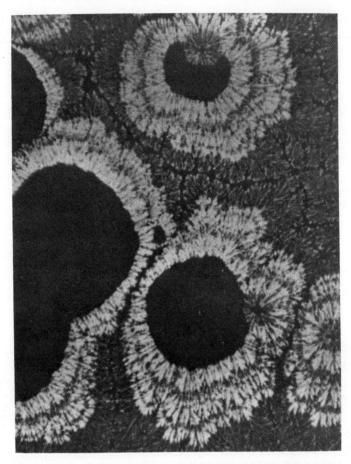

Figure 6.4. Spherulitic structure of a film of a styrene–ethylene oxide block copolymer ($w = 0.40$) obtained on quenching to about 35°C (Kovacs, 1967). Photomicrograph taken with film between crossed Nicols ($\sim 100 \times$).

studies, three casting solvents were used: nitromethane (preferential for ethylene oxide), ethylbenzene (preferential for polystyrene), and chloroform (a good solvent for each component). Employing both optical and electron microscopy, these workers observed that the particular morphology depended on the composition of both the copolymer and the solvent. Such variations are certainly to be expected, in view of the doubly antithetic composition A and B incompatible; B readily crystallizable, A not. Although crystallization of the PEO segments occurred, the nature of the crystallization is affected by the length of the PEO segments and the solvent.

After drying, all films were negatively birefringent at temperatures up to the melting point of the PEO component, at which point the sign of the birefringence quickly changed to positive, with the value of the birefringence being a direct function of the PS content. Interestingly, even at 250°C, birefringence, and hence ordering, persisted.

In general, the films contained spherulitic or pseudospherulitic growth (see Figures 6.5–6.7) (Crystal *et al.*, 1970). The degree of perfection tended to be lower in films cast from ethylbenzene, because the highly swollen PS structures tend to form the more continuous phase, and in films containing more than 40% PS. The spherulitic structure of homopolymer PEO is shown in Figure 6.8 for comparison; relatively perfect structures were obtained using chloroform as the casting solvent.

Films from nitromethane also exhibited rather perfect spherulites of PEO, which contained globules of the insoluble PS (Figure 6.5). On the other hand, films from chloroform, which, being a good solvent for both components, permits more efficient organization during drying, show spherulites in which the fibrils appear to be coated more or less uniformly with polystyrene.

In each case, the basic *crystalline* subunit is the folded-chain lamella, with a thickness of about 180 Å. However, as the content of PS increases,

Figure 6.5. Optical micrograph of a film of PEO/PS block copolymer (19.6%) cast from nitromethane, and observed using crossed polarizers. The white markers are 250 μm apart. (Crystal *et al.*, 1970.)

Figure 6.6. Optical micrograph of a film of **PEO/PS** block copolymer (19.6% PS) cast from chloroform, and observed using crossed polarizers. The white markers are 250 μm apart. (Crystal *et al.*, 1970.)

Figure 6.7. Optical micrograph of a film of PEO/PS block copolymer (19.6% PS) cast from ethylbenzene and observed using crossed polarizers. The white markers are 100 μm apart. (Crystal *et al.*, 1970.)

Figure 6.8. Optical micrograph of a film of PEO homopolymer, 37,000 M.W., cast from chloroform, and observed using crossed polarizers. The white markers are 125 μm apart. (Crystal *et al.*, 1970.)

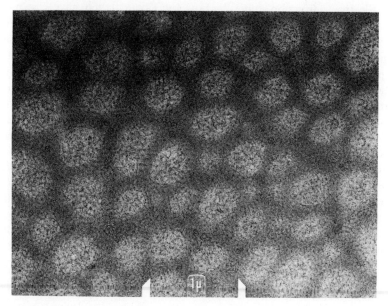

Figure 6.9. Transmission electron micrograph of PEO/PS block copolymer (70.0% PS) cast from ethylbenzene (OsO$_4$ staining). The white markers are 1 μm apart. (Crystal *et al.*, 1970.)

the PS segments play a major role in disturbing the ordering process. In the light of the interpenetrating networks (Chapter 8), it is interesting to note in Figure 6.9 a network consisting of spheres rich in PEO lamellae, but containing some PS segments (dark spots) embedded in a PS honeycomb. The PS obviously forms the more continuous phase in this instance.

6.1.5. Crystalline Forms from Dilute Solution: Single Crystals

During the investigation of mesomorphic gels of PS–PEO block copolymers, it was found that one could obtain not only alternating lamellar structures in ethylbenzene, but also PEO single crystals having exceptionally regular habits (Kovacs, 1967; Kovacs et al., 1966; Lotz, 1963; Lotz and Kovacs, 1966; Lotz et al., 1966; Manson and Kovacs, 1965). Such crystals were obtained by first heating a dilute suspension (0.01 g/ml) of the copolymer in ethylbenzene, amyl acetate, or toluene above the PEO melting point, and then cooling. The most typical form for a crystal was a regular square, though under particular conditions such as higher initial concentration, degenerate forms such as hexagons and "shishkebabs" may be found (Kovacs et al., 1966; Lotz and Kovacs, 1966; Lotz et al., 1966).

Electron diffraction studies by Lotz and Kovacs (1966) have shown that the unit cell of the crystals formed by precipitation is identical to the unit cell of poly(ethylene oxide) homopolymer. As with polymer single crystals in general (Fava, 1969; Geil, 1963), the c axis was found to be approximately perpendicular to the plane of the crystals. Since electron micrographs indicated a thickness of the order of 100 Å for the crystals, it was concluded that the polymer chains in the crystals must be folded, and that an appropriate model would be the sandwich described in Figure 6.10, in which folded, crystallized chains of PEO are sandwiched between layers of uncrystallized amorphous PS (Lotz and Kovacs, 1966; Lotz et al., 1966). Concurrent studies also revealed the existence of double-layered lamellae of PEO (Figure 6.10). In a later investigation, Crystal et al. (1970) found spacings of 95 or 180 Å, which they also ascribed to single- or double-layered lamellar formation, respectively. The extensive morphological studies are interesting in their own right, and the reader is referred to Lotz and Kovacs (1966) and Lotz et al. (1966) for more details.

Although at high concentrations of PS ($w \sim 0.9$), remarkably perfect crystals can still be grown by using a solvent preferential for the uncrystallizable PS, the strong tendency for PEO to crystallize can, in general, only overcome the significant encumbrance of the PS segments at PS concentrations of less than about 40%. In fact, however, the presence of the PS layers

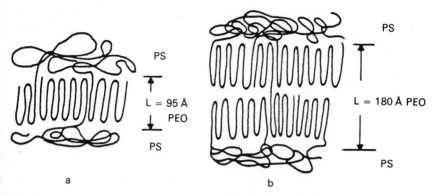

Figure 6.10. Models of (a) single and (b) double folded chain lamellae in PEO/PS block copolymers. (Crystal *et al.*, 1970.)

sometimes inhibits the formation of screw dislocations and other irregularities, and thus confers mechanical stability on the crystals, making them useful for studies of crystallization phenomena per se. It should be stressed that single-crystal formation in dilute solution is favored by the use of good solvents for PS; in contrast, film evaporation techniques demand a good solvent for PEO if good crystallization is to be achieved.

6.1.6. Bulk Thermal and Transition Behavior

The first study of both first- and second-order transitions in PS–PEO block copolymers ($w = 0.3$–0.9) was conducted by Lotz and Kovacs (Kovacs, 1967; Lotz and Kovacs, 1969; Sadron, 1962a), using a sensitive dilatometric technique. In each case, T_m, the melting point of the PEO segments, T_f, the freezing point of the PEO segments, and T_g, the glass temperature of the PS segments, were measured. As a result of these studies (Lotz and Kovacs, 1969), it was possible to construct a phase diagram (Figure 6.11). A clearly defined T_m exists for all samples; as might be expected, the value of T_m, though always in the range of 50–60°C, varies inversely with the fraction of PS. Since in each sample the PEO sequences were identical, the variation in T_m must be due to variations in the perfection of lamellar units. This conclusion is consistent with the notion of an increasing degree of imperfection as styrene content increases, as shown previously (Crystal *et al.*, 1970; Erhardt *et al.*, 1970; O'Malley *et al.*, 1970), and perhaps with the tendency for the triblock polymers to melt at slightly lower temperatures than the diblock polymers.

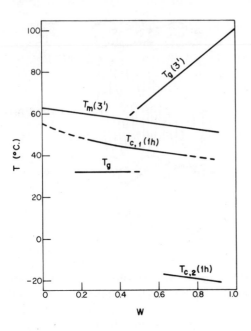

Figure 6.11. Phase diagram for PEO/PS block copolymers in terms of weight fraction polystyrene content w. (Lotz and Kovacs, 1969.)

A more recent study of solvent cast films by O'Malley *et al.* (1970), using scanning calorimetry, includes PEO–PS–PEO triblock polymers as well as the PS–PEO diblock materials. Values of T_g for the PS segments were found over the whole PS concentration range in the DSC study (O'Malley *et al.*, 1970), but only for $w \geq 0.4$ in the dilatometric study. In each case, T_g varied directly with PS content—probably because the molecular weight of the PS segment varied from the rather low value of 3000 to 75,000 g/mol. However, a wide discrepancy exists between the DSC and the dilatometric values, the T_g's measured by DSC being about 30°C higher than obtained by dilatometry; this observation may reflect the faster heating rate of DSC studies.

In any case, the general transition behavior is quite consistent with the notion of separation into two phases: one, crystalline PEO; the other, glassy PS. Further, the separation must be essentially complete, for the glass phase obviously exists in the presence of the molten crystalline phase ($T_g > T_m$). (It should be pointed out that a case in which $T_g > T_m$ has never been observed for homopolymers.) The particular finding here, of course, reflects the existence of two phases, each behaving nearly independently with respect to transition temperatures.

Several other features of the transition behavior are worth noting. First, both the DSC and dilatometric experiments revealed the existence of *two* crystallizing temperatures, about 60°C apart, for copolymers rich in

PS ($w > 0.6$). This phenomenon may be attributed to the occurrence of both heterogeneous ($T_m \cong 50$–$60°$C) and homogeneous ($T_m \cong -20°$C) nucleation. Thus, when domains of PEO are of molecular dimensions, and are dispersed in a matrix of glassy PS, some domains contain heterogeneous nuclei and some do not. Since most crystallization appears to be homogeneously nucleated, the domain size must in fact be small and discontinuous (Kovacs, 1967).

Second, when the PEO segments were crystallized homogeneously in the glassy PS matrix near $-20°$C, a significant volume contraction was observed by Lotz and Kovacs (1969) even though the T_m was about 100° below the T_g of pure PS. Thus the matrix would seem to be more mobile than ordinary glassy PS; indeed the T_g's noted for the PS segments appear to be lower than expected for PS of the same molecular weight.

Third, the values of T_g noted by Lotz and Kovacs may be extrapolated to a limiting value of about 35°C at zero concentration of PS. The authors have suggested that this may correspond to a limiting case of a monolayer of PS which is presumably relatively free of entanglements. In such a state one might expect to satisfy the requirements of the Gibbs–DeMarzio (1958, 1959) theory for a "true" thermodynamic second-order transition, uncomplicated by rate effects. However, since T_g also varies with molecular weight, especially when the molecular weight is low, this question is unresolved. Finally, although the transition temperatures are sensitive to composition, they do not appear to be very sensitive to whether the copolymer is of the A–B or B–A–B type (O'Malley et al., 1969).

In summary, the thermal behavior of the PEO–PS block copolymers can be accounted for in terms of a phase separation into domains of PEO and PS. However, the existence of a matrix and the interconnection between the domains appear to influence the behavior of the domains themselves. The phenomena observed are certainly worth much further study.

6.1.7. Mechanical, Dielectric, and Rheological Behavior

The melt rheology of amorphous block copolymers, e.g., styrene–butadiene block copolymers (Arnold and Meier, 1968; Holden et al., 1969a; Meier, 1969), has been described and interpreted already (Section 4.11). It is interesting to compare the amorphous block copolymers with block copolymers that have the additional feature of crystallizable sequences. A basic study of block copolymer rheology was carried out by Erhardt et al. (1970), who determined the complex modulus and tan δ, and studied melt behavior at temperatures between about 60 and 200°C. A report on dielectric behavior by Pochan (1971) is also significant.

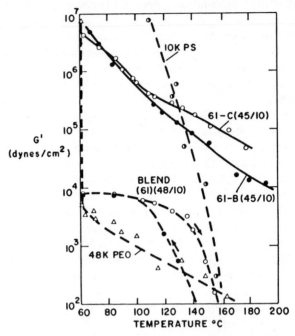

Figure 6.12. $G'(T)$ for the homopolymers PS and PEO, the AB block copolymer 61, and a solution blend of the homopolymers having the composition of copolymer 61. Numbers in parentheses indicate the $\overline{M}_n$ in the order (PEO/PS). The B indicates method of preparation; the C represents preparation by vacuum evaporation from chloroform solution. (Erhardt *et al.*, 1970.)

Erhardt *et al.* noted two striking features: (1) considerable birefringence at temperatures above 250°C—a phenomenon indicative of an appreciable degree of ordering, and (2) in comparison with the corresponding homopolymers, a highly elastic character in the melt. Both observations are consistent with continued phase separation at elevated temperatures. Results are shown in Figure 6.12, in which values of the storage shear modulus G' of PEO-rich specimens are compared as a function of temperature with homopolymers of similar molecular weight and for blends. Below about 130°C, the values of G' for the block copolymer are intermediate between the values for the homopolymers, but higher than the values for the blend of PS and PEO. However, at temperatures above 130–140°C, G' for the block copolymer becomes at least two orders of magnitude higher than for the homopolymers or blend, the latter tending to undergo phase separation uncontrollably. [Polymer blends may have lower viscosities than the corresponding homopolymers (Sections 4.11 and 9.6).] Thus, even at elevated temperatures, at which the PEO phase is molten (and, in this case, continuous) the PS phase appears to reinforce the melt (cf. the observations of birefringence at such temperatures). In other measurements (not shown) the reinforcement at high temperatures was shown to depend on the block length of PS. This behavior is quite consistent with observations on amorphous blocks and blends. Differences become significant when the temperature falls below 60°C, at which temperature G' rises rapidly to crystallization of the PEO blocks. As pointed out in Section 4.11, viscous flow in block copolymers can occur only when the molecules of one block are pulled into the phase of the other block, a phenomenon ordinarily involving a positive change in free energy.

The storage and loss shear modulus behavior of a PS-rich specimen ($w = 0.67$), shown in Figure 6.13, is also interesting (Erhardt *et al.*, 1970). A T_g for the PS sequences is clearly evident from the maximum in the loss modulus. Apparently the strength of the transition observed depends on the PS content, becoming high when PS forms a continuous phase, even though a PS phase may be observed microscopically at low PS concentrations. It was possible to obtain the melt viscosity as a function of temperature,

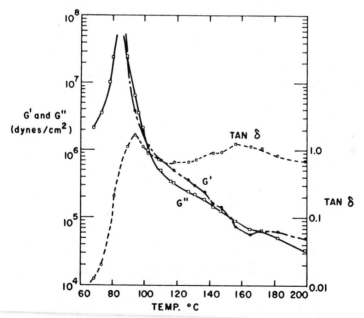

Figure 6.13. G', G'', and tan δ as a function of temperature for the PEO/PS block copolymer 57 (Erhardt *et al.*, 1970). Copolymer type AB; frequency 3.5 cps.

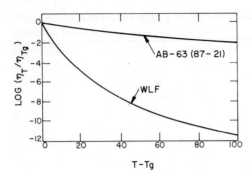

Figure 6.14. Comparison of the temperature dependence of viscosity above the polystyrene T_g for AB block copolymer 63 and that predicted from the WLF expression (Erhardt *et al.*, 1969).

but, as shown in Figure 6.14, values of viscosity did not agree with values calculated using the WLF equation (Section 1.5.7).

In addition to the T_g observed in Figure 6.13, a broad transition was observed in some specimens near 160°C, as indicated by the tan δ curve. This transition, which has an apparent activation energy of about 20 kcal/mol, is related to the phase-separated structure in the melt.

Further evidence for the role of phase separation is given by Pochan (1971). On casting films from a solvent preferential for PEO, a dielectric constant ε' of 2 was noted (Figure 6.15). In contrast, by casting from a mutual

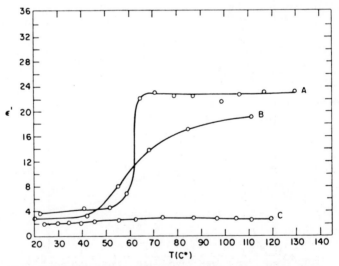

Figure 6.15. Dielectric constant of styrene–ethylene oxide block copolymers A, B, and C cast from solution in solvents preferential for the polystyrene component (C, ethylbenzene) and in a mutual solvent (A, B, chloroform). Molecular weights: A, 6×10^4, B and C, 1×10^5; frequency, 1 Hz. (Pochan, 1971.)

solvent, unusually high values of ε'—up to 25—were noted. These observations would seem to be consistent with a "locking-in" of ordered PEO by a continuous matrix of PS [cf. results given by O'Malley *et al.* (1969)]. Such high values of ε' are quite rare in polymers, and suggest a high degree of correlation of dipole orientations.

In summary, the mechanical and flow behavior in the melt is dominated by the block character and phase relationships between the PS and PEO sequences.

6.2. CRYSTALLIZATION: GENERAL ASPECTS

In this section, we shall consider several aspects of the crystallization of block copolymers (specifically styrene–ethylene oxide blocks) in dilute solution (Lotz, 1963; Lotz and Kovacs, 1966; Manson and Kovacs, 1965): fractionation during crystallization, criteria for crystal stability, and the kinetics of crystallization. It will be seen that the existence of a connected uncrystallizable sequence does markedly affect the crystallization kinetics of a block polymer, and that the course of events is controlled by the length of the uncrystallizable sequence and its interaction with the solvent. Although few quantitative data have been published on the crystallization of block polymers in bulk (Kovacs, 1967), one may expect general similarities. Also, as it turns out, single crystals obtained by crystallizing block copolymers from dilute solution have served as convenient models for the study of nucleation and growth phenomena in general, and have led to seeding techniques useful in non-block systems (Lotz *et al.*, 1966).*

6.2.1. Fractionation

When amorphous homopolymers having a significant distribution of molecular weight are dissolved in a thermodynamically poor solvent, addition of a nonsolvent or cooling may result in phase separation. Usually the molecules of higher molecular weight are thrown down first, thus effecting a fractionation (Holden *et al.*, 1969a). The PS/PEO block copolymers behave similarly, but with several complications (Lotz and Kovacs, 1966; Manson and Kovacs, 1965): (1) The blocks, being of different

---

*Since much of the literature has been published in French rather than English, and is hence unfamiliar to many scientists in this country, we have treated the crystallization studies in considerable detail.

Table 6.1

Degree of Precipitation and Composition of Crystals of PEO/PS Block Copolymers B and E Obtained in Ethylbenzene (EB) and Amyl Acetate (AA) at Different Temperatures[a]

	B ($w = 0.40$)						E ($w = 0.66$)					
	EB			AA			EB			AA		
T, °C:	15	20	25	15	25	35	15	20	25	15	25	35
p	0.91	0.86	0.86	0.96	0.94	0.91	0.23	0.22	0.15	0.52	0.44	0.21
w_p[b]	0.37	0.34	0.34	0.36	0.36	0.37	0.32	0.28	0.22	0.55	0.51	0.35
w_s	0.56	0.16	0.57	—	—	—	0.79	0.79	0.76	0.82	0.84	0.80
w_{p+s}	0.39	0.38	0.37	—	—	—	0.68	0.68	0.68	0.68	0.70	0.70

[a] $c_0 = 0.01$ g/ml. Lotz and Kovacs (1966).
[b] Subscripts p and s refer to crystals and solution, respectively.

chemical composition, have different solubility requirements; and (2) the PEO segments can crystallize.

In the following discussion on fractionation and selective crystallization of the PEO–PS blocks, it should be borne in mind that although the molecular weight distribution attained by means of anionic polymerization is narrow, it is still sufficiently broad to allow for significant fractional separation according to molecular weight.

Convenient measures of the fractionation are the total fraction crystallizable (PEO plus associated PS) p and the weight fraction of PS attached to the PEO in the crystal w_p (see Table 6.1). In general, the crystals have associated with them less PS than is in the parent polymer ($w_p < w$), the difference

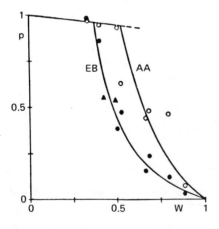

Figure 6.16. Variation of degree of precipitation of single crystals of a PEO/PS block copolymer in two solvents as a function of initial composition of the samples (Lotz and Kovacs, 1966). EB, ethylbenzene; AA, amyl acetate.

$w - w_p$ decreasing with higher original PS contents. Figure 6.16 shows that p remains close to unity until a critical value of PS content is reached, w^*, which depends on the solvent. At PS contents higher than w^*, p decreases rapidly.

6.2.2. Structural Models

Let us return to the two molecular models proposed for PEO/PS crystals* and shown in Figure 6.10; each model is based on actual observations of typical platelets (Lotz *et al.*, 1966). In one model, folded chains of crystallized PEO segments ($L = 100$ Å) are sandwiched between layers of amorphous PS. It is assumed that the block junctions are rejected from the crystalline layer. Segments of PEO that are too short (average length, $L/2$) to complete a fold period are also assumed to be rejected, and are not considered further. The second model is similar except for the assumption of a *double* folded layer of PEO, with the PEO lamellae presumably back-to-back and with the PS layers again forming the outer layers in the sandwich. The role of the solvent and the PS sequence in determining the crystallization behavior was made clear by the following analysis.

First, we define the *total* fold surface S of a PEO sequence:

$$S = 2\tilde{v}_{PEO}/L \tag{6.3}$$

where $\tilde{v}_{PEO}$ is the average molar volume of PEO chains in the crystalline PEO layer of thickness L. Next, we define the *specific* surface of each PS sequence on the fold surface, Σ, which depends on the model. For the single-layer model

$$\Sigma_1 = S \tag{6.4}$$

and for the double-layer model

$$\Sigma_2 = S/2 \tag{6.5}$$

Now the composition of the crystals must depend on the relative thicknesses of the two layers. Specifically, x, the ratio of PS to PEO in the crystal, is given by

$$x = \frac{w_p}{1 - w_p} = \frac{M_n(PS)}{M_n(PEO)} = \rho \frac{\tilde{v}_{PS}}{\tilde{v}_{PEO}} \tag{6.6}$$

where ρ is the ratio of the specific volumes ($\rho = \bar{v}_{PEO}/\bar{v}_{PS}$).

---

* In this context, the term crystal refers to a whole platelet, which in this case contains segments of both crystalline PEO and amorphous PS (see Figure 6.10).

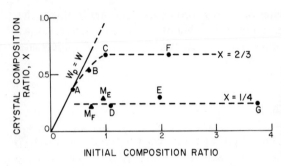

Figure 6.17. Variation of the composition of single crystals of a PEO/PS block copolymer obtained at 25°C in ethylbenzene as a function of initial composition ratio (PS/PEO), $w/(1 - w)$. (Lotz and Kovacs, 1966.)

Combining equations (6.3)–(6.6), and introducing l, the thickness of a PS layer, we then have the following two relationships (subscripts 1 and 2 referring to the single- and double-layer models, respectively):

$$x_1 = 2\rho l_1/L_1, \qquad x_2 = 2\rho l_2/L_2 \tag{6.7}$$

If L is fixed, as appears to be the case, then the variation in x depends on the variation in l, the thickness of the disordered PS layer. Hence

$$x_1/x_2 = 2l_1/l_2 \tag{6.8}$$

Thus the *composition* of the crystals can take up *two* discrete values, depending on whether a single or double layer structure is present, for comparable values of L and l. This prediction was confirmed by experimental results (Lotz *et al.*, 1966).

Indeed, when $w > w^*$, fractionation proceeds according to two different curves, with limiting values of the composition parameter x of 2/3 and 1/4. These two values of x must correspond to *two* critical compositions of PS in the crystals, x_1^* and x_2^*, and imply two different thicknesses of the disordered PS layers, one being 4/3 the other. Thus, if the existence of the two models is assumed, then from equation (6.8) we have

$$l_1^*/l_2^* \sim x_1^*/2x_2^* \sim \tfrac{4}{3} \tag{6.9}$$

a ratio in agreement with the data in Figure 6.17. Hence two critical compositions, x_1^* and x_2^*, in the crystals are possible, depending on the crystal model assumed.

6.2.3. Criterion of Stability

Now let us consider the compromise between the tendency of PEO sequences to crystallize and the tendency for PS sequences to hinder the crystallization. As discussed in Section 6.2.4, it seems reasonable to focus attention on the specific contact surface Σ, and on the tendency of crowding of the PS segments to result in elongation perpendicular to the crystal plane. Since both x^* and l^* depend on the particular crystal structure adopted, a dimensionless parameter λ may be defined in terms of the specific surface and the specific volume of PS:

$$\lambda = \tilde{v}_{PS}/\Sigma^{3/2} = l/\Sigma^{1/2} \tag{6.10}$$

This *elongation parameter* indicates the ratio of the thickness of the PS layer to the *average distance between junctions* $\Sigma^{1/2}$. It is thus related to the crowding of PS on the fold surface (see Figure 6.18). The quantity λ is in a sense analogous to the quantity $\alpha = L/L_0$ of rubber elasticity theory; in both cases we are concerned with molecules having perturbed conformations and reduced entropy.

It is assumed that each critical value of l (l^*) is associated with a critical value of λ (λ^*); then for constant S

$$l_1^*/l_2^* \sim \sqrt{2}\,\lambda_1^*/\lambda_2^* \tag{6.11}$$

Experimentally, $l_1^*/l_2^* \sim \sqrt{2}$ so that $\lambda_1^* \sim \lambda_2^*$. Thus the critical elongation of the volume element pervaded by a PS sequence appears to play the major role in governing the crystal growth. Indeed, one may write as a criterion for fractionation (equivalent to the criterion for crystal stability)

$$l/\Sigma^{1/2} \leq \lambda^* \tag{6.12}$$

If $l/\Sigma^{1/2} < \lambda^*$, then no fractionation occurs and stable crystals can be formed which have compositions equal to that of the parent polymer. Experimental results may be adduced in verification of this prediction, as shown in Table 6.2. Clearly fractionation is occurring in samples B and E, as predicted.

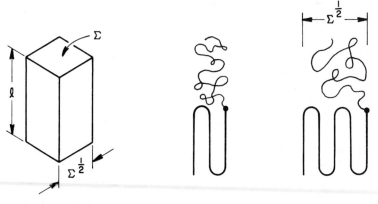

Figure 6.18. Schematic showing rectangular volume element for polystyrene segments in single crystals of styrene–ethylene oxide block copolymers, the cross-sectional area being given as Σ. (Lotz and Kovacs, 1966.)

Table 6.2
Fractionation of PEO/PS Block Copolymers During Crystallization[a]

Sample	Original weight fraction PS, w	Weight fraction PS in crystals, w_p	Critical elongation parameter, λ^*
A	0.28	0.27	$\sim 1\,(\simeq \Sigma^{1/2})$
B	0.40	0.34	~ 1.7
E	0.66	0.22	~ 1.7

[a] Lotz and Kovacs (1966).

Thus when the PS sequence is short enough to be randomly disposed in space (as a cubic volume element), crystallization is possible without fractionation. However, if the sequence is long, crowding at the fold surface occurs, and the volume element must undergo elongation, i.e., the conformation of the PS blocks is perturbed from its equilibrium[†] state. Consequently, the PS sequence, now ordered to some degree, must have a lower entropy, and the free energy of crystallization must compensate for the lower entropy of the PS sequences. The higher the PS content, then, the greater the fractionation tendency, and the lower the stability of the crystals, as is seen in the lower values observed for T_m (Kovacs, 1967).

6.2.4. Energetics

The role of solvent will now be considered. The importance of solvent interactions can be readily visualized, for at constant specific surface Σ any swelling must be anisotropic, and elongation of the PS element must occur in the direction perpendicular to the crystal surface.

If α_l is defined as the linear extension coefficient, and λ_s is the value of λ in the presence of solvent, then

$$\alpha_l = \lambda_s/\lambda = 1/\tilde{v}_2 \tag{6.13}$$

where $\tilde{v}_2$ is the average volume fraction of swollen PS and

$$\lambda_s = \alpha_l \lambda = (l/\tilde{v}_2)\Sigma^{1/2} \tag{6.14}$$

The parameter α_l will vary with PS molecular weight and with the free energy of dilution (Flory, 1953), $kT\psi_2(1 - \theta/T)v_2^2$, where ψ is the entropy of dilution, θ is the Flory theta temperature, and v_2 is the volume fraction of solute.

[†] In this discussion a dry state is assumed. The role of solvents is discussed separately below.

Now λ_s really expresses a state of equilibrium in which all thermodynamic forces balance, for the case in which one end of a PS sequence is fixed and the other end is free. Since one end of each chain is constrained, the neighboring chains must exhibit an excluded volume effect (proportional to $1/\Sigma$) in addition to the excluded volume otherwise characteristic of PS itself in the same solvent. As a result, the chains tend to undergo elongation in a direction at right angles to the plane of the lamella. At equilibrium, the tendency toward randomization will be compensated by an elastic retractive force due to the lower conformational entropy of the elongated PS sequences.

In general, for stable crystallization

$$\Delta G_{PEO} + \Delta G_{PS} \cong \Delta G_{PEO} - T\Delta S_{PS} \leq 0 \qquad (6.15)$$

where ΔG_{PS} represents the difference in free energy between PS in solution and PS in the dry, amorphous state, and is approximately equal to $-T\Delta S_{PS}$ (ΔS_{PS} being the loss of entropy due to extension), assuming the entropy and enthalpy of mixing are small.

To a reasonable approximation,

$$\lambda_s = \alpha_l/\Sigma^{1/2} \leq \lambda_s^* = \alpha_l\lambda^* \qquad (6.16)$$

where λ_s^* (numerically about 7) depends on interaction parameters and on ΔG_{PEO}. Hence, as the degree of solvation is decreased, α_l decreases, λ^* increases, and fractionation becomes progressively less significant. Indeed, as we go from ethyl benzene to amyl acetate, this is the case. At the same time, an increase in supercooling should cause Σ to increase, L to decrease, and hence w_p and p to increase—a prediction also verified by experience. Further confirmation of these considerations was provided by experiments in which the addition of pure PEO to an otherwise uncrystallizable solution stimulated the crystallization of the dissolved polymer. Integration of pure PEO into the crystal lattice relieves the crowding of PS sequences, and thus allows the PS sequences to be accommodated on the surface of the lamellae. By careful purification, crystals were obtained that had compositions predicted by the criterion for stability, equation (6.12).

Thus, in a quantitative manner, the dimensions and composition of crystals of block polymers can be related to the composition of the parent block copolymer and to the thermodynamic goodness of the solvent used as crystallization medium.

6.3. KINETICS OF CRYSTALLIZATION

As has been shown in previous sections, the presence of an uncrystallizable tail may hinder crystallization of an otherwise crystallizable sequence in a block polymer, but does not necessarily prevent the attainment of well-ordered lamellar single crystals provided that crystallization is thermodynamically possible. Indeed, crystallization may be a quite rapid process. In order to determine the rules governing the crystallization and recrystallization of AB block polymers, Kovacs et al. (1966) conducted a study of the nucleation and growth of a typical PEO–PS block copolymer ($w = 0.40$; $M_n(PS) = 7 \times 10^3$) in several typical solvents. Results of this study reveal several aspects of the formation of single crystals* in dilute solution which should be generalizable, at least in some respects, to other systems.

The homogeneity in size and form of the crystals produced by the techniques developed made it possible, for the first time, to relate microscopic observations under isothermal conditions to dilatometric measurements of the partial specific volume of the dissolved polymer $\bar{v}_l$, of the precipitated polymer $\bar{v}_{p\infty}$, and of the partially crystallized polymer $\bar{v}_p$. Thus the crystalline fraction z is given by

$$z = \frac{\bar{v}_l - \bar{v}_p}{\bar{v}_l - \bar{v}_{p\infty}} \tag{6.17}$$

Since the thickness of the PEO lamellae L is essentially constant (see Section 6.2.2), the isothermal contraction ($\bar{v}_l - \bar{v}_p$) is always proportional to the surface area of the crystals S (neglecting the lateral faces and assuming the crystals are two dimensional). Thus

$$z = S/S_\infty \tag{6.18}$$

where S_∞ is the final surface area of the crystal when crystallization has been completed.

For studies of nucleation, the number of nuclei N must be determined. In this case, N must be given by

$$N = \frac{p(1 - w_p)c_0 v_c}{a_\infty^2 L} \tag{6.19}$$

where p is the fraction of copolymer precipitated, $(1 - w_p)$ is the weight fraction of PEO in the crystal, c_0 is the *original* concentration in g/ml of solvent, v_c is the specific volume of the crystal (~ 0.815 ml/g), and a_∞ is the length of the crystal side. If some heterogeneity of size is present, then a_∞ will be an average, defined by

$$\bar{a}_\infty^2 = \frac{1}{n} \sum_{i=1}^{n} a_{i,\infty}^2 \tag{6.20}$$

where n is the number of units in the ensemble to be averaged.

---

* As mentioned above, the PS layers serve to inhibit the growth of abnormalities on the crystals, and also to confer mechanical stability. Hence the PEO–PS crystals are useful as models for the study of nucleation and growth of crystals per se. Some of the techniques developed have been adapted for use with other non-block copolymer systems (Lotz et al., 1966).

6.3.1. Effect of Dissolution Temperature

It was found by Kovacs *et al.* (1966) that at a given temperature the kinetics of crystallization and the size and morphology of the crystals depend on the thermal history of the dilute solutions used and concentration of the polymer (concentrations as low as 0.005 g/ml). In particular, the temperature of dissolution T_s plays a major role, with two types of behavior observed for a typical crystallization temperature T_c of 25°C. (It may be recalled that the temperature of fusion T_m of PEO is 44°C.)

Three groups of isothermal contraction curves are plotted in Figure 6.19. Experiments were performed by dissolving the block copolymer in ethyl-benzene at the value of T_s indicated, quenching to the crystallization temperature T_c, and measuring the contraction $(\bar{v}_l - \bar{v}_p)$ as a function of $(t - t_i)$, where t is the time after quenching, and t_i the time needed for thermal equilibration.

Enormous differences in the curves may be noted, even for differences in T_s of only a few degrees. If T_s was greater than about 45°C (curve *a*), an induction period of about 8 hr was observed before crystallization started; crystallization then proceeded extremely rapidly, on a log time basis. As long as $45 < T_s < 60$°C, excellent reproducibility was found. On the other hand, though the overall *rates* of crystallization were well behaved, the nature of the crystals obtained was complex. Although it had been

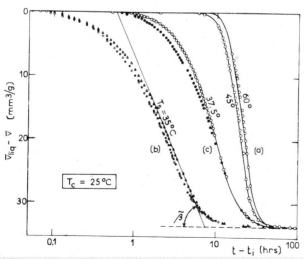

Figure 6.19. Isotherms of original crystallization ($T_s \geq 45$°C) and successive recrystallizations ($T_s < 45$°C) at $T_c = 25$°C. (Kovacs *et al.*, 1966.) The symbol $\bar{v} \equiv v_p$ in the text.

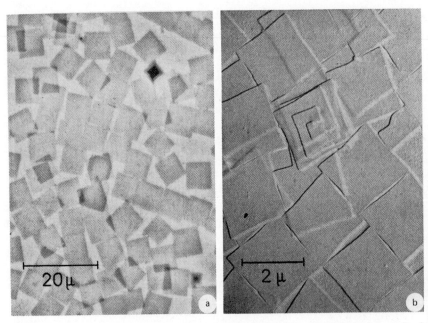

Figure 6.20. Monolamellar crystals of PEO/PS block copolymers: (a) Optical micrograph of crystals obtained by recrystallization at 25°C; (b) electron micrograph of crystals formed by recrystallization at 25°C. (Kovacs et al., 1966.)

thought that heating solutions to T_s would yield a reproducible number of nuclei, complex crystals were found, and the sizes of crystals tended to be polydisperse. Sporadic nucleation could, of course, account for the spread in size, for a spread would be obtained if not all crystals started to grow at the same time. [As will be discussed below, if the dissolution temperature T_s was above T_m, the melting point of pure PEO in the absence of solvent (about 40°C), the crystallization was regular and reproducible; if below T_m, order was retained, nuclei persisted, and recrystallization was relatively rapid.] If, then, the original* crystals were redissolved and recrystallized, the kinetics became complex. With dissolution at $33 \leq T_s < 45°C$, and recrystallization as before at 25°C, quite different kinetics was observed; both the shape and position of the isotherms depended on T_s. Thus for $T_s = 35°C$ (curve b), the induction period noted before disappeared and crystallization was more rapid. Repetition of the dissolution and crystallization cycles

---

* In this and following discussions, the term "original" ["primitive" according to Lotz and Kovacs (1966)] refers to the products of the *first* crystallization from solution, at a temperature T_1, which yields the single crystals in the first place. The symbol T_c is used to refer to the temperature of a subsequent crystallization.

gave very nearly the same isotherm. Under these conditions, examination by optical and electron microscopy revealed the presence of many small crystals which were homogeneous in size and which possessed discernible nuclei at their centers (Figure 6.20b). In contrast to the previous case ($T_s >$ 45°C), few irregular crystals were noted.

Now, if dissolution was effected at an intermediate temperature, 37°C, an isotherm having an intermediate shape and position was obtained (Figure 6.19, curve c). The crystals, while essentially free from degenerate forms, were rather more heterogeneous in size than for $T_s = 35$°C. Repetition of the dissolution–crystallization cycle gave a curve (filled circles) which was consistent with an increase in the concentration of nuclei N; at least the first part of the isotherm was superimposable on curve b by shifting the log t axis.

The results of this study strongly suggest that dissolution well above the melting temperature of PEO is required to destroy the crystallinity of the lamellae, even in dilute solution. However, dissolution at lower temperatures allows ordered entities to persist; these entities can serve as nuclei in subsequent crystallizations.

6.3.2. Effects of the Original Crystallization Temperature

If the nuclei concerned in a recrystallization (see footnote, p. 196) at T_c are derived from crystals that have not been completely dissolved on the molecular scale, one would expect their number to depend on the inherent stability of the original crystals. In the following, T_1 may be taken as the value of the crystallization temperature during the first crystallization. Since the stability varies directly with T_1 and with aging time t_1, one would expect a significant dependence of crystallization kinetics on both T_1 and t_1. Indeed this was observed (Figure 6.21): The higher T_1 and the longer the aging time t_1, the more the curves are shifted to the left, reflecting a progressive increase in the concentration of nuclei. If $T_1 \geq T_c$, successive recrystallizations are, as in Figure 6.21, very little affected. However, if $T_1 < T_c$, the isotherms approach the limiting isotherm given by the right-hand curve in Figure 6.21; in effect the thermal history is erased.

Effects of the original crystallization temperature T_1 and of the dissolution temperature T_s in relationship to the crystalline melting point of PEO (in equilibrium with solvent) T'_m may be summarized in two rules:

(a) If $T_s \geq T'_m$, the *kinetics* of crystallization is independent of the previous thermal history of the sample, and depends only on the temperature T_1 and on the concentration and composition of the copolymer. The crystals, however, contain a high proportion of degenerate forms.

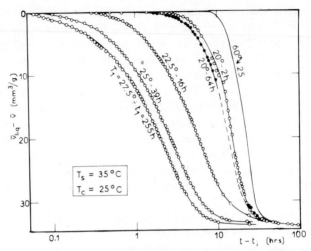

Figure 6.21. Recrystallization isotherms at $T_c = 25°C$ for a PEO/PS block copolymer in ethylbenzene, with original dissolution temperature T_0 of 50°C, original crystallization temperature and time T_1 and t_1, and redissolution temperature and time T_s and t_s. Sequence: $(T_0 = 50°C) \rightarrow (T_1, t_1) \rightarrow (T_s = 35°C, t_s = 0.5 \text{ hr}) \rightarrow (T_c = 25°C)$. The curve at the far right corresponds to the case $(T_0 = 60°C) \rightarrow (T_c = 25°C)$. (Kovacs et al., 1966.)

(b) If $T_s < T'_m$, the undissolved fraction of the copolymer, which serves to provide seed nuclei for subsequent crystallization, depends on the stability of the crystals first obtained. The crystals formed tend to be homogeneous in size and relatively perfect.

Thus the thermal history of a copolymer even in a dilute suspension determines the number of nuclei available for recrystallization. In addition, the *nature* of the nuclei is evidently also affected.

6.3.3. Number and Nature of Nuclei

It is instructive to consider the *shape* of the isotherms in order to gain some insight into the crystallization process. For this purpose, a shape parameter $\tilde{n}$ may be defined:

$$\tilde{n} = \frac{1.18\tilde{\beta}}{\bar{v}_l - \bar{v}_{p\infty}} \tag{6.21}$$

where $\tilde{\beta}$ is the experimental slope at the inflection of the sigmoidal crystallization curve (Figure 6.21). Clearly $\tilde{n}$, which corresponds to the Avrami dimensionality parameter (see below) varies significantly as a function of the time scale (Figure 6.22). The parameter $(t - t_1)_{20}$, the crystallization time corresponding to a value of $(\bar{v}_l - \bar{v}_p)$ of 20 mm^3/g, is used to characterize the position on the time scale.

The *shape* of a curve is closely related to its position on the log t axis. Two general classes are found (Figure 6.21):

(1)   When $1 \leq (t - t_1)_{20} < 10$ hr, isotherms are similar in shape ($\tilde{n} \sim$ 1.0–1.2), and are superimposable by shifting.

(2)   When $(t - t_1)_{20} > 10$ hr, $\tilde{n}$ is strongly dependent on $(t - t_1)_{20}$.

Evidently, the existence of these classes reflects variations in size and morphology, which in turn reflect variations in the number and nature of the nuclei. With type 1 crystallization it was shown that the size varied inversely with the concentration of nuclei and it was concluded that the slight increase in $\tilde{n}$ was due to a slight increase in the number of defective crystals, e.g., those having screw dislocations. Such crystals, having greater surface areas for growth, would increase the observed rate of crystallization, as is observed.

The situation is much more complex in crystallization of type 2. Here it appears that nuclei are effectively destroyed in heating to T_s, and that a relatively few new nuclei form sporadically with time. These nuclei tend to be complex in form, and induce secondary crystallization. Such a conclusion is in accord with microscopic observation of complexity and heterogeneity. Once the nuclei initiate crystallization, many growth surfaces are involved, and the rate (and $\tilde{n}$) increases very rapidly. On the other hand, seeding of a type 2 system during its incubation period results in kinetics

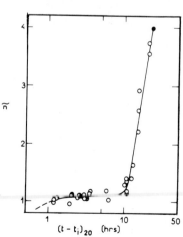

Figure 6.22. Variation of the shape of isotherms for the crystallization of a PEO/PS block copolymer. Isotherms characterized by $\tilde{n} = 1.180\beta/(\bar{v}_l - \bar{v}_{p\infty})$ as a function of their position along the log time axis. Position defined by $(t - t_1)_{20}$; $T_c = 25°C$. (Kovacs *et al.*, 1966.)

like that initiated by self-seeding, with a reversion of $\tilde{n}$ to the range 1.0–1.2, and rather regular crystals which are relatively homogeneous in size.

6.3.4. Effect of Final Crystallization Temperature

As T_c is increased, the isotherms change their slope as well as their position (Figure 6.23), as also noted by Mandelkern (1964) for the crystallization of dilute solutions of polyethylene.

When T_c is low, $\tilde{n} \sim 1.0$–1.2, and crystals tend to be monolamellar and homogeneous in size. At higher values of T_c, curves are no longer superimposable, $\tilde{n}$ increases rapidly with T_c, and the crystals again become more complex. (Kovacs et al., 1966.)

To a good approximation, isotherms were consistent with Avrami's equation (Avrami, 1939, 1940, 1941) relating z, the degree of crystallization [see equation (6.17)] to time and to the geometry of the nuclei:

$$z = 1 - \exp(-kt^n) \qquad (6.22)$$

where k is a rate constant and n characterizes the frequency of nucleation and the geometry of growth. To a further good approximation (at least for $0.4 < z < 0.8$), $n \sim \tilde{n}$.

A plot of $\tilde{n}$ as a function of T_c (Figure 6.24) reveals a significant difference in kinetics (and, specifically, in the Avrami coefficient n) depending on whether or not T_s is greater or less than 45°C. For $T_s = 35$°C, $\tilde{n}$ is relatively independent of T_c, and close to unity. For $T_s = 50$°C (original crystallization),

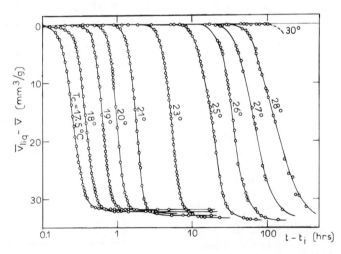

Figure 6.23. Isotherms for original crystallization of a PEO/PS block copolymer. (Kovacs et al., 1966.)

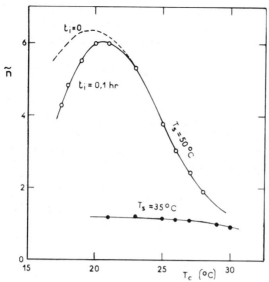

Figure 6.24. Variation of $\tilde{n}$ as a function of T_c in isotherms for original crystallization and recrystallization of a PEO/PS block copolymer. (Kovacs et al., 1966.)

$\tilde{n}$ passes through a maximum, reaching a value of ~ 6 at the peak. Again, these results are consistent with the earlier conclusion that sporadic nuclei tend to be complex and to initiate growth on many surfaces, while predetermined nuclei tend to be more regular.

So far the kinetics may not appear to reflect specifically the existence of a noncrystallizable sequence. It is worth noting, however, the growth law deduced from geometrical considerations:

$$\bar{a} \cong \bar{a}_\infty [1 - \exp(-k/t^{1/2})] \qquad (6.23)$$

where $\bar{a}$ is the length of a single crystal side; this expression corresponds to the case in which $n = 1$. Since for growth in two directions from predetermined nuclei Avrami's derivation predicts a value of 2, it seems likely that the slower overall rate reflects the fact that fractionation is occurring. Thus molecules containing long PS sequences are rejected, and hence the overall rate is diminished.

6.3.5. Rate of Crystallization

While the exponent n characterizes the *shape* of crystallization curves, in equation (6.22), the rate parameter k characterizes the *position* on the time scale. As shown earlier, k depends not only on T_c, but also on the thermal history. It is interesting to examine the effect of both T_c and the solvent on k. In order to do so as simply as possible, we may use a characteristic time θ defined as follows:

$$\theta = k^{-1/n} \tag{6.24}$$

Thus θ represents the time corresponding to the inflection of the isotherms, and hence to a constant fraction crystallized (regardless of n). Values are given in Figure 6.25 for original crystallization ($T_s = 50°C$) and for the first recrystallization ($T_s = 35°C$). The data were derived from dilatometric and microscopic experiments.

For self-seeded systems ($T_s = 35°C$), the values of θ at any given T_c

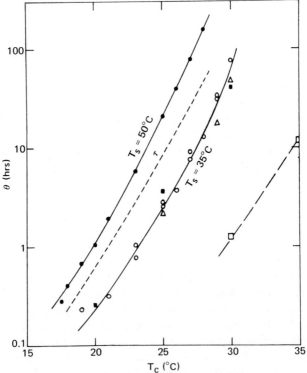

Figure 6.25. Variation of the rate parameter θ as a function of T_c for isotherms for original crystallization ($T_s = 50°C$) and recrystallization ($T_s = 35°C$). Solvent used: ($\square$) amyl acetate; ($\triangle$) xylene; and ($\bullet$, $\bigcirc$, $\blacksquare$) ethylbenzene. (Kovacs et al., 1966.)

are invariably lower than for the non-self-seeded systems ($T_s = 50°C$). The effect of solvent is interesting (see also Figure 6.26). In amyl acetate, a poor solvent for PS, the rate for a given value of T_c is 50 times higher than in ethyl benzene or xylene—good solvents for PS. Thus the rate of crystallization and the fraction precipitated are higher, the poorer the solvent for the PS, and the smaller the value of $\bar{v}_{PS}$ in solution. One may estimate a value of T'_m in amyl acetate about 7°C higher than in ethyl benzene, reflecting the higher thermodynamic stability of crystals in the former solvent.

Although this study did not include the effect of copolymer composition, it seems reasonable to expect that the rate of crystallization will vary inversely with the PS content in the copolymer.

6.3.6. Variation in Partial Volumes

In solution, the values of calculated partial volume $\bar{v}_l$, assuming additivity of polymer specific volumes, are higher than observed (Kovacs et al., 1966). However, after crystallization, the difference becomes very small, and values calculated for PS and PEO are close to values observed independently for crystalline PEO and PS dissolved in toluene. Thus in the liquid state a specific interaction between PEO and solvent must exist. In the crystalline state, however, the specific volumes become more nearly additive; the specific interaction must then have essentially disappeared on crystallization.

Thus the presence of the uncrystallizable PS does not appreciably affect the specific volume of the PEO once the latter crystallizes. Again, if a PS sequence is too long to be accommodated, the crystal will not form.

6.4. SILOXANE BLOCK COPOLYMERS

In the preceding sections of this chapter the properties of poly(ethylene oxide-b-styrene) were considered. This block copolymer was seen to differ structurally from other materials, such as poly(styrene-b-butadiene) (see Chapter 4), because the poly(ethylene oxide) portion crystallized, thus profoundly affecting morphology and physical behavior. In this section we shall consider a novel inorganic block copolymer poly(diphenylsiloxane-b-dimethylsiloxane-b-diphenylsiloxane). The poly(diphenylsiloxanes) form a high-melting crystalline phase, while the poly(dimethylsiloxanes) form an elastomeric phase that retains useful rubbery properties from -50 to $+250°C$ (Fordham, 1960; Lee et al., 1966; Meals and Lewis, 1959; Sperling et al., 1966; Stone and Graham, 1962).

6.4.1. Synthesis

Block copolymers with known block lengths were prepared by Bostick (1970) using organocyclosiloxanes with lithium-based catalysts. For example, when hexamethylcyclotrisiloxane is reacted with the lithium salt of diphenylsiloxane diol, a multistep addition polymerization occurs:

$$n\left(\begin{matrix} CH_3 \\ | \\ -Si-O- \\ | \\ CH_3 \end{matrix}\right)_3 \quad + \quad Li-OSi-OLi$$

n ≥ 1

$$\xrightarrow[\text{room temperature}]{\text{tetrahydrofuran}} \quad Li\left(\begin{matrix} CH_3 \\ | \\ -O-Si- \\ | \\ CH_3 \end{matrix}\right)_{3n/2} -O-Si-O\left(\begin{matrix} CH_3 \\ | \\ Si-O \\ | \\ CH_3 \end{matrix}\right)_{3n/2} -Li \quad (6.25)$$

Further, when another cyclosiloxane is added to this reaction system, polymerization resumes to produce block poly(organosiloxanes) of the ABBA type. Because the B sections are identical, we may treat the product as a triblock copolymer. For example, when hexaphenylcyclotrisiloxane is added to the product of reaction (6.25), the reaction may be represented as follows:

$$Li\left(\begin{matrix} CH_3 \\ | \\ O-Si \\ | \\ CH_3 \end{matrix}\right)_{3n/2} -O-Si-O\left(\begin{matrix} CH_3 \\ | \\ Si-O \\ | \\ CH_3 \end{matrix}\right)_{3n/2} -Li \; + \; m\left(\begin{matrix} \emptyset \\ | \\ -Si-O- \\ | \\ \emptyset \end{matrix}\right)_{3n} \xrightarrow[\text{or anisole}]{\text{THF/C}_6\text{H}_6}$$

$$(6.26)$$

$$Li\left(\begin{matrix} \emptyset \\ | \\ O-Si \\ | \\ \emptyset \end{matrix}\right)_{3m/2}\left(\begin{matrix} CH_3 \\ | \\ O-Si \\ | \\ CH_3 \end{matrix}\right)_{3n/2} -O-Si-O\left(\begin{matrix} CH_3 \\ | \\ Si-O \\ | \\ CH_3 \end{matrix}\right)_{3n/2}\left(\begin{matrix} \emptyset \\ | \\ Si-O \\ | \\ \emptyset \end{matrix}\right)_{3m/2} -Li$$

Table 6.3
Composition of ABA Siloxane Block Copolymers[a]

Polymer	A = $(\emptyset_2SiO)_x$, x	B = $(Me_2SiO)_y$ y	DP = 2x + y	% $\emptyset_2SiO$
BC 1[b]	500	4000	5000	20
BC 2	96	2150	2340	8
BC 3	150	2200	2500	12
BC 4	190	2170	2450	16

[a] Tobolsky (1960).
[b] Values for this polymer are approximate.

Typical compositions of the materials prepared are shown in Table 6.3 for triblock copolymers (Fritzsche and Price, 1970).

6.4.2. Morphology and Properties

When a crosslinked or otherwise insoluble polymer is swollen with a series of solvents of differing solubility parameters δ, a maximum in swelling will occur when the solubility parameter of the solvent roughly equals that of the polymer (Tobolsky, 1960, pp. 64–66). Fritzsche and Price (1971) reasoned that a two-phase insoluble block copolymer should display two independent peaks, and used the swelling method to examine this aspect of their material. Typical results are shown in Figure 6.26. The peak

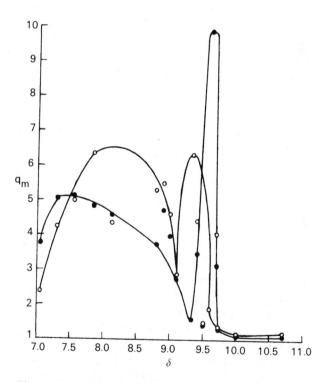

Figure 6.26. Swelling as a function of solubility parameter of the swellant at 25°C of ABA siloxane block copolymers cast from two different solvents at 25°C. (O) BC-2 ($\delta = 9.1$) from o-chlorohexahydrotoluene; (●) BC-2 ($\delta = 9.6$) from o-dichlorocyclohexane. (Fritzsche and Price, 1971.)

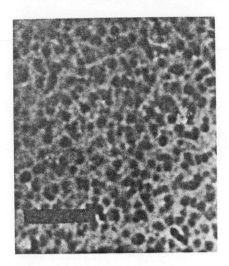

Figure 6.27. Optical micrograph for a thin film of ABA siloxane block copolymer BC-1 after heating for 45 min at 500°C. Marker equals 0.5 μm. (Fritzsche and Price, 1970.)

near $\delta = 7.8$ was believed due to the poly(dimethylsiloxane) component, and the peak near 9.5 due to the poly(diphenylsiloxane) component.

Electron microscopy, even in the absence of staining, yielded surprisingly good contrast in thin sections of these polymers (Fritzsche and Price, 1970). Casting of BC-1, followed by heating at 500°C, produced the structure shown in Figure 6.27. It was established that the dark, discontinuous phase was due to the poly(diphenylsiloxane) component. Depending on composition, precipitation conditions, and annealing, either tough, crystalline elastomers or higher modulus materials could be obtained (Bostick, 1970).

6.5. POLYALLOMERS

The polyallomers constitute the class of block copolymers where both components are capable of crystallizing independently (Coover et al., 1966; Hagenmeyer and Edwards, 1966, 1970; Eastman Chemical Products, n.d.). The most important member of this family contains crystalline, stereoregular polypropylene as the major component and polyethylene as the minor component. As expected for a block copolymer, these products differ greatly in behavior from mechanical blends of polyethylene and polypropylene, and also from their random copolymers, poly(propylene-co-ethylene). When crosslinked with a diene monomer, the latter copolymers are known as EPDM rubbers (Lee et al., 1966; Rodriguez, 1970, Chapter 13), while the former blends are of apparently little interest. In Figure 6.28 and 6.29 the

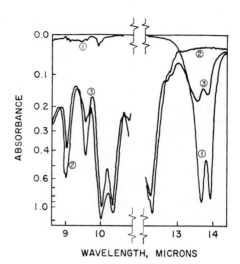

Figure 6.28. Infrared curves of (1) linear polyethylene, (2) polypropylene, (3) propylene–ethylene polyallomer. (Hagenmeyer and Edwards, 1966.)

infrared spectra for compositions representing the several ways of combining the two polymers are presented. Infrared bands for the polyallomer resemble bands for the blend, but the spectrum differs greatly from the spectrum for the corresponding random copolymer, which tends to exhibit much weaker absorptions.

From a practical point of view, incorporation of a few percent of polyethylene in block form as a tail on the polypropylene chain lowers the material's brittle temperature and increases impact strength. Table 6.4

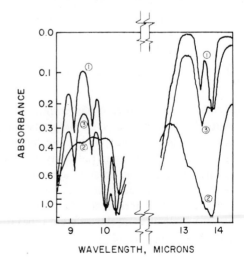

Figure 6.29. Infrared curves of (1) polypropylene–linear polyethylene blend, (2) ethylene–propylene copolymer rubber, (3) propylene ethylene polyallomer. (Hagenmeyer and Edwards, 1966.)

Table 6.4

Comparison of Properties of Propylene–Ethylene Polyallomers and Polypropylene[a]

	Run 11	Run 70	Run 71	Run 79
Ethylene, %	0.0	0.6	2.0	3.0
Flow rate at 230°C, deg/min	2.4	2.8	2.9	2.5
Density (annealed), g/ml	0.9100	0.9093	0.9044	0.9010
Brittleness temperature, °C	+8	−5	−22	−35
Tensile strength at yield, psi	4700	3870	3560	3050
Elongation, %	360	550	500	>650
Stiffness in flexure, psi	142,000	99,500	92,400	80,000
Hardness, Rockwell R scale	93	87	68	60
Vicat softening point, °C	145.0	140.3	127.2	124.5
Notched Izod impact strength (23°C), ft-lb/in	0.5	0.9	1.9	3.5
Unnotched Izod impact strength (23°C), ft-lb/in	24	No break	No break	No break
Tensile impact strength, ft-lb/in^2	32	56	79	90

[a] Hagenmeyer and Edwards (1966).

compares polypropylene homopolymer with several polyallomers of increasing polyethylene content. As shown, 3% of polyethylene decreases the brittle temperature from +8 to −35°C and increases the notched impact strength from 0.5 to 3.5 ft-lb/in., with only modest changes in other properties.

While these materials do contain two crystalline phases, it appears that small quantities of amorphous polyethylene are important in the mechanical property improvements noted above. It should be noted that the amorphous portions of polyethylene have a glass temperature near −80°C, compared to −10°C (Nielsen, 1962, p. 16) for polypropylene.

6.6. SUMMARY OF BLOCK COPOLYMER SYSTEMS

Most of Chapter 4 has emphasized block copolymers based on, for example, combinations of polybutadiene with polystyrene, where both components are amorphous. The multiblock copolymers were treated in Chapter 5. The system poly(ethylene oxide)–polystyrene, containing one amorphous and one crystalline component, and the crystalline–crystalline block copolymer polyethylene–polypropylene were treated in this chapter. These three chapters illustrate some of the differences that can be brought about by considering the different types of polymers that can be attached to form block copolymers.

Table 6.5
Typical Block Copolymer Compositions

Component 1	Component 2	Phases crystalline	References
Polystyrene	Polybutadiene	0	Holden *et al.* (1969a)
Polystyrene	Polyisoprene	0	Prud'homme and Bywater (1970)
Polystyrene	Poly(ethylene oxide)	1	Sadron (1962)[a]
Polystyrene	Poly(dimethyl siloxane)	0	Saam *et al.* (1971a,b)
Polystyrene	Polyester	0	Tobolsky and Rembaum (1964)
Cellulose triacetate	Polyethers and polyesters	1	Steinman (1970)
Polycarbonate	Poly(dimethylsiloxane)	0	Kambour (1970)
Poly(diphenylsiloxane)	Poly(dimethylsiloxane)	1	Fritzche and Price (1970)
Polypropylene	Polyethylene	2	Battaerd and Tregear (1967, p. 239), Vermillion (1968),[b] Yakobson *et al.* (1970) Hagenmeyer and Edwards (1966)
Poly(ethylene terephthalate)	Poly(ethylene oxide)	2(?)	Burlant and Hoffman (1960, pp. 75–77)

[a] Also see other references in Section 6.1.
[b] Part of the class of materials known as polyallomers.

In summary, Table 6.5 contains a list of selected block copolymers containing either two or three blocks; also shown are the number of crystalline phases encountered. Although we have chosen to discuss selected systems, Table 6.5 shows that, in fact, many block copolymer pairs have been synthesized.

With triblock copolymers containing either crystallizable or plastic-forming end blocks as the dispersed phase and an elastomer midblock as the continuous phase, thermoplastic elastomers that are dimensionally stable to T_m or T_g of the hard block are achieved. In some cases the crystallinity also imparts special strength and stability.

7

Miscellaneous Grafted Copolymers

7.1. GENERAL CONSIDERATIONS

The first graft copolymers were synthesized in the 1930's, and since then several different classes of materials have been developed. Chapter 3 considered graft copolymers designed for toughness and impact resistance. The present discussion centers on the synthesis, morphology, and physical and mechanical behavior of miscellaneous graft copolymers, with some emphasis on heavily grafted materials. In general, two major routes to grafting have been examined. The first involves starting the graft at a specific active site on the backbone chain, for example, treating nylon with ethylene oxide (Battaerd and Tregear, 1967, Chapter 3), so that the amide hydrogen reacts with the primary hydroxyl group:

$$
\underset{\substack{\displaystyle \\ }}{-\overset{\displaystyle O}{\overset{\|}{C}}-NH\!+\!CH_2\!\!\xrightarrow{}_x\!\!N-\overset{\displaystyle O}{\overset{\|}{C}}\!+\!CH_2\!\xrightarrow{}_y} \tag{7.1}
$$

$$
\begin{array}{l} \quad\quad\quad\quad\quad\quad CH_2 \\ \quad\quad\quad\quad\quad\quad | \\ \quad\quad\quad\quad\quad CH_2\!-\!O\!+\!CH_2\!-\!CH_2\!-\!O\!\xrightarrow{}_n\!H \end{array}
$$

However, the most important method of producing heavily grafted materials utilizes radiation techniques, usually using either ^{60}Co gamma radiation or Van de Graaff-type electron accelerators. According to Chapiro (1962, Chapter 12), there are four major types of radiation grafting:

1. The direct irradiation of a polymer I containing a second monomer II.
2. Grafting onto radiation-peroxidized polymers.
3. Grafting initiated by trapped radicals.
4. The intercrosslinking of two different polymers.

Of course, the first is the most important, having attained a considerable and growing commercial usage. Usually the major component in such systems is polymer I, less than 50 % by weight of monomer II being swollen in. The less monomer contained in the polymer, the more efficient the grafting, in terms of number of grafts per unit weight of the second polymer. When polymerization is complete, grafting may continue by route 4 if the radiation treatment is continued, as in the normal mode of preparing heavily grafted materials.

Graft polymerizations have a long history (Burlant and Hoffman, 1960; Ceresa, 1962), and in the opinion of the present authors, one of the major shortcomings in understanding them is still in the area of morphology and phase separation. Historically, early workers in the field of grafting tended to be either synthetic organic chemists or radiation specialists, and their research programs were naturally oriented about their interests. Thus, following synthesis, the most common method of analysis involved measurement of swelling and gel fraction, which tended to be significantly decreased and increased, respectively. As a result of such changes, it was often concluded that the material was richly grafted. Those who carried their investigations further often proceeded to investigate mechanical properties, such as dimensional stability, found improvements, and ascribed such improvements to the supposedly extensive grafting. Especially before the development of osmium tetroxide staining by Kato (1966, 1968) in 1964, many investigators were understandably unaware of phase separation in their materials. Others noted the presence of two phases (Wellons et al., 1967), but, lacking the tools to examine the structure on the submicroscopic level, could go no further.

Polymerization of monomer II in the presence of polymer I definitely leads to increases in molecular weight, increased viscosity, and often true gel formation. This latter is actually a form of joined IPN or AB crosslinked copolymer. Using the cellular model of phase separation, we may predict that gelation, if it occurs, will be most extensive near the cell wall. In Section 3.1.1.3, we concluded that polymer II was unable to pass easily through polymer I even in the highly swollen state, because of polymer incompatibility. Obviously, this transport difficulty is augmented by a surrounding gel network.

Most important, we may add that extraction tends to remove coarse, noncellular material, and that true grafts and finely divided cellular material together remain unextracted. As brought out in Chapter 3, it is the latter that especially promotes toughening. So, regardless of molecular interpretation, we conclude that the original extraction experiments did and still do provide an important chemical guide for estimating toughening.

The specialized materials exemplified by HiPS and ABS have already been considered in Chapter 3. The electron microscope reveals two distinct phases for these impact-resistant materials, which should be called lightly grafted polymers in the present context. (The distance between phase domain boundaries, compared to molecular dimensions, suggests that the extent of true grafting must be limited to the neighborhoods of domain boundaries.) In the light of the work by Kato (1966, 1968) and the important studies of the morphology of blocks, blends, and grafts by Matsuo (1968), Molau (1971), Aggarwal (1970), and others, it is becoming widely accepted that most, if not all, of the materials do undergo phase separation. In particular, the grafted copolymers tend to form domains with the second component within discrete cellular structures, sometimes exhibiting a phase-within-a-phase-within-a-phase; see Chapter 3. It may be suspected that more heavily grafted materials exhibit similar, but probably finer (smaller) phase domains. An important hint that the heavily grafted materials do undergo phase separation was given a decade ago by Katz and Tobolsky (1964), who reported that unsaturated polyesters grafted with polystyrene were often opaque, and displayed broadened glass transition regions.

In summary, two points can be made: (1) Regardless of the true extent of grafting, little material is extractable by classical techniques because polymer II, within the cell walls of polymer I, cannot pass through the barrier and escape (because of thermodynamic incompatibility), even for high levels of swelling. Thus the cellular morphology offers an attractive alternative explanation of the extraction experiments, and raises anew the question of the exact extent of grafting. (2) The very presence of the two phases may be instrumental in attaining the improved physical and mechanical properties noted. When, for example, the added polymer has a higher (or lower) glass temperature, or develops crystallinity, real property improvements may be noted, due, in part, to an appropriate degree of phase separation. The point is that while true grafting does occur, the importance of phase separation in governing the extent of grafting (vs. crosslinking) on the one hand, and influencing mechanical behavior on the other hand, appears to have been under-investigated, even in recent years. (Undoubtedly, one reason is the lack of suitable staining techniques for systems not containing double bonds.) In the following discussions, reinterpretations of older data in the light of present knowledge will be made wherever possible.

7.2. SURFACE GRAFT COPOLYMERS

Broadly speaking, three different kinds of polymer grafts may be distinguished: surface grafts, heterogeneous grafts, and homogeneous grafts. Surface grafting occurs when monomer II contacts, but does not appreciably swell polymer I (Battaerd and Tregear, 1967, pp. 44, 216, 224; Ceresa, 1962,

p. 55); when the polymer I surface is activated, grafting will occur. In hetero-geneous grafts, monomer II enters and swells portions of polymer I, but some important portions of polymer I remain unswollen. One example would be the case in which monomer II swells only the amorphous portion of a partly crystalline polymer, or enters pore spaces. Homogeneous grafting may be defined as the uniform swelling of polymer I by monomer II to form one phase prior to polymerization. The ABS and HiPS plastics (Chapter 3) are examples of this type of graft.

7.2.1. Uses of Surface Grafts

Surface grafts are usually employed to modify the surface of another polymer for specific property improvements. For example, improvements in dyeability, weathering, microbiological resistance, water resistance, and adhesive and wetting properties can often be attained. Interfacial polymeriza-tion on wool (Battaerd and Tregear, 1967) improves shrink and matting resistance. Wash-and-wear finishes based on melamine and urea-resins gave cotton and rayon new life (Rodriguez, 1970, pp. 337, 346, 449). Surface modifications by means of irradiation of polymer I include polystyrene onto polyethylene and poly(vinyl acetate) onto poly(tetrafluoroethylene). These materials are obviously phase-separated, often in the sense of the bicom-ponent fibers treated in Section 9.2.1. The cellular model of phase separation would not be expected to hold in true surface grafts.

7.2.2. Surface Grafts Involving Proteins

Two additional application areas of surface graft copolymers include: (1) the use of heparinized surfaces to reduce thrombosis in surgical implants, and (2) the use of immobilized enzymes as catalysts.

Heparin, a natural protein, is an important constituent of blood and contributes to blood clotting. When circulatory assist devices were first proposed, it was found that the artificial surfaces promoted unnatural thrombosis. An important proposed solution to the problem involved grafting heparin to the surfaces of the materials involved, including such polymers as poly(vinyl chloride) and poly(dimethyl siloxane) (Artificial Heart Program, 1968; Lyman, 1966; Sears, 1965).

The grafting of several enzymes to surfaces has also found application in the form of biologically based catalysts (Carbonell and Kostin, 1972;

Kay, 1968; Messing and Weetall, 1970; Silman and Katchalski, 1966). In this application the enzymes are immobilized on polymeric or inorganic solid support surfaces. The substrate to be reacted then is made to flow over the catalytic surface. The principal advantages include: (1) The enzymes do not remain in the product as a difficultly separable impurity, (2) the enzymes may be reused, reducing cost, and (3) under certain conditions the useful life of the enzyme is prolonged. (Hasselberger *et al.*, 1974.)

Immobilized enzymes have many uses. For example, Hustad *et al.* (1973) bonded *β*-galactosidase to Teflon stirring rods in connection with milk processing; Capet-Antonini *et al.* (1973) grafted urokinase to water-insoluble polymers for application in blood chemistry purification. Shuler *et al.* (1973) examined the possibility of living cells having immobilized enzymes. A common commercial method of initiating grafting involves radiation techniques. Unfortunately, detailed treatments of these interesting topics are beyond the scope of this monograph.

7.3. DEGRADATION AND CROSSLINKING REACTIONS

When polymers are subjected to high-energy radiation or free-radical-producing conditions generally, a combination of degradation and cross-linking reactions also ensues (Chapiro, 1962). These reactions occur with or without the presence of the second monomer, but if the second monomer is present, obviously grafting and other reactions are also involved. As one might imagine, these reactions tend to be complex, and many products are formed. For example, polyisobutylene may be degraded by means of disproportionation reactions:

$$
\sim CH_2 - \underset{\underset{CH_3}{|}}{\overset{\overset{CH_3}{|}}{C}} - CH_2 \sim \xrightarrow{h\nu}
\begin{cases}
\sim CH_2 - C \underset{CH_3}{\overset{\parallel CH_2}{\diagdown}} + CH_3 \sim \\[2em]
\sim CH_2 = C \underset{CH_3}{\overset{CH_3}{\diagup}} + CH_3 \sim
\end{cases}
\tag{7.2}
$$

Poly(vinyl chloride), on the other hand, undergoes crosslinking following the elimination of HCl,

$$
\begin{array}{ccc}
\text{---CH---} & & \text{---CH---} \\
\left[\begin{array}{c} \text{Cl} \\ | \\ \text{H} \end{array}\right] & \rightarrow & | \\
\text{---CH---} & & \text{---CH---}
\end{array}
\quad + \text{HCl} \qquad (7.3)
$$

Thus, while polyethylene can be crosslinked relatively efficiently and cheaply, especially by use of electron radiation techniques, serious problems have existed with other polymers. Some polymers, such as polyisobutylene, tend to undergo chain scission predominantly on irradiation; some, such as poly(vinyl chloride), do undergo crosslinking but also degrade to an unacceptable extent; others, such as polypropylene, undergo crosslinking, but not very efficiently. As a general rule, when a polymer contains α-hydrogen groups, crosslinking tends to predominate, whereas degradation tends to predominate when α-substituents other than hydrogen are present.

7.4. GRAFTING ONTO CELLULOSIC MATERIALS

As a natural polymeric product, cellulose has been used by man since the beginning of civilization. Cellulose derivatives in the modern sense date from the middle of the 19th century. Older uses of cellulosic materials range from cotton dresses (natural cellulose), to cellophane (regenerated from cellulose zanthate), to photographic film (a composite based on cellulose acetate). More recent uses for cellulosic products include such items as desalination membranes (Turbak, 1970). Although much of the following discussion on grafting is quite general, and has been applied widely to other polymeric systems, the importance of cellulose and its singularity among polymeric materials as a crystalline, natural product warrant special attention.

There are two general methods for grafting a polymer onto cellulose. The first involves a heterogeneous grafting technique, in which the cellulose is reacted with a second monomer with a slight ability to swell the cellulose. The second method involves swelling (or dissolving) the first polymer with (or into) the second monomer, often in the presence of a simple solvent. The resulting graft copolymer then more closely resembles the type of graft polymer exemplified by HiPS or ABS resins. This type of graft is sometimes important with cellulose esters, such as secondary cellulose acetate. Of course, a broad range of incompletely swollen or partially heterogeneous parent polymers exists, wherein most of the cellulose grafts appear to lie. One major difference from ABS-type graft copolymers discussed previously should be emphasized: most of the cellulose–synthetic polymer grafts contain sidechains that have specifically undergone initiation, transfer, or termination on the cellulosic backbone chain (Stannett and Hopfenberg, 1971). The materials in this section should be compared instead with polymer-impregnated wood (see Section 11.2); the cellular structure of cotton cellulose is similar to the structure of wood tracheids (see Figure 9.19 and Section 9.8).

7.4.1. Synthetic Methods

Most of the important syntheses of cellulosic graft polymers involve irradiation with high-energy beams. Ultraviolet and x-ray sources are sometimes used, but ^{60}Co radiation is used most often (Chapiro, 1962). In an oversimplified manner, the synthesis can be described as follows (Chapiro, 1962, p. 600):

$$
A \underset{\searrow}{\overset{\nearrow}{}} \qquad
\begin{matrix}
A \\ \{ \\ \cdot
\end{matrix} +
\begin{matrix}
\cdot \\ \{ \\ A
\end{matrix} \overset{+2nB}{\rightarrow}
\begin{matrix}
A \\ \{ \\ B_n
\end{matrix} +
\begin{matrix}
B_n \\ \{ \\ A
\end{matrix} \qquad \text{(a)}
$$

$$
A \overset{A}{\underset{A}{\{}} \cdot + R\cdot \overset{+nB}{\rightarrow}
\overset{A}{\underset{A}{\{}}\!\!\sim\!\!\!\sim B_{n-q} + B_q \qquad \text{(b)}
$$

(7.4)

where n mers of monomer B are added to each chain of polymer A. Route (a) involves nominal block copolymerization, and route (b) yields true grafts. B_q represents homopolymer containing q mers formed. Sometimes the free radicals are introduced chemically, as by direct oxidation with ceric ions onto cellulose:

$$
\text{R cell–H} + \text{Ce}^{4+} \rightarrow \text{R cell}\cdot + \text{Ce}^{3+} + \text{H}^+
$$

$$
\text{R cell}\cdot + n\text{M} \rightarrow \text{graft copolymer}
$$

(7.5)

where M represents monomer II, and R cell—H represents cellulose.

7.4.2. More-Heterogeneous-Type Grafts

This and the following subsection will compare heterogeneous and homogeneous synthetic methods, with emphasis on cellulosics. Fibrous cellulosic graft copolymers, prepared by radiation-induced free-radical copolymerization reactions of vinyl monomers with cellulose, retain some cellulosic properties, and acquire new ones that increase their usefulness. For example, grafting of poly(acrylonitrile-co-butyl methacrylate) onto cotton cellulose greatly reduces the extent of rupture of fibers during abrasion (Arthur, 1971). Figures 7.1 and 7.2 compare results of flex abrasion tests, utilizing low-power scanning electron microscopy. The untreated fabric was found to rupture after 2000 cycles, while the grafted material was still unbroken even after 24,000 cycles. Depending on the surface properties of the grafted polymer, the soil release properties of these fibers may be improved. Because synthetic polymers are generally resistant to

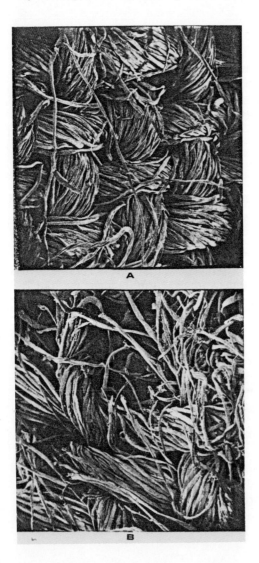

Figure 7.1. Scanning electron microphotograph of cellulose fabric sample before (A) and after (B) flex abrasion (2000 cycles).

attack by bacteria and fungus, the resistance of cellulose to rot is increased by grafting. Also, grafted cellulosic fabrics have higher wash-and-wear appearance ratings because they retain creases better. Today, nearly all wash-and-wear fabrics are either crosslinked, grafted, or both. The presence of distinct second-order transitions in the grafted materials strongly suggests their phase-separated nature.

As noted above, the fine structure of cotton fibers is similar to that of the wood tracheid cells discussed in Section 9.8. Thus, grafting can occur on the outer and inner surfaces, as well as in the amorphous portions of the fiber.

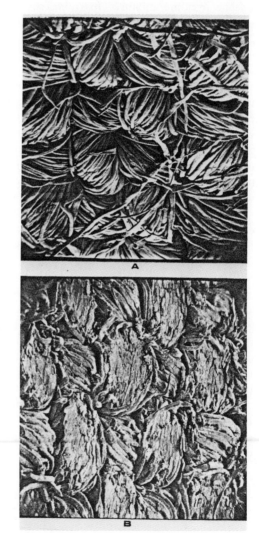

Figure 7.2. Scanning electron microphotograph of cellulose–polyacrylonitrile–poly(butyl methacrylate) copolymer fabric sample before (A) and after (B) flex abrasion (24,000 cycles).

7.4.3. More-Homogeneous-Type Grafts

In order to accomplish the synthesis of homogeneous grafts, the backbone polymer must be swelled by or dissolved into the second monomer. Grafting may be initiated by irradiation with ^{60}Co sources or Van de Graaff-type electron accelerators. The ratio of graft to homopolymer formed will depend on the relative reactivities of the backbone chain atoms vs. the corresponding monomer atoms. Hopfenberg et al. (1969) prepared a homogeneous-type graft of polystyrene onto amorphous secondary (~ 2.5 degree of substitution)* cellulose acetate film by swelling the polymer with styrene–pyridine mixtures (pyridine serving as a solvent), followed by irradiation with a ^{60}Co source. After homopolymer extraction, nominally 20–40% of grafted polystyrene remained. The effects of total dose and film thickness are shown in Figure 7.3. The films studied by Hopfenberg et al. were intended for use in the desalination of sea water by reverse osmosis. The grafted material had a tensile creep rate several times lower than that of ungrafted cellulose acetate, without significant sacrifice in transport properties. Although the phase separation characteristics of this material were not discussed in detail (Wellons et al., 1967), it seems probable that the polystyrene forms the dispersed phase, perhaps in the form of spheroidal or cylindrical shapes (see Section 3.1), binding the cellulose ester chains together (Wellons et al., 1967). If so, the continuous cellulose acetate phase would be free for water permeation.

It is interesting to note that homogeneous grafting procedures can be employed to produce cellulose-based elastomers (Nakamura et al., 1968). In this case rayon fibers were first crosslinked with formaldehyde and then swelled with aqueous zinc chloride. The graft copolymerization was carried out in an aqueous emulsion system by the ceric ion method, using ethyl acrylate as monomer. These materials are semi-IPN's of the first kind (Section 8.6). Materials containing several hundred percent or more of grafted material revealed typical rubber elastic behavior. One may speculate that the cellulose and poly(ethyl acrylate) undergo phase separation. If so, either the cellulose may form the stiff phase dispersed in the softer acrylate matrix or two continuous phases may be present. At lower acrylate levels, the cellulose probably would retain phase continuity.

It should be noted that wool, the other major fibrous polymer of natural origin, has also been used as a substrate for grafting. Indeed, materials of commerce are often grafted in a manner analogous to that used for the cellulosic fibers (Rebenfeld, 1971).

---

* The degree of substitution in cellulose is based on the three hydroxyl groups on each glucose residue (see Table 1.1).

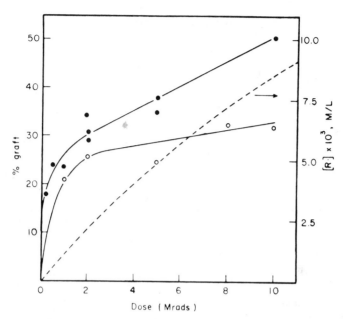

Figure 7.3. Effect of film thickness on grafting kinetics of styrene to cellulose acetate: (●) 0.030-mm cellulose acetate film; (○) 0.15-mm cellulose acetate film; monomer solution 80 styrene: 20 pyridine. Dashed line shows estimated radical buildup during grafting process. (Hopfenberg *et al.*, 1969.)

7.5. NYLON GRAFT COPOLYMERS

Graft copolymers may be composed of two amorphous polymers, two crystalline polymers, or one crystalline and one amorphous polymer. Generally speaking, the crystalline component of a crystalline–amorphous pair such as a nylon–amorphous polymer graft may serve as either the backbone or the side chain. (The natural cellulose polymers discussed in the previous section are, however, exceptions, and, not possessing an accessible monomeric base, serve only as the backbone.)

For example, Matzner *et al.* (1973) succeeded in grafting side chains of nylon 6 onto a polystyrene backbone by means of anionic polymerization of ε-caprolactam. To facilitate grafting, polystyrenes containing up to 5% of ethyl acrylate or methyl methacrylate as comonomer were synthesized. The ester group served as the initiator for the lactam polymerization, thus permanently attaching the side chain to the backbone. The mechanism of the reaction is as follows:

$$^{\ominus}N\underset{(CH_2)_5}{\diagdown\diagup}CO + RCOOR' \xrightarrow{\text{initiator}} RCON\underset{(CH_2)_5}{\diagdown\diagup}CO + {}^{\ominus}OR' \qquad (7.6a)$$

$$RCON\underset{(CH_2)_5}{\diagdown\diagup}CO + {}^{\ominus}N\underset{(CH_2)_5}{\diagdown\diagup}CO$$

$$\xrightarrow{\text{fast}} RCO-{}^{\ominus}N-(CH_2)_5-CO-N\underset{(CH_2)_5}{\diagdown\diagup}CO \qquad (7.6b)$$

$$RCO-{}^{\ominus}N-(CH_2)_5-CO-N\underset{(CH_2)_5}{\diagdown\diagup}CO + HN\underset{(CH_2)_5}{\diagdown\diagup}CO \qquad (7.6c)$$

$$\xrightarrow{\text{fast}} RCON\overset{H}{\overset{|}{}}-(CH_2)_5-CO-N\underset{(CH_2)_5}{\diagdown\diagup}CO + {}^{\ominus}N\underset{(CH_2)_5}{\diagdown\diagup}CO \qquad (7.6d)$$

$$\xrightarrow{\text{etc.}} RCO-[-\overset{H}{\overset{|}{N}}-(CH_2)_5-CO-]_n-N\underset{(CH_2)_5}{\diagdown\diagup}CO \qquad (7.6e)$$

When ethyl acrylate was used at a concentration in excess of 1%, a gelled material was obtained. The crosslinking reaction thus revealed was tentatively identified as arising from the presence of tertiary hydrogen atoms in the α position with respect to the ester group. The crosslinked materials must topologically resemble the structures discussed in Section 8.7.

As can be seen in Figure 7.4, the mechanical properties of the system show the glass and melting transitions of the nylon 6, the T_g of the polystyrene, and strong indications of a two-phase nature.

The inverse synthesis (of polymers grafted to nylon) was considered by Huglin and Johnson (1972). Employing γ-irradiation of nylon 6 films in the presence of acrylic acid, Huglin and Johnson found that copper salts suppressed the homopolymer reaction in favor of graft copolymerization.

The component polymerized first usually constitutes the more continuous phase in graft copolymers (Section 3.1), unless the reacting mass is stirred. An unusual test of this principle was accomplished by Korshak and co-workers (1968) during an investigation of the coefficient of friction of systematically prepared graft copolymers of styrene on nylon 6 (see Table

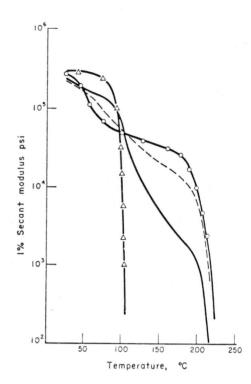

Figure 7.4. Modulus–temperature behavior of nylon 6/polystyrene graft copolymers. —, 50% N_6; --, 75% N_6; ○, Nylon 6; △, poly-styrene. (Matzner *et al.*, 1973.)

Table 7.1
Coefficient of Friction and Wear Rate of Styrene–ε-Caprolactam Copolymers of Varying Composition

Copolymer composition		Coefficient of friction				Wear rate	
		in atmosphere at sliding velocities of (m/sec)		in vacuum, sliding velocity 0.2 m/sec	in vacuum with irradiation, sliding velocity 0.2 m/sec	in atmosphere	in vacuum with irradiation
Styrene	ε-Capro-lactam	0.2	2.0				
—	100	0.23	0.23	0.23	0.51	1.0×10^{-10}	8×10^{-10}
10	90	0.23	0.24	0.23	0.44	1.1×10^{-10}	5×10^{-10}
20	80	0.23	0.23	0.23	0.41	1.4×10^{-10}	4×10^{-10}
33	67	0.24	0.24	0.24	0.40	2.0×10^{-10}	3×10^{-10}
100	—	0.48	0.55	0.55	0.62	5×10^{-9}	8×10^{-9}
$\Phi_{\text{Teflon-4}}$	—	0.17	0.23	0.17	0.36	10^{-9}	6×10^{-9}

[a] The values of the coefficients of friction and the wear rate were determined as the arithmetic means of ten experiments, the mean-square error for the coefficients of friction being ±0.003. Korshak *et al.* (1968).

7.1.). Nylon 6 is known to have a low coefficient of friction and high wear resistance, while polystyrene, in contrast, has poor antifriction properties. The experimental data suggest that, since the coefficient of friction of the grafts (at least for polystyrene contents up to 33 %) is essentially equal to that of the nylon, nylon 6 must make up the continuous phase.

In work reported in a recent paper, Khanderia and Sperling (1974) employed a suspension-modified interfacial polymerization route to form a graft copolymer of nylon 6,10 and polystyrene. Optical microscopy revealed that most of the nylon was on the surface of the suspension-sized particles; hence this product actually is very heterogeneous, resembling the surface graft copolymers. Khanderia and Sperling found that when the material was molded below the melting point of the nylon, the nylon minor component retained phase continuity and imparted nylon mechanical properties to the bulk material. Such compositions serve as a prototype "poor man's nylon" (Khanderia and Sperling, 1974; Sperling, 1974b). Materials molded at higher temperatures, with nylon forming the discontinuous phase, serve only as a "rich man's polystyrene."

7.6. GRAFTING WITH POLYFUNCTIONAL MONOMERS

The crosslinking of polymers by irradiation has been studied for many years (Chapiro, 1962) as a means for improving physical and mechanical properties. Indeed, radiation-crosslinked polyethylene has been a commercial product for quite some time. As mentioned above, irradiation of a polymer in the presence of a monomer (either swollen into the polymer or not; see Sections 7.2 and 7.4) has also been a standard technique for the preparation of graft polymers (Charlesby, 1960; Harmer, 1967; Odian and Bernstein, 1964; Odian and Kruse, 1968). An interesting class of polymer–polymer systems has resulted from attempts to enhance crosslinking and grafting efficiency by combining the two approaches, i.e., by irradiating polymers that have been compounded with polyfunctional monomers capable of undergoing crosslinking.

Thus a number of studies have been concerned with the irradiation of polyfunctional monomers, such as diacrylates, which have been incorporated in polymers such as polyethylene (Odian and Bernstein, 1963), polypropylene (Odian and Bernstein, 1964), polyisobutylene (Odian and Bernstein, 1963), unsaturated polyesters (Charlesby and Wycherly, 1957), and, especially, poly(vinyl chloride) (Bonvicini et al., 1963; Gladstone et al., 1971; Hammon, 1958; Harmer, 1967; Izumi et al., 1965; Koozu et al., 1963; Miller, 1959; Nicholl, 1969; Pinner, 1960; Stanton and Traylor, 1962; Szymczak, 1970;

Szymczak and Manson, 1974a,b; White and Mann, 1967). As with electron-irradiated polyethylene, crosslinked poly(vinyl chloride) formulations (usually plasticized) are now of commercial interest as wire and cable coatings and insulation [see, for example, Nicholl (1969)]. In the following sections, typical properties will be described and discussed; details of other standard polymer–monomer graft polymerizations are given by, e.g., Chapiro (1962) and Charlesby (1960).

7.6.1. Polyolefins

Odian and Bernstein (1963) and Lyons (1960) have reported results of an extensive study of the crosslinking of polyethylene using gamma irradiation in the presence of polyfunctional monomers. The monomers, which included a variety of diallyl and triallyl compounds and polyfunctional acrylates, were incorporated by swelling into the polyethylene matrix until equilibrium had been attained (corresponding to monomer concentrations between 1 and 13%). Two of the most efficient monomers in terms of gel formation were allyl methacrylate (I) and allyl acrylate (II).

$$CH_2{=}\overset{\overset{\displaystyle H}{|}}{C}{-}CH_2{-}O{-}\overset{\overset{\displaystyle O}{\|}}{C}{-}\overset{\overset{\displaystyle CH_3}{|}}{C}{=}CH_2 \qquad CH_2{=}\overset{\overset{\displaystyle H}{|}}{C}{-}CH_2{-}O{-}\overset{\overset{\displaystyle O}{\|}}{C}{-}\overset{\overset{\displaystyle H}{|}}{C}{=}CH_2$$

$$\text{(I)} \qquad\qquad\qquad\qquad\qquad \text{(II)}$$

Using allyl methacrylate as a model, Odian and Bernstein (1964) found that the dose of radiation needed to induce a detectable amount of gelation in polyethylene (as determined by extraction) was, in fact, markedly lowered—by an order of magnitude—from about 0.5 to 0.05 Mrad by the incorporation of polyfunctional monomers. This tendency was maintained over the whole range of gel contents obtained (Figure 7.5). It was possible to improve efficiency still further, by partially crosslinking, reswelling, and repeating the cycle several times (cf. swelling of crosslinked networks by Millar (1960), Donatelli et al. (1974), and Sperling et al. (1971); see Chapter 8 for further details).

Physical properties of the polyethylene were, in general, improved. The use of the difunctional monomer resulted in better dimensional stability at 185°C (a combination of 1.2 Mrad with 4% monomer being equivalent to a dose of 30 Mrad without monomer addition) and in improved tensile strength.

The same two monomers were also studied in combination with poly-propylene. Although polypropylene requires about 50 Mrad for incipient gelation, the use of allyl methacrylate, at a concentration of 5.5%, resulted

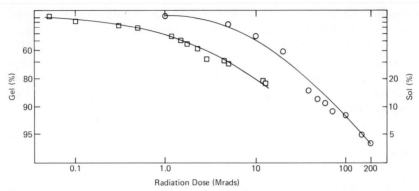

Figure 7.5. Radiation crosslinking is accomplished with a lower radiation dose in the presence of a monomer such as polyfunctional allyl methacrylate, 3.7 % monomer incorporated. (○) Polyethylene; (□) polyethylene-allyl methacrylate. (Odian and Bernstein, 1964.)

in a gel fraction of 0.70 after exposure to only 5 Mrad. Only preliminary results were reported for polyisobutylene. In contrast to the degradation normally observed in polyisobutylene on irradiation, crosslinking was noted instead (a gel fraction of 0.6) when 14 % allyl acrylate was incorporated.

Thus the incorporation of polyfunctional monomers in polyolefins, followed by radiation polymerization, can improve properties, lower the dose for gelation, or minimize degradation, depending on the polymer. It should be pointed out that a major use of the materials described here is as insulators for electrical wires and cables (Nicholl, 1969).

7.6.2. Poly(vinyl chloride)

Most of the literature on polymer–polyfunctional monomer systems has been concerned with poly(vinyl chloride). Radiation crosslinking of poly(vinyl chloride) (PVC) has long been of interest as a means for increasing the modulus at elevated temperatures and inhibiting plastic flow. Since the T_g of rigid PVC is only 80°C, and is much lower for the plasticized resin— say − 20°C—the onset of plastic flow has restricted maximum use temperature to about 100°C. However, another limiting factor is the tendency to undergo dehydrochlorination, which results in embrittlement and loss of other properties, at relatively low temperatures; unfortunately, irradiation by itself induces degradation as well as crosslinking.

In 1959, Miller (1959a,b) reported that radiation polymerization of a plasticized PVC containing a divinyl monomer could be used to crosslink PVC effectively without seriously affecting thermal stability. Three monomers were used: poly(ethylene glycol dimethacrylate) (PEGDMA), ethylene glycol diacrylate (EGDA), and divinylbenzene (DVB). Most results were based on milled specimens containing 10 or 20% (based on PVC) of PEGDMA.

Gelation was induced at much lower doses than in the absence of the monomer—at about 0.1 Mrad, compared to about 20 Mrad. The higher the percentage of monomer, the higher the degree of gelation at all stages after initiation of gelation; about 70% of the ultimate gelation was reached after only 1 Mrad. At equivalent gel fractions, the degree of swelling was less, the higher the monomer concentration, as would be expected if a tighter gel was formed (Figure 7.6). Curiously, however, the degree of swelling was found to *increase* as the dose and gel yield were increased. This was attributed to the concurrence of some degradation of radiation-sensitive methacrylate crosslinks along with crosslinking elsewhere.

Both tensile strength and modulus were affected to an insignificant degree by exposure of untreated specimens to a dose of 1 Mrad. In contrast, tensile strength and modulus were increased by about 90% in the presence of 20% PEGDMA at the same dose level (Figure 7.7); further irradiation resulted in only slight additional increases. The addition of carbon black was also examined; both modulus and tensile strength were increased still further—to 4100 and 3200 psi, respectively, at 1 Mrad.

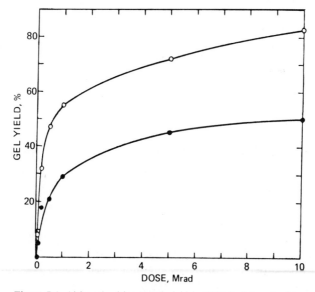

Figure 7.6. Although ultimate gel yield in PVC is determined by the amount of added dimethacrylate, most of the gel is formed at a dose of only 1 Mrad. (○) 20% monomer; (●) 10% monomer. (Miller, 1959b.) [Reprinted with permission from *Ind. Eng. Chem.* **51**, 1271 (1959). Copyright by The American Chemical Society.]

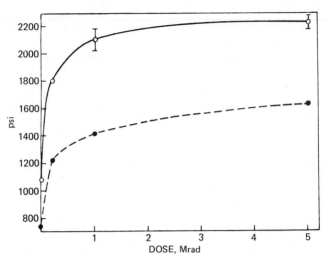

Figure 7.7. With 20 % added dimethacrylate, irradiation of PVC causes
large increases in tensile strength (upper curve) and modulus (lower
curve) up to a dose of 1 Mrad. (Miller, 1959b.) [Reprinted with permission from *Ind. Eng. Chem.* **51**, 1271 (1959). Copyright by the American
Chemical Society.]

All these results may well be expected, but would be of no practical
interest unless the thermal stability is improved, or at least not adversely
affected. Miller observed that as long as the total dose did not exceed about
1 Mrad, the discoloration on aging at 150°C characteristic of irradiated
PVC was suppressed by incorporation of the divinyl monomer before
irradiation. The enhanced stability was attributed to the ability of the added
monomer to scavenge propagating free radicals that would otherwise
facilitate dehydrochlorination.

In the decade following publication of Miller's papers, a diverse group
of papers appeared on the same general subject; several typical ones are
noted here. Wippler (1959) investigated the irradiation of PVC alone and
as a plasticized gel; he found that the crosslinking behavior was dependent
on the plasticizer (or other additive) present. Pinner (1960) investigated the
irradiation of PVC containing dioctyl fumarate (a monovinyl compound),
diallyl sebacate (a divinyl compound), and triallyl cyanurate (a trivinyl
compound). In an absolute sense, i.e., at the same concentration, he found that
the higher the functionality, the higher the modulus and tensile strength.
As noted previously by Miller, the thermal stability was also improved by
incorporation of the monomers. Izumi *et al.* (1965) also reported that
polyfunctional monomers were more efficient. While most studies have been
concerned with identifying potentially useful monomers, Pinner (1960) and
Izumi *et al.* (1965) did try to determine the effect of monomer structure and
functionality. However, since the structures were of different types and had

different concentrations of reactive groups, a comparison of intrinsic reactivity is not possible.

Other monomers have been used. Koozu *et al.* (1963) studied the effects of monomers, including diallyl phthalate, and ethylene glycol dimethacrylate (EGDM). Gladstone *et al.* (1971) used monomers such as triallyl phosphate, acenaphthalene, and crotonic acid. Trimethylolpropane trimethacrylate has been reported to give a tough, heat-resistant PVC after irradiation (White and Mann, 1967).

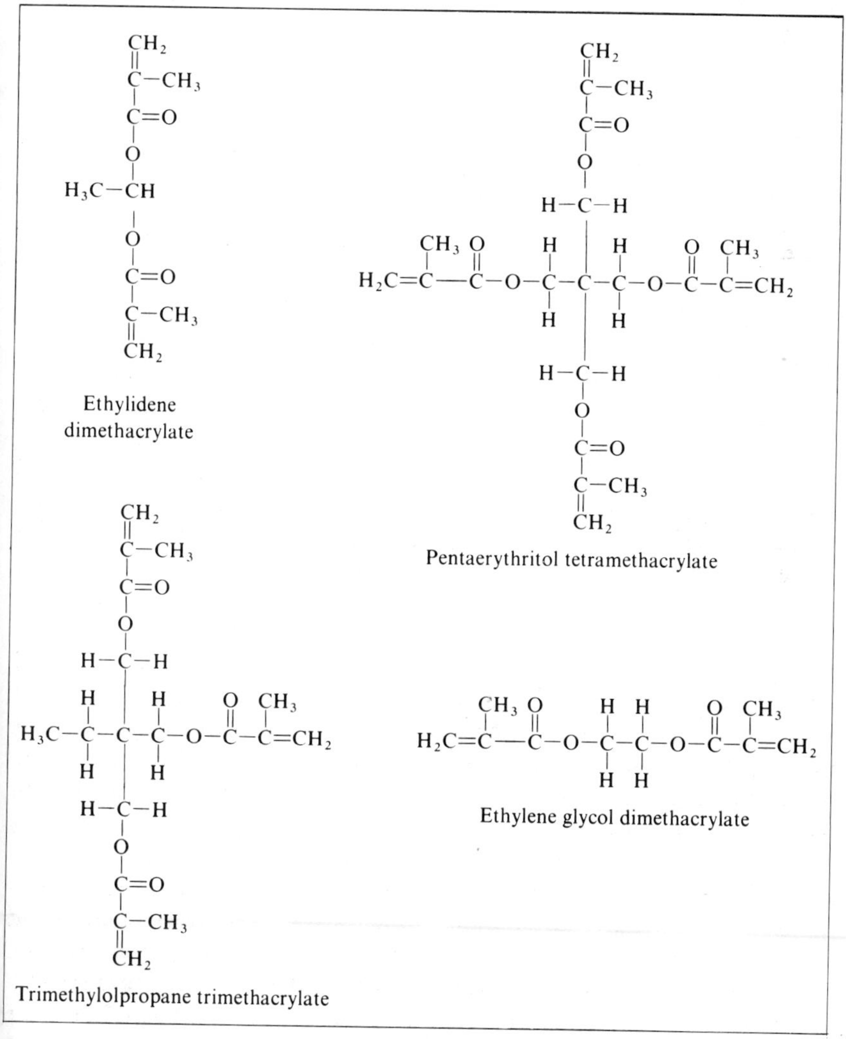

Figure 7.8. Monomer structures. Monomers of this general type have found application in reducing creep in PVC.

Table 7.2[a]
Actual Composition of PVC–Monomer Mixtures

| | Actual composition, % | | | | | | | |
	A_1	A_2	B_1	B_2	C	D	E	F
Component								
PVC	68.5	66.7	66.1	64.7	64.6	64.6	71.1	64.6
DOP	25.7	25.7	25.7	25.7	25.7	25.7	25.7	32.2
Monomer								
(a) EDMA	2.5	4.3	—	—	—	—	—	—
(b) EGDMA	—	—	4.9	6.4	—	—	—	—
(c) TMPTMA	—	—	—	—	6.5	—	—	—
(d) PTMA	—	—	—	—	—	6.5	—	—
Stabilizer	3.4	3.3	3.3	3.2	3.2	3.2	3.2	3.2

[a] Szymczak (1970).

More recently, Szymczak (1970) and Szymczak and Manson (1974a,b) have reported a comparison of reactivities in a series of acrylates and methacrylates selected to have approximately equal concentrations of double bonds per mole. In this study, four monomers were used: ethylidene dimethacrylate (EDMA) (divinyl); ethylene glycol dimethacrylate (EGDMA) (divinyl, isomeric with EDMA); trimethylolpropane trimethacrylate (TMPTMA) (trivinyl); and pentaerythritol tetramethacrylate (PTMA) (tetravinyl); see Figure 7.8. Concentrations up to 10% (based on PVC) were examined; concentrations of double bonds (based on PVC) were kept equal to within ±10% (see Table 7.2). In order to induce flexibility into the PVC resins, approximately 25% of DOP plasticizer was added before irradiation.

Of these monomers, EDMA appeared to be quite ineffective, even though it is isomeric with EGDMA; discussion of EDMA is therefore omitted.

In the other monomer-containing specimens, the hardness and specific gravity were increased consistently by irradiation. As noted earlier by Miller (1959a,b), the major increase occurred after only a very small dose—about 0.2 Mrad. Effects of monomer concentration and functionality were slight. On the other hand, the gel fraction was affected not only by the dose, but also by the functionality. Although irradiated control formulations remained soluble, at least up to a dose of 11 Mrad, gel fractions increased with the level of monomer, with the total dose, and with the functionality. Thus, at 11 Mrad, the gel fractions increased from 0.43 to 0.57 as the functionality was changed (at the same concentration) from 2 to 4 (Table 7.3). Since even at 0.2 Mrad the gel fraction as determined by extraction was higher than the fraction of monomer present, clearly PVC was being bound into the

Table 7.3[a]
Results of Gel Fraction Determinations

Dose, Mrad	Gel fraction, wt %							
	A_1	A_2	B_1	B_2	C	D	E	F
0	0	0	0	0	0	0	0	0
0.19	1.2	0	7.5	9.1[b]	10.9	10.9	0	0
1.2	0.9	0	19.8	25.0[b]	27.7	29.1	0	0
7.3	0.7	0	36.2	39.6	46.8	47.4	0	0
10.7	0.7	0	39.0[a]	42.7	53.8	56.7	0	0

[a] Szymczak (1970).
[b] Average of two readings.

three-dimensional network. Again, the gelation process is relatively more effective at the lower doses.

As shown in Figure 7.9, the curve of Young's modulus vs. dose is sensitive to the concentration of EGDMA [cf. Miller's (1959a,b) results]. At equal concentrations (specimens B2, C, and D), however, both the shape and position of the modulus curves depend on the functionality. Clearly (Figure 7.10) the trivinyl and tetravinyl monomers gave higher moduli than the divinyl monomer (about 14,000 psi at 7 Mrad, compared to 9000 psi). At higher doses, however, the modulus begins to decrease with the trifunctional and tetrafunctional monomers. In comparison, moduli for control formulations were independent of dose, and relatively low, between 1000 and 5000 psi. Thus, as indicated by previous studies, the presence of a polyfunctional monomer results in a dramatic increase in modulus. The functionality is also important, even at the same concentration of double

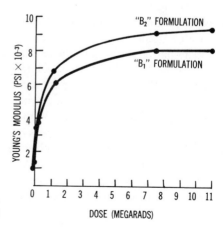

Figure 7.9. Modulus vs. dose for EGDMA formulations. (Szymczak and Manson, 1974a.)

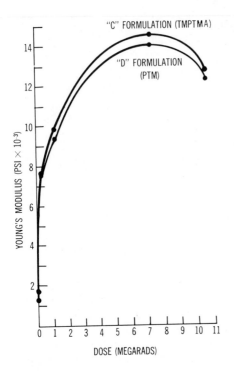

Figure 7.10. Modulus vs. dose for TMPTMA and PTM formulations. (Szymczak and Manson, 1974a.)

bonds, with trivinyl and tetravinyl monomers being the most effective in forming tight networks.

Results of thermal stability tests form a rather complex pattern. On aging at 100°C, all formulations, though less discolored than monomer-free specimens, exhibited maxima in color development after an hour or so, indicating that irradiation in the presence of monomer produces transient color-producing species, which later decay, perhaps by prolonged exposure to air. Again, a dependence on functionality was noted—the higher the functionality, the greater the resistance to discoloration.

A more detailed study of thermal behavior using both differential thermal (DTA) and thermogravimetric (TGA) analysis gave results that indicate a different aspect of thermal stability (Table 7.4). In ordinary PVC, HCl is eliminated at about 300°C, with depolymerization at about 450–460°C. The latter endotherm was found to be unchanged by the presence of monomer, stabilizer, or plasticizer or by irradiation. However, the dehydrochlorination endotherm (T_{HCl}) was quite sensitive to all these factors. The addition of plasticizer and stabilizer alone *raised* T_{HCl}; irradiation of such controls had no further effect. In the presence of polyfunctional monomer

Table 7.4[a]

Temperatures (°C) of Endothermic Reactions—DTA[b]

Dose, Mrad	A_2		B_2		C		D		E		F		PVC resin	
	1	2	1	2	1	2	1	2	1	2	1	2	1	2
0	331	463	330	463	330	458	330	460	329	460	330	463	300	460
0.19	330	463	327	460	325	457	325	460	331	460	331	463	—	—
1.2	334[c]	461	323	458	320	458	320	460	330	460	330	463	—	—
7.3	331	460	320	458	318	458	317	460	328	460	332[c]	463	—	—
10.7	330	460	317	460	316	459	310	460	326[c]	458	332[c]	463	—	—

[a] Szymczak (1970).
[b] 1, first reaction; 2, second reaction.
[c] Questionable result.

however, T_{HCl} was *reduced*, though not lower than 310°C. The sensitization may be due to the transfer of a free radical with the PVC chain, and subsequent formation of a double bond, which would be a potential source of instability. On the other hand, at T_{HCl}, the weight loss on prolonged heating was consistently lower for the monomer-containing specimens, and approximately independent of functionality.

Further evidence of the tightness of the network developed was provided by thermomechanical studies using a Brabender plastograph. The higher the functionality, the greater the resistance to shearing and the shorter the time to reach maximum torque.

A little speculation on a possible explanation for the behavior noted is in order.

Simple calculations based on Figure 7.9 yield insight as to the probable supermolecular structure of the multifunctional crosslinked materials. Let us assume first a simple, one-phase semi-IPN of the second kind (Chapter 8) (Szymczak, 1970) and apply the theory of rubber elasticity to the modulus increase of material B_2. Young's modulus increases from 6.9×10^7 dyn/cm^2 (a stiff rubber) to 6.2×10^8 dyn/cm^2 (a leather) during the polymerization of the second component. Calculations based on $E = 3nRT$ yield 7.4×10^{-3} mol/cm^3 of new network chains introduced by means of the second polymerization. From the actual amount of EGDMA introduced (6.4%), one may estimate that some 23 equivalent new network chains were introduced per monomer unit. In spite of the probable high degree of true grafting and crosslinking introduced by the radiation, this number seems unrealistic, thus casting serious doubt upon the applicability of rubber elasticity theory to the problem.

Let us assume instead a two-phase material, with the second component being glassy, having $E = 3 \times 10^{10}$ dyn/cm^2. Elsewhere in the text (Chapters 3

Table 7.5
Modulus Prediction for Material B_2

Relationship	Modulus, dyn/cm^2
Upper bound	2×10^9
Lower bound	7.5×10^7
Experimental	6.2×10^8

and 8) it was shown that the second component polymerized tends to form cellular domain structures, suggesting the applicability of the upper and lower bound composite relationships (see Section 12.1.1). Results are tabulated in Table 7.5. It is seen that the experimental value for the fully polymerized material lies part way between the upper and lower bounds, suggesting some phase continuity and some cellular domain structure. Neglecting truly grafted material, the material may be depicted as bundles of straw bound together by a three-dimensional chicken-wire. Thus creep is reduced and mechanical properties generally are improved.

While no detailed studies of the network characteristics of mechanical behavior or phase relationships appear to have been made, one may speculate on possible morphologies. First, all studies with multifunctional monomers were done with very low monomer II/polymer I ratios. Although it is not clear whether one ought to expect phase separation when only 5% or so of monomer II has been added, the multifunctional monomers should form rather dense networks in addition to true grafts. Such networks would tend to entrap mechanically polymer I chains, lowering creep rates below those expected for grafting only (a desirable result). If phase separation does occur, experience with other systems (for example, ABS polymers; Section 3.1.2) suggests that polymer II would form domains within a cellular structure. However, due to the low concentrations and high crosslink density, the phase separation would probably remain very incomplete. Finally, from a topological point of view, the use of multifunctional monomers results in a semi-IPN of the second kind (see Section 8.6) since the first component is essentially linear, while the second forms a continuous network (Donatelli et al., 1974).

7.7. MULTIPOLYMER GRAFTS

Nearly all of the polymer blends, grafts, and blocks considered in this monograph are concerned with combinations of two polymers. Some materials, however, contain three or even four identifiable, distinct polymer components. Each, presumably, has its own phase; thus multiphase structures are indicated. This section will briefly cover the literature focused on multipolymer grafts.

A ternary graft copolymer semi-IPN is described by Rogers and Ostler (1973) and Rogers *et al.* (1974), where an original crosslinked polyethylene-graft-poly(potassium acrylate) was swelled with styrene and radiation-polymerized. The authors comment that most of the grafting of polymer III was on polymer II, the poly(potassium acrylate).

Some rather interesting, if complex, materials have been prepared by Vollmert (1962). For instance, in his example 5, Vollmert emulsion polymerized *n*-butyl acrylate, styrene, 1,4-butane-diol monoacrylate, and 1,4-butane-diol diacrylate, forming polymer I. Separately, *n*-butyl acrylate, styrene, and acrylic acid were emulsion-polymerized, forming polymer II. Polymers I and II were mixed, precipitated, washed, dried, and dissolved in styrene and acrylic acid, followed by polymerization, forming polymer III. The material was then heated to induce grafting between polymers I and II, and I and III.

In a more recent patent, Ryan (1972) reported poly(*n*-butyl acrylate)/ poly(methyl methacrylate) latex semi-IPN's of the first kind. These materials were then blended into linear poly(vinyl chloride), resulting in a ternary polymer mixture.

In an interesting related patent, Amagi *et al.* (1972) described a triple latex IPN (three crosslinked polymers). In brief, polymer I was a crosslinked SBR, polymer II was a crosslinked styrene–methyl methacrylate copolymer, and polymer III was a crosslinked poly(methyl methacrylate), all sequentially synthesized on the same latex particle. The latex material was then mechanically blended with linear poly(vinyl chloride). Finally, Torvik (1971) blended four polymers together, each with a different glass transition temperature.

7.8. CATIONIC GRAFT COPOLYMERS

The previous types of graft copolymers discussed in this monograph, all synthesized by free radical techniques, contain a very large proportion of homopolymer, perhaps as little as 1–5% of the molecules actually being grafted by statistical chain transfer or by reaction with the double bond in a diene mer. Recently Kennedy and co-workers (Kennedy, 1971; Kennedy and Baldwin, 1969; Kennedy and Smith, 1974) reported that true graft copolymers could be obtained by means of cationic polymerization techniques that ideally allow only graft-type copolymers to be formed. Grafting efficiencies of up to 90% were reported, where the grafting efficiency G.E. was defined by

$$G.E. = \frac{(\text{component } 2)_g}{(\text{component } 2)_g + (\text{component } 2)_h} \tag{7.7}$$

where (component 2) represents the weight of second polymer synthesized, and the subscripts g and h represent graft and homopolymer, respectively.

The mechanism proposed by Kennedy requires that allylic or tertiary chlorines be attached to the first polymer chain. Chlorobutyl rubber, PVC, and other chlorine-containing polymers usually have 1–2% of mers with chlorines in the required reactive positions. The halogen-containing polymer is dissolved in an inert but polar solvent in the presence of an organoaluminum catalyst and a cationically polymerizable monomer, such as styrene, and polymerization is effected, usually at about $-50°C$. The major point is that the second component chains can be initiated only at an active chlorine site on the first polymer.

Use of a chemically saturated rubber such as chlorobutyl, together with styrene as the monomer, with the rubber the predominant component by weight, results in a material similar in behavior to the thermoplastic elastomers (see Chapter 4). In all cases two glass transitions and other evidence suggested phase separation.

8

Interpenetrating Polymer Networks

Interpenetrating polymer networks (IPN's) are a unique type of polyblend, synthesized by swelling a crosslinked polymer (I) with a second monomer (II), together with crosslinking and activating agents, and polymerizing monomer II *in situ* (Sperling, 1974–1975; Sperling and Friedman, 1969). The term IPN was adopted because, in the limiting case of high compatibility between crosslinked polymers I and II, both networks could be visualized as being interpenetrating and continuous throughout the entire macroscopic sample.* As with other types of polyblends, if components I and II consist of chemically distinct polymers, incompatibility and some degree of phase separation usually result (Sperling, 1974–1975; Sperling and Friedman, 1969; Sperling *et al.*, 1970a,b,c; 1971). Even under these conditions, the two components remain intimately mixed, the phase domain dimensions being on the order of hundreds of angstroms. If one polymer is elastomeric and one polymer is plastic at the use temperature, the combination tends to behave synergistically, and either reinforced rubber or impact-resistant plastics result, depending upon which phase predominates (Curtius *et al.*, 1972; Sperling and Mihalakis, 1973; Sperling *et al.*, 1971; Huelck *et al.*, 1972). Among the other kinds of polymer blends discussed in this monograph, the graft-type copolymers are the ones most closely related to the IPN's.

This chapter will discuss the synthesis, morphology, and mechanical behavior of IPN's. Related materials, in particular the interpenetrating elastomeric networks (IEN's) of Frisch (Frisch and Klempner, 1970; Klempner *et al.*, 1970; Matsuo *et al.*, 1970b) will be included for comparison.

---

* An interesting analog of an IPN is given by Gamow (1967, p. 56): Two worms eat out independent tunnels in an apple. Each worm may cross its own path but not that of its neighbor. If the tunnels so formed are considered polymer chains, then one has IPN's.

8.1. SYNTHESIS

The synthesis of an IPN first requires a crosslinked polymer I, which may be either the elastomer or plastic component. In the first papers on IPN's, the elastomer was the first component. When the plastic was component I, the term "inverse" IPN was employed on an arbitrary basis.

Network I may be synthesized and crosslinked simultaneously, for example, by using ethyl acrylate (EA) and tetraethylene glycol dimethacrylate (TEGDM). This network may be photopolymerized using ultraviolet light and benzoin as an activator (Sperling, 1974–1975; Sperling and Friedman, 1969; Sperling et al., 1970a,b,c; 1971). Alternatively, the linear polymer may be prepared first, and crosslinked by subsequent reaction, as, for example, by subjecting polybutadiene films containing dicumyl peroxide to heat and pressure.

The second network was synthesized in all cases by swelling in a controlled amount of monomer (containing dissolved benzoin and TEGDM), allowing ample time for diffusion so that the swelling was uniform, and photopolymerizing, or thermally initiating with peroxide species. An example of the former would be styrene (S) monomer containing 0.5% TEGDM and 0.3% benzoin, swelled into poly(ethyl acrylate) (PEA) to equal weight. The resulting 50/50 PEA/PS is opaque, white, and leathery in texture. This synthesis produces a material presently called a "sequential IPN." Other types will be discussed in later sections of this chapter. Two types of IPN's differing in compatibility of the respective components have been synthesized. The first type includes such normally incompatible polymer pairs as polybutadiene/polystyrene, PB/PS, or PEA/PS. The second type includes the semicompatible pair poly(ethyl acrylate)/poly(methyl methacrylate),* PEA/PMMA. Films of the latter type are clear in comparison to those of the first type. A series of IPN's having intermediate morphologies and properties was prepared by random copolymerization of the MMA and S monomers to produce PEA/P(S-co-MMA).†

---

* PEA and PMMA are chemically isomeric, having structures

$$-CH_2-\underset{\underset{\displaystyle O-CH_2-CH_3}{\overset{\displaystyle |}{\underset{\displaystyle |}{C=O}}}}{\overset{\overset{\displaystyle H}{\displaystyle |}}{\underset{\displaystyle |}{C}}}- \qquad \text{and} \qquad -CH_2-\underset{\underset{\displaystyle O-CH_3}{\overset{\displaystyle |}{\underset{\displaystyle |}{C=O}}}}{\overset{\overset{\displaystyle CH_3}{\displaystyle |}}{\underset{\displaystyle |}{C}}}- \qquad (8.1)$$

respectively. This structural similarity reduces the (positive) heat of mixing, and greater mutual solubility results.

† The glass–rubber transition temperatures T_g of PS and PMMA are 100 and 105°C, respectively. Consequently, their random copolymers have essentially the same T_g. In contrast, the T_g's of PEA and PB are −22 and −90°C, respectively.

With both the PEA/P(S-co-MMA) and PB/PS IPN's, an important variable is the ratio of elastomer to plastic in the final material. When the plastic component predominates, a type of impact-resistant plastic results. In this manner the PB/PS IPN's are analogous to the impact-resistant graft copolymers. When the elastomer component predominates, a self-reinforced elastomer results, the behavior resembling that of the ABA-type block copolymers (thermoplastic elastomers) described in Section 4.4. When the overall compositions of both the PB/PS and the PEA/P(S-co-MMA) series are close to 50/50, the materials behave like leathers.

8.2. MORPHOLOGY OF IPN's

Although both polymeric networks in IPN's may be visualized as continuous, their milky appearance hints at a more complex morphology. In order to characterize the morphology, the elastomeric portion of the IPN's was stained selectively using osmium tetroxide (see Section 2.4); to facilitate staining, about 1% of butadiene was incorporated as a comonomer in the PEA. This special product was designated PEAB.

8.2.1. General Features

The IPN's were found to exhibit a characteristic cellular structure, where the first component made up the cell walls and the second component

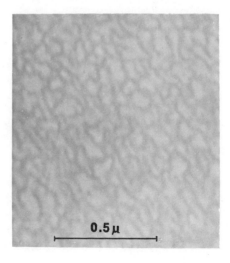

Figure 8.1. Electron micrograph of a *cis*-PB/PS IPN containing 76% PS. The dark portion is the PB phase stained with osmium tetroxide. Although the cell shape is irregular, the cell wall thickness appears nearly constant. Curtius *et al.*, 1972.)

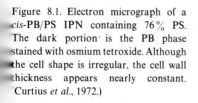

0.5 μ

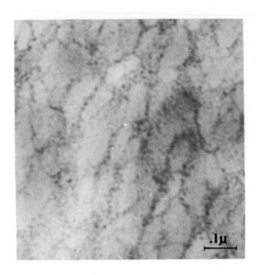

Figure 8.2. Electron micrograph of a 50/50 PEAB/PS IPN. The residual double bonds in the subsequent elastomeric component allow staining with osmium tetroxide (dark phase). Note the characteristic two-phase morphology with fine structure within the cell walls. (Huelck et al., 1972.)

the contents of the cells, as shown in Figures 8.1 and 8.2, for *cis*-PB/PS and PEAB/PS IPN's, respectively. Cellular structures have already been encountered with graft copolymers (see Figure 3.2). The major difference between IPN's and graft copolymers is that the cellular structure in IPN's pervades the entire macroscopic sample rather than just discrete regions.

The actual size of the cellular structures depends on the crosslink density of the two components, varying from about 1000 to 500 Å. In Figures 8.3 and 8.4, for example, the crosslink density of the random PB

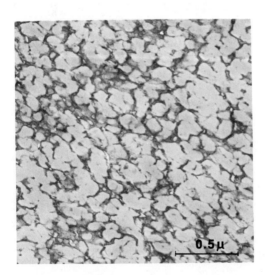

Figure 8.3. Electron micrograph of an 85/15 R-PB/PS IPN showing a two-phase, cellular structure. The symbol R denotes random (mixed *cis* and *trans*). This particular composition had an impact strength of over 5 ft-lb/in. of notch. (Curtius et al., 1972.)

Figure 8.4. Electron micrograph of a PB/PS IPN containing 72% PS (Curtius *et al.*, 1972). This structure is much finer, and more suggestive of molecular interpenetration, than the structure shown in Figure 8.3. The major difference is the level of crosslinking, being greater by a factor of three in the more finely divided material.

differs by a factor of about three, while the crosslink density of the styrene remains the same. The phase domain sizes also depend greatly on the compatibility of the two polymers. Thus, cellular structures of the PEAB/PS IPN pair are larger than those of the corresponding PEAB/P(S-co-MMA) polymer pair.

Close examination of Figure 8.2 reveals additional structure at an exceedingly fine level (the order of 100–200 Å), especially in the PEAB/P(S-co-MMA) materials, in which the staining is necessarily lighter.* The fine structure has the appearance of dots or specks, and in Figure 8.2, appears primarily in the cell walls. Fine structure on such a scale is indicative of interpenetration. When phase separation occurs initially, much monomer remains dissolved in both phases. On continued polymerization, a second, finer phase separation of polymer II occurs within polymer I. Since the chains are crosslinked to each other, the opportunity exists to form a continuous network throughout the cell walls, which connects the cell contents. The proposed structure of the fine, dispersed phase within the cell walls is shown schematically in Figure 8.5.

A somewhat different explanation is required for the small quantities of cell wall material sometimes seen to be dispersed within the cell structure (see Figure 8.2). Here we may imagine that some chains of polymer I remain entrapped within polymerizing monomer II. As polymerization proceeds,

---

* The appearance of such a fine structure in the walls of solution-graft copolymers was predicted and recently observed in our laboratories. Perhaps the heavy staining characteristic of the diene rubbers masked this feature for previous investigators.

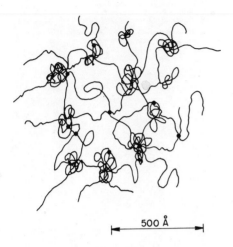

Figure 8.5. Schematic showing fine structure within domains for an IPN. The solid black circles represent crosslinks. The distance between phase domains is less than half the chain contour length between crosslinks in the network, thus permitting phase separation to occur. At the same time, some network segments provide the requisite interconnection between the phase domains.

500 Å

these chains tend to undergo phase separation, resulting in a second distinct fine structure.

8.2.2. Effect of Polymerization Sequence

When the IPN's are synthesized in inverse order, i.e., the plastic component first, the gross features are roughly inverted with respect to the "normal" preparations. Figure 8.6, showing a midrange PS/PEAB composition, should be compared to Figure 8.2. An important observation is that in both Figures 8.2 and 8.6 the first component forms the cell wall phase and the second component forms the cell contents phase.

Let us now introduce the concept of degree of continuity of a phase. In the beginning of the IPN synthesis, polymer network I obviously exhibits continuity of both the network structure and its phase. When monomer II is uniformly swollen in, before polymerization of II, one phase also exists. Polymer network I is continuous (the sample is usually a swollen elastomer), and the monomer II is also distributed everywhere. Upon polymerization of II, phase separation takes place. Polymer network I is still continuous, but is partially or wholly excluded from some regions of space. Assuming the previous even distribution of monomer II, we have reason to believe polymer network II will exhibit some degree of chain continuity. Sometimes polymer network II also appears to exhibit a degree of phase continuity Usually, polymer network II has less continuity than polymer network I A simple example of greater and lesser phase continuity in everyday life i chicken-wire in air.

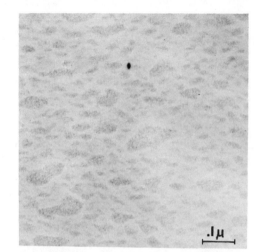

Figure 8.6. Electron micrograph of IPN 12: 50.7 PS/49.3 PEAB. This structure is roughly the inverse of the structure shown in Figure 8.2. Note the greater continuity of the plastic phase in this case. (Huelck *et al.*, 1972.)

Thus, both components (networks) are assumed to be continuous in both the normal and inverse compositions, but obviously one phase is more continuous than the other.* As shown in Figures 8.1 and 8.2 the network formed first appears to have the greater degree of continuity, even though it may be the minor component by weight. In Section 8.3 we shall see that the network formed first also tends to control such features as the stiffness of the product.

8.2.3. Compatibility and Interpenetration

The system PEA/PS, exemplified by Figure 8.2, is clearly incompatible. As MMA mers replace S mers in the plastic component, the compatibility is increased, because PEA and PMMA are isomeric. As shown in the series: Figures 8.2, 8.7, 8.8, and 8.9, the cellular structure becomes progressively smaller in size and less distinct. When MMA has completely replaced S, the cell structure has almost entirely disintegrated and has been replaced by a structure more obviously interpenetrating.

The probable reason for the cell shrinkage and disintegration can be outlined as follows. As compatibility is increased, initial phase separation takes place later during the polymerization of the second component. As a consequence, greater mixing of the two networks takes place, and more interpenetration on a molecular scale may be found in the final product.

In Orwell's *Animal Farm*, all animals are equal, but some are more equal than others. The meaning is different, but the analogy, we hope, is clear.

However, as seen in Figure 8.9, despite the facts that PEA and PMMA are chemically isomeric and the resultant IPN is optically clear, full thermodynamic compatibility and mutual solubility have not been achieved. The blurring of the remaining phase boundaries suggests that phases of distinct compositions in the classical sense probably do not exist in PEA/PMMA IPN's. Perhaps all compositions, as illustrated in Figure 8.10, exist side by side (Sperling *et al.*, 1970a,b).

The most incompatible compositions are the 50/50 series examined above. IPN's containing 25/75 PEA/PMMA, the most compatible of the

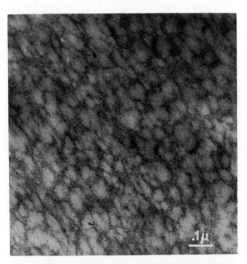

Figure 8.7. Electron micrograph of IPN 51.2 PEAB/P(25.4 MMA-co-23.4 S). On replacement of half the S by MMA, the cellular structure is much finer than that shown in Figure 8.2. (Huelck *et al.*, 1972.)

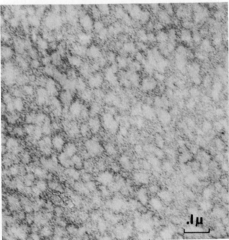

Figure 8.8. Electron micrograph of IPN 48.4 PEAB/P(38.0 MMA-co-13.6 S). Phase domains are finer and less distinct than in Figure 8. (Huelck *et al.*, 1972.)

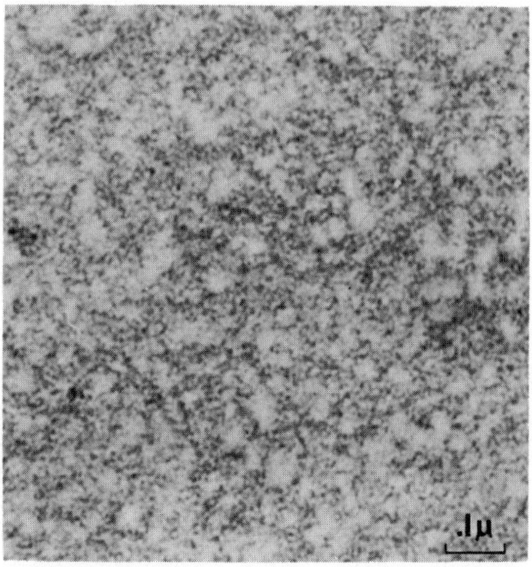

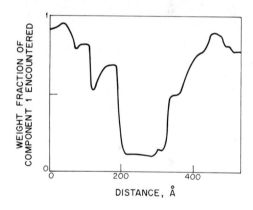

Figure 8.9. Electron micrograph of IPN L6: 52.9 PEAB/47.1 PMMA. The cellular structure has been replaced by a finer, more interpenetrating structure. (Huelck *et al.*, 1972.)

Figure 8.10. Schematic representation of a concentration profile as found by a Maxwell demon traveling through a midrange composition, semicompatible IPN.

structures studied by Sperling and co-workers (Huelck *et al.*, 1972), have structures with sizes entirely in the 100-Å range, as shown in Figure 8.11. Since simple calculation shows that a phase domain of this size can be stained by only about a dozen osmium atoms, the details must be, and are, quite indistinct. Although mixing takes place at the finest level and network interpenetration is extensive, true solution apparently never occurs.

One point further remains to be considered. Even if the two networks were completely compatible, so that only one phase was formed in the classical sense, an important topological difference between the networks still remains, which would permit the polymerization sequence to be established. We refer here to the fact that the first network formed is swollen, so that its chains have an extended and less probable conformation. Network

Figure 8.11. Electron micrograph of IPN P4: 23.3 PEAB/76.7 PMMA. Phase domains are of the order of 100 Å, with both phases continuous. (Huelck et al., 1972.)

II chains would be expected to be found in a much more nearly relaxed conformation.

8.3. PHYSICAL AND MECHANICAL BEHAVIOR OF IPN's

This section will examine some of the characteristic features of IPN's from a physical and mechanical point of view. Emphasis will be on relating the glass transition behavior to corresponding aspects of morphology. The principal instrumentation employed in the studies discussed here includes a torsional tester for creep-type studies (Section 8.3.1) and a fixed-frequency vibrating unit for dynamic mechanical spectroscopy (see Section 8.3.2). In addition, stress–strain, tensile, and Charpy impact strength values will be briefly discussed.

8.3.1. Static Relaxation in IPN's

While simple homopolymers and random copolymers exhibit single, sharp glass transitions, polymer blends in general, and IPN's in particular, show two such transitions, one for each phase. The intensity of each transition is clearly related to the overall composition and phase continuity, while shifts and broadening of the transition indicate the extent of molecular mixing. In contrast, electron microscopy (previous section) shows phase size

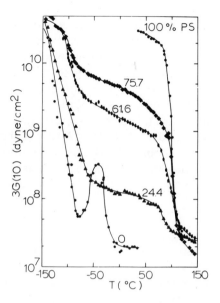

Figure 8.12. Modulus–temperature behavior of *cis*-PB/PS IPN's (Curtius *et al.*, 1972). Two transitions are observed for all IPN compositions. The IPN's with mid-range compositions behave in a leathery manner at room temperature. The sharp rise at −80°C for the pure *cis*-PB is due to crystallization. The fact that none of the IPN's shows PB crystallization indicates the existence of molecular mixing. (Modulus taken at 10 sec, using a Gehman torsional tester.)

and shape, but gives only slight indication of the extent of true molecular mixing.

Figure 8.12 shows the behavior typical of an incompatible polymer pair, *cis*-PB/PS (Curtius *et al.*, 1972). The shift and broadening of the transitions are at a minimum. The morphology of compositions shown in Figure 8.12 was already examined (see Figure 8.1) and found to indicate relatively sharp phase domain separation. The mechanical and morphological observations tend to confirm each other.

Semicompatible IPN's, on the other hand, exhibit one broad transition (see Figure 8.13). This broad transition may be considered to result from an extensive overlap of the two primary transitions, or, more probably, of transitions of all possible compositions that can contribute independent

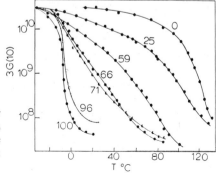

Figure 8.13. Modulus–temperature behavior of PEA/PMMA IPN's. Numerical values indicate wt % PEA. All compositions were found to exhibit only one broad transition region. (Sperling *et al.*, 1970a.)

transitions (see Figures 8.9 and 8.10 and Section 8.2). No discrete, sharp transitions were obtained for any IPN composition, thus confirming the conclusion (based on morphology) that no IPN pair studied was truly compatible and mutually soluble.

The apparent breadth of the glass transition can be examined further with the aid of time-dependent creep experiments. A master curve, composed from data for several creep experiments at different temperatures performed with a 50/50 PEA/PMMA IPN, is shown in Figure 8.14.

The time dependence of the transition in Figure 8.14 was subjected to mathematical analysis, based on the use of a modified double Rouse function (Sperling *et al.*, 1970*b*; Tobolsky and Aklonis, 1964):

$$E_r(t) = \frac{E_1 \tau_{\min}^{1/2}}{\tau_{\min}^{1/2} + t^{1/2}} + R_2 \tag{8.2}$$

where $E_r(t)$ represents the time-dependent relaxation modulus at time t for a sample having a minimum relaxation time of τ minutes. The factor R_2 represents the modulus in the rubbery plateau region; it has a value of approximately 4×10^7 dyn/cm^2 for this sample. Equation (8.2), which predicts a maximum slope of $-\frac{1}{2}$ in a log E vs. log t plot, is shown as curve 2 in Figure 8.14. Equation (8.2) represents the behavior expected of a homopolymer or random copolymer, and predicts a relatively narrow transition having a breadth of some ten decades of time.

If, instead of a single-phase composition, a series of phase compositions is envisioned, as discussed above, then equation (8.2) may be modified further to read:

$$E_r(t) = E_1 \cdot \sum_{i=1}^{n} \frac{W_i \tau_{\min,i}^{1/2}}{\tau_{\min,i}^{1/2} + t^{1/2}} + R_2 \tag{8.3}$$

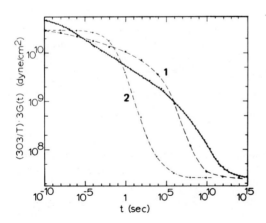

Figure 8.14. Master relaxation curve for a 50/50 PEA/PMMA IPN composition at 30°C. The transition covers 23 or 24 decades of time. Curve 1 represents equation (8.3) as discussed in the text. Curve 2 represents equation (8.2). The original points were taken as shear creep data, and converted to relaxation data for plotting purposes. (Sperling *et al.*, 1970*b*.)

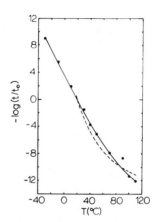

Figure 8.15. Comparison of experimental shift factors vs. those predicted by the classical WLF formulation. Although agreement is good over the portion covered by theory, this result may be fortuitous. $T_{ref} = 20°C$; 49.4% PEA. (Sperling et al., 1970b.)

where W_i is the weight fraction of composition i, normalized to satisfy

$$\sum_{i=1}^{n} W_i = 1 \qquad (8.4)$$

Equation (8.3) is plotted as curve 1 in Figure 8.14, with the assumption that all possible phase compositions are equally likely. While equation (8.3) fits much better than equation (8.2), neither fits the data well.

The empirical shift factors obtained in the formulation of Figure 8.14 are compared to the theoretical WLF (Section 1.5.7) values in Figure 8.15, 20°C being chosen as approximating the classical glass transition temperature. Surprisingly, the WLF equation fits the center and lower portions of the data quite well. The reason probably lies in the value of the derivative $-d[\log(t/t_0)]/dT$, which is the slope of the transition. The slopes are nearly equal because the IPN transition covers both a broader time scale and a broader temperature range. The increased breadth of the IPN transition becomes apparent on observing that the upper portion of Figure 8.15 lacks an analog in the WLF formulation. Note also the lack of fit of the WLF equation to relaxation data for polyblends and block copolymers discussed elsewhere, particularly with respect to PEO/PS block polymers (Chapter 6).

8.3.2. Dynamic Mechanical Spectroscopy (DMS)

DMS techniques yield the loss modulus E'' in addition to the storage modulus E'. (See Section 1.5.5.) As is the case with the static mechanical tests discussed above, DMS and electron microscopy techniques provide complementary information on two-phase materials: The former is a sensitive indicator of the extent of molecular mixing, while the latter shows the size and shapes of the phase domains. In the following sections, loss moduli were calculated from E' and tan δ as follows:

$$E'' = E' \tan \delta \qquad (8.5)$$

In all cases a frequency of 110 Hz was used; such a frequency raises the glass transition of the individual components to about 18–20°C above the 10-sec modulus reported in the previous subsection.

The advantage of the dynamic method (DMS) over the static tests (10-sec modulus) is that the loss modulus E'' is obtained. Some materials show important transitions with DMS that are missed with the static test; for example, when a minor, discontinuous phase has a transition, the Young's modulus (and also E') may not change appreciably, while E'' may exhibit a significant maximum.

Figures 8.16 and 8.17 show E' and E'' as a function of temperature for PEA/S and PEA/PMMA IPN's (Huelck et al., 1972). Figure 8.16 shows results typical of an incompatible system, with two distinct transitions and loss peaks. When PMMA is substituted for PS, a single, very broad transition appears. Since PS and PMMA are very nearly an iso-T_g pair and hence the copolymer T_g's are essentially invariant with respect to composition, the shifting and broadening of the higher transition may be attributed to mixing with the PEA component. Likewise, changes in the lower temperature

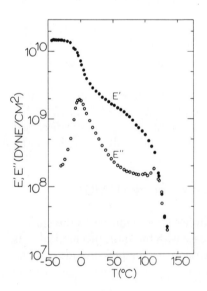

Figure 8.16. Temperature dependence of E' and E'' for IPN L1. Note two sharp transitions and two correspondingly sharp loss peaks. 48.8 PEAB/51.2 PS; 110 cps. (Huelck et al. 1972.)

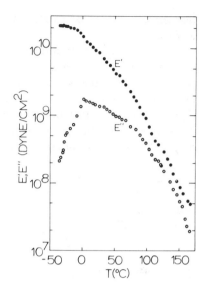

Figure 8.17. Temperature dependence of E' and E'' for IPN L4. The two transitions have broadened and merged sufficiently that E'' appears as two shoulders with a single broad peak. 47.1 PEAB/52.9 PMMA; 110 cps. (Huelck et al., 1972.)

transition may be caused by entrapped or dissolved* P(S-co-MMA) chain segments.

The extent and type of molecular mixing depends on the method of preparation. Even for nominally similar overall compositions, quite different mechanical behavior may be obtained by altering synthetic procedures, particularly the order of preparation of the two networks. Figure 8.18 shows sets of data for: (1) the 50/50 PEA/PS IPN concerned in Figure 8.10, (2) The IPN having the *inverse* composition 50/50 PS/PEA, and (3) a mechanical blend of PEA and PS (50/50) prepared by simultaneously coprecipitating their dilute solutions. The "inverse" IPN is seen to be much stiffer than the "normal," thus confirming the observations (based on electron microscopy) that PS has greater continuity in the inverse material (Figures 8.2 and 8.6). As discussed elsewhere, the first component of an IPN tends to form the more continuous phase. Surprisingly, the solution blend has the highest modulus between the individual component T_g's, and the lowest E'' peak for PS. The high modulus in the intermediate temperature range suggests that in this case PS forms the most continuous phase of the three materials; in contrast, the lower modulus of the IPN phases suggests less continuity, which can be explained by assuming interpenetration with elastomer. One anomaly remains: The PS transition in the PS blend is shifted to a lower

---

* The mechanical measurements do not distinguish among molecules visiting because of true thermodynamic solubility or because of forced mechanical chain entanglements caused by the presence of crosslinks.

temperature compared to PS itself; perhaps some unusual type of mixing is involved.

It is of interest that both the PEA and PS transitions are shifted inward (PEA to a higher temperature and PS to a lower one), relative to the homo-polymers under the same experimental conditions. This result can be explained in terms of a limited extent of molecular mixing (mechanical entrapment, thermodynamic equilibrium, or both).

Surprisingly, the peaks in the more compatible PEA/PMMA IPN's (Figure 8.17) are not shifted inward further; rather, the value of E'' in between

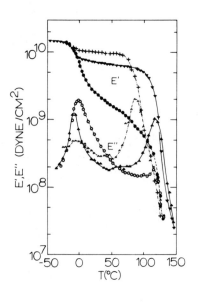

Figure 8.18. Temperature dependence of E' and E'' for two IPN's (I2 and L1) and for a blend having a similar mer composition. Three quite different mechanical spectra result from the same nominal composition. ($\triangle$, $\blacktriangledown$) 50.7 PS/49.3 PEAB; (O, ●) 48.8 PEAB/51.2 PS; ($\wedge$, $+$) blend 50 PEA/50 PS; 110 cps. (Huelck et al., 1972.)

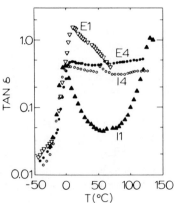

Figure 8.19. Tan δ vs. temperature for several IPN's. The loss tangent of semicompatible compositions remains high over a broad temperature range. (Huelck et al., 1972.)

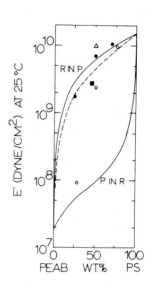

Figure 8.20. Study of phase continuity with the aid of mechanical models. Storage moduli at 25°C for PEAB/PS IPN's. (○) Normal IPN; (●) inverse IPN; (△) blend; (■) graft. (Huelck *et al.*, 1972.)

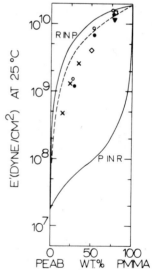

Figure 8.21. Study of phase continuity with the aid of mechanical models. Storage moduli at 25°C for PEAB/PMMA IPN's. (○) Normal IPN; (●) inverse IPN; (×) blend; (◇) mixed latex; (□) CHCl₃; (▼) graft. (Huelck *et al.*, 1972.)

the peaks is increased enormously. The increased damping in the region between the two transitions is most clearly shown with the aid of tan δ vs. temperature graphs, as shown in Figure 8.19. Semicompatible compositions E4 and I4 exhibit tan δ values that remain nearly invariant over a 100°C range. The increased damping will be discussed further (see Sections 8.8.2 and 13.5).

Studies of storage moduli and composition studies can yield further insight into the question of the relative continuity of the two phases. By applying the theories of Bauer *et al.* (1970) and Fujino *et al.* (1964) [which are essentially refinements of the earlier theory of Takayanagi *et al.* (1963)], we may judge whether the plastic (or rubber) phase is continuous or dispersed, or whether both phases are continuous (see also Section 2.6.4). Figures 8.20 and 8.21 present storage moduli as functions of composition for PEAB/PS and PEAB/PMMA, respectively. Also shown are data for a solution blend, a graft copolymer (Hughes and Brown, 1963), and a mixed latex (Hughes and Brown, 1961) system. The upper solid lines represent Bauer's relationship for a rubber in a plastic matrix, while the lower solid lines represent a plastic in a rubber matrix. The dashed lines represent the results for two continuous phases. Allowing for the relative oversimplification of the Bauer–Takayanagi–Fujino model, the results tend to indicate the existence of two continuous phases for most of the IPN data.

8.4. ULTIMATE MECHANICAL BEHAVIOR OF IPN's

The ultimate utility of any material lies in its performance. Let us examine now the types of reinforcement obtained with IPN's. Stress–strain curves for random (R) PB/PS IPN's are shown in Figure 8.22. As is well known, random (*cis–trans* mixture) polybutadiene homopolymer is very weak, breaking at a rather low elongation. Both ultimate elongation and stress to break are increased by addition of the polystyrene network; the work required to break, as measured by the area under the curves, is vastly increased. Further, the shapes of the curves are affected by the presence

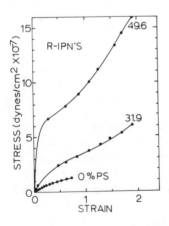

Figure 8.22. Stress–strain behavior of random PB/PS IPN's. Both ultimate elongation and stress to break are significantly increased. (Curtius *et al.*, 1972.)

Table 8.1
Tensile Data for R-IPN's[a]

Styrene content, %	Tensile, psi	Elongation, %
0.0	190	70
20.5	490	140
25.6	540	140
31.9	870	190
42.9	1500	180
49.6	2400	190

[a] Curtius et al. (1972).

of PS; both curves for the IPN's exhibit an upturn in the stress–strain relationship which is characteristic of a reinforced elastomer. All the tensile data for these random, R-IPN's are summarized in Table 8.1.

Both the modulus–temperature relationships presented in the preceding sections and the tensile data presented above are strikingly similar to those demonstrated for other rubber–plastic combinations, such as the thermoplastic elastomers (see Chapter 4 and the model system presented in Section 10.13) and the impact-resistant plastics (Chapter 3). The IPN's constitute another example of the simple requirement of needing only a hard or plastic phase sufficiently finely dispersed in an elastomer to yield significant reinforcement. Direct covalent chemical bonds between the phases are few in number in both the model system (Section 10.13) and present IPN materials. Also, as indicated in Chapter 10, finely divided carbon black and silicas greatly toughen elastomers, sometimes without the development of many covalent bonds between the polymer and the filler.

Some of the plastic IPN's exhibit very high Charpy impact resistance; impact strengths for these IPN's, as well as for ordinary polystyrene and a graft-type high-impact polystyrene (HiPS), are compared in Table 8.2.

Table 8.2
Impact Resistance Data[a]

Material	Sample thickness, in.	Styrene content, %	Impact resistance, ft-lb/in.
PS	0.166	100	0.34
HiPS	0.252	—	1.29
R-IPN	0.147	68.8	4.90
R-IPN	0.177	85.6	5.06

[a] Curtius et al. (1972).

The HiPS typically is about four times as tough as the unmodified plastic. The IPN's concerned exhibit values of impact strength about four times as high as the HiPS. The 15/85 composition of the random PB/PS is similar to the overall composition of the HiPS, but contains a few percent more rubber, which may contribute to the observed toughness.

In contrast to the PB/PS compositions, the PEA/PS and PEA/PMMA IPN'S exhibit only marginal increases in impact resistance values in comparison to the homopolymer PS and PMMA; values of impact strength are typically below 1.0 ft-lb/in. of notch. This relatively poor performance may have two explanations: (1) the domain sizes may be too small, and (2) the T_g of the PEA ($-22°C$) may be too high. According to the theory developed in Section 3.2, a T_g of about $-40°C$ of the rubber phase is required for impact resistance, suggesting that explanation 2 may be correct.

8.5. INTERPENETRATING ELASTOMERIC NETWORKS (IEN's)

This interesting class of polymers is prepared by first mixing emulsions of elastomeric polymer I with emulsions of elastomeric polymer II, followed by film formation and subsequent crosslinking.* Frisch et al. (Frisch and Klempner, 1970; Frisch et al., 1972; Klempner et al., 1970; Matsuo et al., 1970b) employed latexes of a urethane-urea (U) and a polyacrylate (A) over the composition range of A/U = 90/10 to 10/90 by weight. The latexes of the A component were crosslinked together by reacting the double bonds present in the polymer with sulfur, the U polymer being self-crosslinking upon the application of heat because of the presence of free hydroxyl groups.

8.5.1. Morphology of IEN's

The morphology of the IEN's is shown in Figure 8.23. The dark portion represents the U phase, which has been selectively stained by osmium tetroxide. At a composition of A/U = 70/30 and 50/50, U particles of 1–5 μm are dispersed in the A matrix. At the composition of A/U = 30/70, however, the U particles touch, two continuous phases exist, and the system

---

* An IEN is a special case of an IPN; the term IEN will be used to designate the formation of an IPN from two distinguishable latexes subsequently mixed, coagulated, and crosslinked. Except for the crosslinking reactions, the IEN's are topologically very similar to the mixed latexes. See Section 3.1.5.

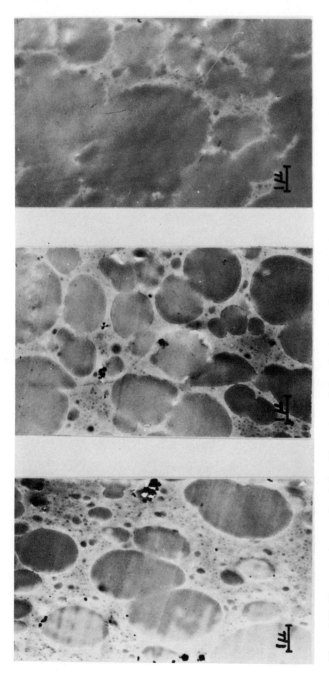

Figure 8.23. Electron micrographs of ultrathin sections of three interpenetrating elastomeric networks. A/U = 70/30 (left), 50/50 (middle), and 30/70 (right). (Matsuo et al., 1970b.)

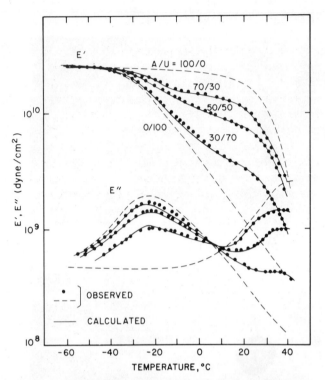

Figure 8.24. Temperature dependences of the dynamic storage modulus E' and the loss modulus E'' of the interpenetrating elastomeric networks. Filled circles: observed values. Solid lines: calculated for Takayanagi's model 2. Dashed lines: component homopolymers. (Klempner *et al.*, 1970.)

may be visualized as interpenetrating. As a result of the crosslinking reactions, each phase consists of only a single molecule of infinite molecular weight. Because the polymers were synthesized separately, intermolecular grafting may be entirely absent.

The domain sizes shown in Figure 8.23 are from one to two orders of magnitude larger than polymer molecules, typical dimensions being 1–4 μm, and considerably larger than reported by Sperling and co-workers for their IPN's (Section 8.2). Although some molecular interpenetration must exist at the phase boundaries, the domains in Figure 8.23 remain sharply defined regardless of composition. Thus the morphological evidence suggests that only very limited true molecular interpenetration probably takes place with IEN's, and that limited to the phase boundaries. The main interpenetration is of the phases, rather than the molecules.

8.5.2. Mechanical Behavior of IEN's

The above picture is further supported by studies of dynamic mechanical spectroscopy. Thus Matsuo *et al.* (1970*b*) measured E' and E'' as a function of temperature for a number of compositions, and compared the results to Takayanagi's mechanical model rubber–plastic phase continuity (Takayanagi *et al.*, 1963) (see Section 2.6.4); results are shown in Figure 8.24.

Values of storage modulus at 23°C are plotted as a function of composition in Figure 8.25. Takayanagi's parallel model for the mechanical behavior of a two-component system (Takayanagi *et al.*, 1963) corresponds to the case in which the stiffer component is continuous, while his series model corresponds to the case in which the softer component is continuous. Clearly the experimental results agree best with the parallel model, and hence confirm the observations (based on electron microscopy) that a phase inversion takes place at about 30% acrylic component. It should be pointed out that a 70% concentration (of the U phase) corresponds approximately to the volume fraction for the closest packing of spheres. Phase inversion of IEN's should be compared to the corresponding behavior of IPN's (Section 8.2.2). Also, Figures 8.24 and 8.25 for the IEN's should be compared with Figures 8.16, 8.17, 8.19, and 8.21 for the IPN's. Both sets show two transitions

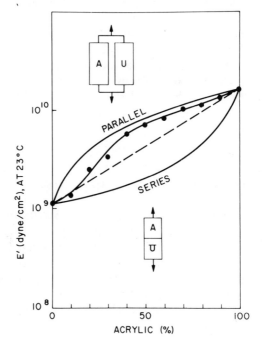

Figure 8.25. Plots of the dynamic modulus E' at 23°C against the polyacrylate content. The upper and the lower solid lines are calculated with a parallel and a series model, respectively, using following equations: $E^*(\text{parallel}) = (1 - \lambda)E_A^* + \lambda E_U^*$; $E(\text{series}) = [(1 - \psi)/E_A^*] + (\psi^{-1}/E_U^*)$. (Klempner *et al.*, 1970.)

indicating two distinct phases, except for the semicompatible PEA/PMMA IPN, which shows only one broadened transition. The dual phase continuity indicated is striking, perhaps greater than justified by the corresponding electron micrographs. Because of their latex origin, the domains of the IEN's are much larger than those of the IPN's. Interpenetration of the former should be expressed in terms of phases and the latter in terms of both interpenetrating molecules and interpenetrating molecular aggregates.

8.6. SEMI-IPN STRUCTURE AND PROPERTIES

In the above sections, the behavior of the IPN's and IEN's was discussed. In this section, we return to the IPN's and discuss their relationship to materials having one linear and one crosslinked polymer, the semi-IPN's. Succeeding sections will discuss other types of IPN's and related materials.

In what sense shall the IPN's be considered a new class of polymer mixture parallel to blends, blocks, and grafts? The synthetic method used clearly places the IPN's as a subclass of the graft copolymers (Donatelli *et al.*, 1974*a*). Indeed, although grafting in the normally defined sense does indeed take place, the crosslinking of both polymer I and polymer II introduces a new structure-influencing element. In most IPN's of the type discussed here, the deliberate crosslinks outnumber the accidental grafts. Thus new morphologies and properties arise because of the controlling effects of the more numerous crosslinks. As we will show below, the ordinary usage of the term graft copolymer becomes more suitable as the number of crosslinks is reduced. As commonly employed, the term "graft copolymer" means the polymerization of monomer II in the presence of polymer I, regardless of the true extent of chemical graft formation (see Sections 2.1 and 3.1.1). The relationships among the blends, grafts, and IPN's are considered in more detail elsewhere (Sperling, 1974*c*).

8.6.1. Definition of Semi-IPN's

At this point, it is useful to recognize the concept of a semi-IPN. A semi-IPN may be defined as a graft copolymer in which one of the polymers is crosslinked and the other essentially is linear. Two semi-IPN's (one the inverse of the other) may be distinguished: (1) a semi-IPN of the first kind, with polymer I in network form and polymer II linear; and (2) a semi-IPN of the second kind, with polymer I linear and polymer II crosslinked. As further discussed by Sperling (1974*c*), many related systems contain one or both polymers in network form.

According to the above definition, a graft copolymer may be considered a limiting form with neither polymer crosslinked.

8.6.2. Morphology

In a recent paper, Donatelli *et al.* (1974*a*) explored the morphological features of IPN's, semi-IPN's, and graft copolymers prepared from poly-(butadiene-co-styrene)/polystyrene combinations (Figure 8.26). As before, the diene phase is stained dark with osmium tetroxide. In Figure 8.26a a micrograph of a commercial graft copolymer (high-impact polystyrene) is depicted. Figure 8.26b illustrates SBR–polystyrene (also a high-impact polystyrene) made without stirring and subsequent phase inversion. Hence, the SBR phase is continuous. The semi-IPN of the first kind is shown in Figure 8.26c, which depicts a far finer state of phase separation than the semi-IPN of the second kind, Figure 8.26d. Finally, Figure 8.26e shows a (full) IPN, with both components crosslinked.

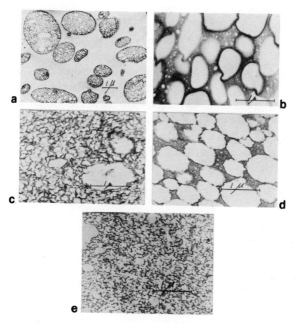

Figure 8.26. Electron micrographs of IPN's and related materials.
(Donatelli *et al.*, 1974*a*.)

The semi-IPN of the second kind resembles the graft copolymer much more than it does the semi-IPN of the first kind. In both the graft copolymer and the semi-IPN of the second kind, phase domain size is large compared to molecular dimensions.

On the other hand, the phase domains of the semi-IPN of the first kind average 1000 Å in size, only slightly larger than the 800-Å domains of the (full) IPN; in each case, the domains are of the order of molecular dimensions. Crosslinking of polymer II obviously reduces the domain size somewhat, but only if polymer I is already crosslinked.

8.6.3. Mechanical Behavior

As shown above (Section 8.4), the IPN's prepared from polybutadiene/polystyrene combinations are much tougher than the corresponding graft copolymers (Curtius *et al.*, 1972). More recently, the mechanical behavior of the IPN's and semi-IPN's was compared (Donatelli *et al.*, 1974*b*) (see Figure 8.27). While all materials exhibited a yield point, the (full) IPN exhibited slightly better mechanical properties than the semi-IPN of the first kind, and much better properties than the semi-IPN of the second kind. The IPN's exhibited Charpy impact strengths of 5–6 ft-lb/in. of notch.

Donatelli *et al.* (1974*b*) and Sperling (1974*c*) studied the effect of crosslink density in the IPN's and semi-IPN's on morphology and mechanical behavior. In general, they concluded that increasing crosslink density of polymer I produces finer phase domains. It is possible to obtain domains that are too small to impart maximum toughness. Crosslinking in polymer I is more important than crosslinking in polymer II for improved mechanical behavior. The (full) IPN, whose micrograph best suggests the existence of two continuous phases (Figure 8.26), yields the best properties.

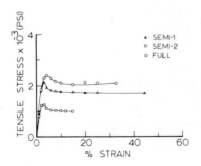

Figure 8.27. Stress–strain studies on an IPN and its two semi-IPN's. SBR (5 % S)/PS; 20/80. (Donatelli *et al.*, 1974*b*.)

8.7. AB CROSSLINKED COPOLYMERS

In this special kind of graft copolymer, *both* ends of polymer II are grafted to different polymer I molecules, causing one crosslinked (grafted) network to arise. Thus, ideally, one molecule is generated for AB crosslinked copolymers, compared to two molecules ideally generated for the IPN's and IEN's. While polymer I is grafted primarily to polymer II, it is not crosslinked to itself. As previously discussed (Section 2.3), the castable polyester resins, synthesized by polymerizing styrene monomer in an unsaturated polyester solution, which long ago achieved commercial importance (Bartkus and Kroekel, 1970; Fekete and McNally, 1969; Kallaur, 1969; Sorenson and Cambell, 1961, pp. 286–289), are AB crosslinked copolymers. These polyester materials are generally characterized by rather short chains (1–10 mers) between grafts (Learmonth and Pritchard, 1969). Another AB crosslinked copolymer of considerable importance is formed when epoxy resin is cured by reaction with a polyamide.

A new synthetic technique for making AB crosslinked copolymers systematically was reported by Bamford *et al.* (1971, 1974). His synthesis involves crosslinking of a suitable polymer (I) with chains of vinyl polymer (II) by reaction of halogen-containing groups in I with an organometallic derivative, e.g., $Mo(CO)_6$ or $Mn_2(CO)_{10}$, in the presence of monomer II. The reaction scheme may be written as

$$(8.6)$$

in which the two free radical chains terminate by combination. Those chains terminating by disproportionation form simple grafts. The crosslinked networks so produced exhibited two glass transitions, indicating phase separation. Dilatometric results for poly(vinyl trichloracetate) crosslinked by methyl methacrylate are shown in Figure 8.28.

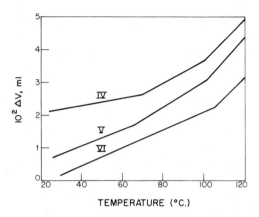

Figure 8.28. Dilatometric data for poly(vinyl trichloroacetate)/poly-(methyl methacrylate) AB crosslinked copolymers. Networks IV and V show two second-order transitions, while network VI shows only one transition. (Bamford et al., 1971.)

Research on a random type of AB crosslinked copolymer, or joined IPN,* is also of interest. Poly(dimethyl siloxane) containing various quantities of vinyl groups substituted for methyl groups serves as polymer I, and methyl methacrylate is polymerized to form polymer II. When about 40% MMA is present, a tough, opaque white elastomer is formed (Sperling and Sarge, 1972). These materials are the analog of the castable polyesters (Katz and Tobolsky, 1964), except that polymer I is a high-molecular-weight elastomer-forming material, and polymer II forms relatively longer chains. The structure of these materials also very closely resembles the structure of alkyd resins, which are important coating materials (Martens, 1968). Attention should also be called to a joined IPN formed by solution-blending a mixture of poly-(acrylic acid) and poly(vinyl alcohol) in water, followed by film formation (Eisenburg and King, 1972; Kuhn et al., 1960) (see Section 9.9). The joining is due to later graft formation between the two polymers. Note also a series of joined IPN's, semi-IPN's (of the first kind; Sections 2.3 and 8.6), and combinations of joined and semi-IPN's reported by Vollmert (1962). The latter materials, prepared from styrenes and acrylates, form a series of tough, slow aging, impact-resistant plastics.

8.8. TOPOLOGY OF IPN's

Let us consider polymer structure for a moment and ask: In how many distinct ways can two kinds of network chains be put together so that they interpenetrate? The original sequential IPN's and the IEN's have

---

* Although the term joined IPN was introduced by Sperling and Sarge (1972), the term AB crosslinked copolymer is now preferred.

already been described, along with the semi-IPN's and AB crosslinked copolymers. The objective of this section is to bring together a series of materials and molecular architectures, and to show their correlation.

8.8.1. Simultaneous Interpenetrating Networks (SIN's)

A very simple jump involves going from a sequential synthesis to a simultaneous synthesis of both networks. From a practical viewpoint, the most vital chemical requirement involves finding independent, noninterfering polymerization reactions that can be run simultaneously under the same overall conditions. A general solution to this requirement entails the use of condensation and addition reactions. Indeed, the art of simultaneously synthesizing two polymers has already received some attention (Rumsheidt and Bruin, 1960) and excellent mechanical properties can be obtained (Tiffan and Shank, 1967). An example of a simple SIN employed an epoxy resin (Kenyon and Nielsen, 1969) (step growth mechanism) with an ethyl acrylate formulation (free radical chain addition). When a tertiary amine was used to cure the epoxy, minimal interference occurred between the two reactions (Sperling and Arnts, 1971).

If the addition reaction was photochemically initiated and the condensation reaction was thermally controlled, the two reaction rates could be adjusted at will over a wide range. Starting with a mixture of the two monomers, application of either heat or light followed by application of the other resulted in one polymer or the other reacting first. Simultaneous application of heat and light resulted in both networks forming together. The tensile strengths of SIN's are compared in Table 8.3. The SIN's reacted simultaneously, had a controlled degree of phase separation, as indicated by intermediate turbidity levels, and were apparently superior in tensile strength to those reacted sequentially.

Table 8.3
Tensile Strengths of 50/50 Epoxy/Ethyl Acrylate SIN's[a]

Synthesis method	Tensile strength, lb/in.^{2}
Light first, then heat	940
Heat first, then light	1400
Light and heat together[b]	1480
Light and heat together[b]	2050

[a] Sperling and Arnts (1971).
[b] Two different light intensities employed.

In a more recent study, Touhsaent *et al.* (1974*a*) investigated the morphology and mechanical behavior of SIN's prepared from 80/20 epoxy/poly(*n*-butyl acrylate) combinations. When simultaneous gelation of both networks was achieved, a minimum in gel size was obtained. The gel size obtained was too small for optimum toughness, and less simultaneously reacted compositions were tougher than the more simultaneous ones (Touhsaent *et al.*, 1974*b*). Frisch *et al.* (1974*a,b*) prepared SIN's based on polyurethanes and acrylics, and found improved toughness. Allen *et al.* (1973*a,b*; 1974*a,b,c*) and Blundell *et al.* (1974) investigated the behavior of semi-SIN's, based on polyurethane (crosslinked) and poly(methyl methacrylate) (linear). A semi-SIN (like a semi-IPN) is defined as containing one linear and one crosslinked polymer.

8.8.2. Latex IPN's

A latex IPN is prepared through emulsion polymerization by taking a crosslinked seed latex of polymer I, adding monomer II, together with crosslinker and activator (but no new soap), and polymerizing monomer II on the original particles. As few as possible new particles should be generated in the second polymerization. Both networks are on each particle, forming an array of micro-IPN's. Due to the thermoset nature of sequential (bulk) IPN's, they have one important drawback, lack of processibility. IPN's prepared by means of emulsion polymerization techniques may offer considerable processing advantages because the thermoset characteristics are limited to individual submicroscopic particles. Advantages may include the use of injection molding techniques, or film formation by painting or casting.

Only a few latex IPN's appear to have been made. Among these are an impact toughening additive prepared by Ryan and Crochowski (1969) and a semicompatible sound-damping material prepared by Sperling *et al.* (1972) (see Figure 8.29). Latex IPN's prepared from methacrylics and acrylics display high values of tan δ over broad temperature ranges, the exact values depending upon the choice of ester groups and overall composition (Sperling *et al.*, 1972, 1973). The latex IPN's are to be distinguished from the ABS-type latexes (Section 3.15) because of the deliberate introduction of crosslinking agents into both polymeric components. In this context it should be mentioned that semicompatible blends (Mizumachi, 1970) and grafts (Allen *et al.*, 1974*a,b*) also form useful damping materials (Warson, 1972, p. 979).

The structure of the latex IPN's is imagined to be complex, especially in light of the Williams shell–core work discussed in Sections 3.1.2.2 and 13.4

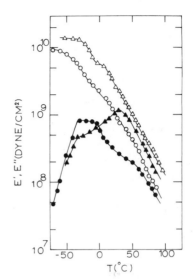

Figure 8.29. DMS of poly(*n*-butyl acrylate)/poly-
(ethyl methacrylate), PnBA/PEMA, latex 50/50
IPN and inverse composition. Dynamic mechanical spectroscopy at 110 Hz shows the two
structures to be different. (△, ▲) 50 PnBA/50
PEMA; (○, ●) 50 PEMA/50 PnBA. (Sperling
et al., 1972.)

Important variables include overall composition, order or preparation, crosslink density, and the size of the final latex particle. With respect to the latter, it should be noted that the phase domain dimensions expected are of the same order of magnitude as the particle diameters.

As with IPN's, semi-IPN's may also be prepared in latex form, latex semi-IPN's of both the first and second kind having been investigated. Employing poly(*n*-butyl acrylate)/poly(methyl methacrylate) combinations, Dickie and Cheung (1973) and Dickie *et al.* (1973) found that when the second component polymerized was PMMA, the material was stiffer than when the order of polymerization was reversed. Interestingly enough, these acrylic/methacrylic semi-IPN's, with their greater molecular mobility, appear to have better defined (sharper) glass transitions than the chemically similar pairs illustrated in Figure 8.29. Again, crosslinking of both polymers is important in the locking of the molecules in close juxtaposition. As the particle diameter increases through the phase domain size obtained with the equivalent sequential (bulk) IPN's, a change in morphological detail may ensue (Figure 8.30). The point of onset of multiple domains may result in interesting mechanical property changes. Alternatively, if the diameter of the particle is kept constant, the same morphological transition may be encountered on systematically increasing the crosslink densities of polymer I, or of both components.

Let us examine some of the effects of order of preparation, assuming that one component is elastomeric and the other plastic (see Figure 8.29). The glass temperature of the outer structures must be below about 50°C to

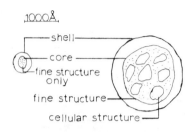

Figure 8.30. Model of predicted latex IPN morphology, showing cellular structures, fine structures, and a shell–core morphology. (Sperling *et al.*, 1972.)

allow simple film formation. If the polymer is to be coagulated and molded, the above condition is ameliorated, but the overall modulus and toughness will still depend on the fine structure. It is thought that the use of a rubbery seed latex and a plastic outer layer should give the best impact properties because the plastic portion will tend to form the matrix material. This is true of the ABS latexes (Chapter 3). [Recall that in sequential (bulk) IPN's the first network formed becomes the more continuous phase.]

Broad temperature damping characteristics can be encouraged by controlling the degree of molecular mixing. For example, the amount of core–shell structure can be reduced or replaced by more intimate molecular contact by the use of swelling agents. To accomplish this, a solvent for the polymer would be added to the emulsion after the first stage of polymerization is complete. The swelling would allow the chains of the first component to assume a more open conformation, and allow improved penetration by the second monomer component.

Other types of IPN's exist, of course. For example, Johnson and Labana (1972) recently synthesized a modified type of latex IPN as follows: A crosslinked polymer network I prepared by emulsion polymerization served as a seed latex to linear polymer II. The resulting semi-IPN exhibited the usual core–shell morphology. After suitable coagulation and molding steps, polymer II was selectively crosslinked to form a macroscopic network, resulting in a thermoset material. The topology of this IPN therefore involves microscopic network islands of polymer I embedded in a continuous network of polymer II.

8.9. DUAL PHASE CONTINUITY IN IPN's

IPN's have two continuous networks. In the case of latex IPN's, each particle ideally is composed of two network molecules. In bulk-prepared materials, the two networks are often presumed to be continuous on a macroscopic scale. However, the domains that result from the ultimate phase separation may or may not be continuous. A case where there need not be two continuous phases would be when visiting chains of polymer II are mechanically trapped in the polymer I phase.

Some of the electron micrographs obtained on IPN's (see Figures 8.3 and 8.26e for example) do suggest dual phase continuity. The effect of composition on modulus in such systems, for example, has been treated by Davies (1971a,b). The Davies equation, analytically developed for dual phase continuity, may be expressed as

$$E = v_1 E_1^{1/5} + v_2 E_2^{1/5} \tag{8.7}$$

where E, E_1, and E_2 are the composite modulus and the moduli of phases 1 and 2, respectively, and v_1 and v_2 represent the volume fractions of phases 1 and 2, respectively.

In Figure 8.31, modulus–composition data for full IPN's and semi-1 compositions based on SBR/PS are shown (Donatelli et al., 1975). Corresponding data for semi-2 compositions are shown in Figure 8.32 (Donatelli, Thomas, and Sperling, 1975). The Davies equation is contrasted in each case with the Kerner equation (see Equation 12.3). The upper bound corresponds to choosing the plastic phase as continuous, and the lower bound results from choosing the elastomer phase as continuous. The full IPN's and the semi-1 compositions in Figure 8.31 fit the Davies equation quite well, while the semi-2 compositions in Figure 8.32 seem to exhibit greater

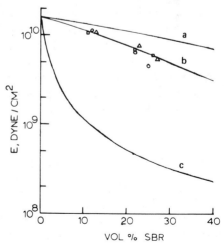

Figure 8.31. Modulus–composition data for SBR/PS full IPN's and semi-1 materials compared with theory: (○) series 3 (semi-1); (△) series 4 (full); (□) series 5 (full). (a) Kerner's equation (upper bound); (b) Davies equation; (c) Kerner's equation (lower bound). (Donatelli et al., 1975.)

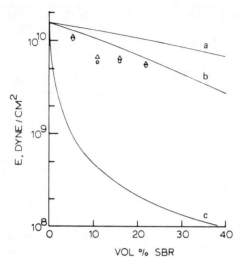

Figure 8.32. Modulus–composition data for SBR/PS semi-2 materials compared with theory: (○) Series 1 (semi-2); (△) series 2 (semi-2). (a) Kerner's equation (upper bound); (b) Davies equation; (c) Kerner's equation (lower bound). (Donatelli *et al.*, 1975.)

continuity for the elastomer phase, according to these relationships. In general, the results are supported by the morphological evidence shown previously in Figure 8.26. While the evidence presented above is clearly suggestive of two continuous phases, neither the electron microscopy nor the mechanical data constitute a proof.

9

Miscellaneous Polymer Blends

This chapter concerns a diverse group of polymeric blends whose constituents may be either compatible or incompatible with each other. While the blends discussed in Chapter 3 are of interest because of their high impact strengths, the materials covered in this chapter are of interest because of a variety of other physical properties.

First, a series of incompatible systems is discussed, including blends of different elastomers, two-component fibers and films, blends having paperlike characteristics, two-component membranes having highly ordered structures, and wood. Next, some aspects of the flow behavior of blends are considered, with emphasis on the effects of flow on morphology. Finally, the behavior of compatible blends, including isomorphic composition, is described.

9.1. RUBBER–RUBBER POLYBLENDS

Blends of two more or less incompatible rubbers are commonly used in the automobile tire industry in order to improve processability or properties. Indeed, as with the high-impact-strength plastics, the improved behavior of such blends depends in part on the limited degree of mixing obtained. As shown in Figure 9.1, either component may serve as the matrix, depending on its relative concentration in the blend.

Several examples may be given (Boonstra, 1970; Callan *et al.*, 1971).

1. The use of dispersions of saturated rubbers, such as ethylene–propylene diene terpolymer (EPDM), in polybutadiene or natural rubber has become important in the manufacture of elastomers that resist cracking due to ozone* attack (O'Mahoney, 1970). The latter two elastomers contain

---

* Ozone occurs naturally at a level of about one part per 100 million. However, as a product of atmospheric reactions in regions polluted by hydrocarbons, it may reach levels 25–30 times normal. Under these conditions, objects such as automobile tires and rubber belts may be severely attacked.

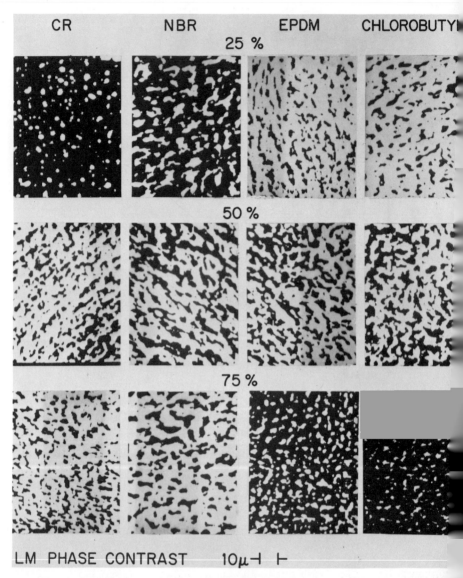

Figure 9.1. Phase-contrast micrographs of blends of chloroprene (CR), nitrile rubber (NBR), ethy‐
propylene terpolymer (EPDM), and chlorobutyl rubber with styrene–butadiene rubber (SBR). The
phase appears white for the blends with CR and NBR, and dark for the blends with EPDM and chloro‐
rubber. At low concentrations, the admixed rubber is the dispersed phase; at higher concentratic‐
phase inversion occurs and the admixed rubber becomes the matrix. (Callan *et al.*, 1971.)

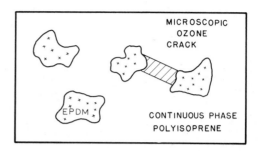

Figure 9.2. When dispersed in an unsaturated, ozone-susceptible rubber, a saturated rubber, such as one based on ethylene and propylene monomers, may stop cracks initiated by the attack of ozone on the unsaturated rubber.

carbon–carbon double bonds in their backbone chain, which are highly susceptible to scission by ozone (Lake *et al.*, 1969). On the other hand, the saturated elastomers, which contain no such double bond groups, resist attack by ozone, and can in fact serve to stop cracks developed in the less resistant phase. Thus EPDM has no effect on the formation of ozone cracks in the continuous phase, but does keep them from spreading (see Figure 9.2).

2. Polybutadiene having a high content of *cis* 1,4 groups is difficult to process and also has a low coefficient of friction. Blending with an oil-extended styrene–butadiene rubber (SBR) overcomes these disadvantages, without adversely affecting resilience or chemical stability.

3. The blending of natural rubber with SBR imparts increased resilience and tack. Increased resilience arises because of the ability of natural rubber to crystallize on deformation.

The field of rubber–rubber blends was recently reviewed by Cornish and Powell (1974), who pointed out that the continuous phase in the blend usually has the higher concentration of polymer or the lower viscosity or both (see Section 9.6).

9.2. BICOMPONENT AND BICONSTITUENT FIBERS

Blends of two fiber-forming polymers can be used to spin several types of two-component fibers (Allied Chemical Corp., n.d.; Buckley and Phillips, 1969; Cresentini, 1971; Fukuma, 1971; Hayes, 1969; Mumford and Nevin, 1967; Papero *et al.*, 1967; Pollack, 1971). As shown in Figure 9.3, the two components may be arranged as mated half-cylinders, in a skin–core configuration, or in a matrix–fibril configuration. The first two configurations are referred to as "bicomponent," the last as "biconstituent" fibers.

9.2.1. Bicomponent Fiber Systems

Strictly speaking, the bicomponent systems are not polyblends in the classic sense, since the dimensions along the fiber axis have macroscopic proportions. However, such fibers commonly have diameters under 100 μm and exhibit the synergistic effects characteristic of other classes of polyblends. It is reasonable, therefore, to include bicomponent fibers within the general class of blends.

Typical bicomponent fibers are melt-spun, as illustrated schematically in Figure 9.4. The dimensions required for the homopolymer capillaries are determined by the melt viscosities of the two polymers and by the composition ratio desired in the final product. In order to achieve satisfactory adhesion between the two components, the two polymers must be sufficiently compatible to at least wet each other. Since the polymers in these as well as other bicomponent fibers are more or less incompatible, it would be interesting to examine the interfacial microstructure in the light of Voyutskii's adhesion theory. (See Section 13.4.)

One interesting, simple bicomponent fiber is made from two types of nylon. If the polymers are selected to have different coefficients of expansion, the resultant fiber will curl or crimp on cooling. This characteristic imparts bulk and stretchiness to the yarn made from the fibers.

A common example of a skin–core fiber is wool, which has a complex inner core with a serrated outer skin. This configuration imparts elasticity and stretchability to fabrics constructed from wool.

Although bicomponent fibers are usually based upon crystalline polymers, crystallinity is not required in both components. For example, in a recent study, Han (1973) coextruded several sheath–core bicomponent systems based on polyethylene and polystyrene, respectively. Han found that the interfacial curvature in the sheath–core fibers became more circular

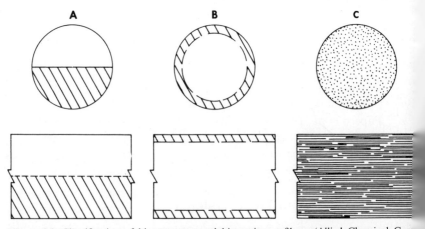

Figure 9.3. Classification of bicomponent and biconstituent fibers. (Allied Chemical Cor advertising literature) (A) Bicomponent system; (B) skin–core system; (C) matrix–fibril syste

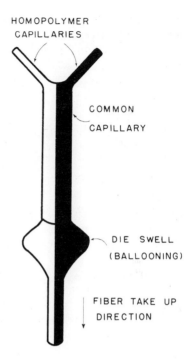

HOMOPOLYMER
CAPILLARIES

COMMON
CAPILLARY

DIE SWELL
(BALLOONING)

FIBER TAKE UP
DIRECTION

Figure 9.4. Results of melt-spinning a simple bicomponent fiber. Light and dark portions represent different polymer materials. Note the ballooning effect (the "die-swell" phenomenon) as the blend leaves the common capillary. Since the pressure drop in the common capillary must be the same for each component, careful regulation of the homopolymer capillary diameters is necessary to obtain the desired result.

with increasing capillary length, a phenomenon due to the intrinsic viscoelastic nature of polymer.

9.2.2. Biconstituent Fiber Systems

These systems, whose phase characteristics resemble those of the polyblends discussed in Chapter 3, can be prepared by first blending the molten polymers together until the minor component is dispersed in the form of droplets that are small in comparison to the fiber diameter desired (Allied Chemical Corp., n.d.; Buckley and Phillips, 1969; Hayes, 1969; Mumford and Nevin, 1967; Papero et al.,1967). The material is then melt-spun and drawn in order to orient both constituents and cause the dispersed phase to form elongated cylinders or fibrils. For satisfactory dispersion, the viscosities of both components must be comparable (for a discussion of rheological effects in molten polymer blends, see Section 9.6). An important biconstituent system is based on a combination of nylon 6 with a linear polyester poly(ethylene terephthalate), with nylon 6 as the continuous phase (Buckley and Phillips, 1969). As shown in Figure 9.5, fibrils of polyester

Figure 9.5. Micrograph of a drawn biconstituent fiber containing 30 % poly(ethylene terephthalate) and 70 % nylon 6. The fibers form elongated, tapered cylinders about 100–200 μm in length. (Buckley and Phillips, 1969.)

dispersed in nylon are in fact observed in conformity with the schematic diagram in Figure 9.3. Dissolution of the nylon component with formic acid yields an interesting proof of structure. Figure 9.6 shows that discrete polyester fibrils remain after dissolution of the nylon phase, thus illustrating that the system was in fact a mechanical blend and not a true solution of one polymer in the other. This conclusion can be confirmed independently by differential thermal analysis (Figure 9.7), which indicates the presence of two separate constituent phases.

The major advantage gained by combining polyester and nylon homopolymers in a fiber is a higher initial modulus, especially just above room temperature, due to the fact that polyethylene terephthalate has a higher glass transition temperature than nylon 6. This feature is important in overcoming the tendency of nylon tire cords to exhibit "flat-spotting" due to creep on standing.

A second important type of biconstituent fiber is a semicompatible blend on nylon 66 with nylon 6I (I referring to isophthalic acid), with nylon 6

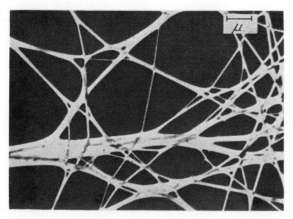

Figure 9.6. Micrograph showing polyester fibrils isolated from a polyester–nylon (30/70) biconstituent fiber after treatment with formic acid. (Buckley and Phillips, 1969.)

as the minor constituent (Zimmerman, 1968; Zimmerman *et al.*, 1973). The presence of the nylon 6I component reduces the "flat-spotting" tendency in tires by imparting a significantly higher modulus and T_g to the blend. A degree of phase continuity of the minor component may be imagined to be present, thus reducing creep.

The crystalline bicomponent and biconstituent blends considered above may also be compared to the polyallomer block copolymers (Section 6.8), which consist of two crystalline components linked together by chemical bonds rather than by mechanical mixing. In all cases, the constituents are independently crystallizable. Melt-blending theory is considered in Section 9.6.

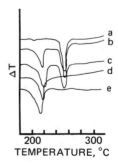

Figure 9.7. Differential thermal analysis of nylon 6, polyethylene terephthalate (PET), and some biconstituent compositions: (a) PET; (b) 45% PET/55% nylon; (c) 30% PET/70% nylon; (d) 10% PET/90% nylon; (e) nylon. (Buckley and Phillips, 1969.)

9.3. MULTILAYER FILMS BY MELT COEXTRUSION

For many years, polymer films have been coated to produce laminar composites having properties unobtainable with any single homopolymer. For example, polyethylene-coated poly(vinylidene chloride)-based films combine the values of low vapor transmission with the good heat-sealing behavior characteristic of poly(vinylidene chloride) and polyethylene, respectively. While most such commercial materials have between two and seven such layers, laminated films containing hundreds of layers have recently been developed (Alfrey *et al.*, 1969; Schrenk, 1973; Schrenk and Alfrey, 1969, 1970; Thomka and Schrenk, 1972). By means of special co-extrusion techniques, as many as 250 distinct layers can be packed into a 1-mil-thick film! There is, however, a lower limit to layer thickness, about 100 Å, below which layers tend to rupture. This last arises because of the dimensions of the polymer chains themselves. Other requirements include the proper matching of melt rheologies, the ability to obtain all components in the melt at a nondegrading temperature, and reasonable adhesion between the layers. Also, with very thin layers, morphological characteristics such as spherulite formation may be altered.

By alternating layers of hard and soft materials, such as polystyrene and polypropylene, it is possible to achieve synergistic effects in properties, similar to polyblending. For example, a better combination of stiffness, ductility, and toughness can be obtained in polystyrene–polypropylene multilayer films than in either component alone. Such materials also display unusual optical behavior, such as total reflection, at certain wavelengths (Alfrey *et al.*, 1969), because the films tend to reflect light in a manner similar to the reflection of x-rays by crystals.

9.4. SYNTHETIC PAPER POLYBLENDS

Traditionally, paper is prepared by blending together cellulose fibers, fillers, and binders. (As such, paper is a composite in its own right, but outside of the scope of this monograph.) Although cheap and useful, cellulose fiber-based papers have several problems, especially sensitivity to water and relative weakness even when dry. In addition, as the level of the world's technological development increases, more and more paper is demanded, at the expense of limited forest resources.

To overcome some of those difficulties, synthetic paper materials have been developed. Two approaches are as follows (International Conference on Plastics and Paper, 1973).

1. Polyblending of cellulose and polyethylene fibers (Morgan, 1961). The resultant paper, which comprises long, soft polyethylene fibers intertwined with the traditional cellulose, can be heat-sealed to a substrate (such as a steel can), and has improved properties, particularly wet strength. Alternatively, a sandwich of polyethylene fibers layered between two cellulose mats may be used; such a paper exhibits considerable water resistance.

The polyethylene fibers may be made by a spin-bonding process, which involves rapid spinning and solvent evaporation. Although the fibers are not as uniform as required for textile uses, they are much more economical when produced in this way.

2. Formation of biaxially oriented films composed of polyethylene–polystyrene blends* with polyethylene as the continuous phase and impact-resistant polystyrene as the reinforcing phase. Such films form tough, paperlike sheets that are impervious to water and are inherently opaque due to the existence of two phases of different refractive indexes.

9.5. CHARGE-MOSAIC MEMBRANES

An important method for recovering fresh water from saline or brackish waters is based upon the phenomenon of reverse osmosis, in which a suitable membrane such as cellulose diacetate permits purified water to pass through but retains most of the salt. In 1963, however, Kedem and Katchalsky (1963) predicted that in certain two-component ionic polymer materials, salt might be induced to diffuse faster than the water, thus allowing an alternate mode of product recovery. This process, operated by means of an externally applied pressure, is usually referred to as "piezodialysis," or sometimes as "negative reverse osmosis." Piezodialysis is believed to possess great potential in desalination, one of its greater advantages being that it is the by-product (salt) and not the product (water) that must be transported through the membrane (Dresner, 1972; Leitz and McRay, 1972).

As a model, Kedem and Katchalsky (1963) assumed alternating parallel arrays of elements of the two polymers, the elements being perpendicular to the membrane surface. To satisfy the requirements for ion transport, one polymer must be negatively and the other positively charged.

The efficiency of the process may be characterized by means of a reflection coefficient σ, which embodies the various transport coefficients concerned. When $\sigma = 0$, no separation of salt from water can occur. A value

---

* "Polyart," Neusiedler AG für Papierfabrikation, Vienna, Austria; "Acroart," Mead Papers, Dayton, Ohio; "Q'Kote," Japan Synthetic Paper Co., Ltd.

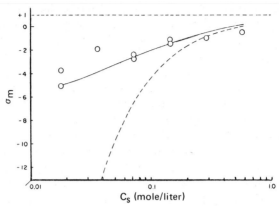

Figure 9.8. The reflection coefficient σ for a charge-mosaic membrane measured at a series of 2:1 concentration ratios. The solid curve was calculated from theory. The dashed curve was calculated from the unmodified Kedem–Katchalsky treatment. (Weinstein et al., 1973.)

of $\sigma = +1$ indicates that only water can flow through the membrane, a case equivalent to reverse osmosis. On the other hand, negative values of σ indicate that salt can permeate faster than water. It was found experimentally that with an appropriate parallel combination of positive and negative elements, σ may assume values of -8 or greater, thus confirming the occurrence of piezodialysis.

Recently, several investigators have extended the original theory and developed several experimental approaches (Leitz and Shorr, 1972; Weinstein et al., 1972, 1973). To verify the essential features of the developing theory, Weinstein et al. (1973) prepared "charge-mosaic" particles of ion-permeable cationic and anionic ion-exchange resins in an impermeable matrix of a flexible silicone resin. A square array was formed with nearest neighbors having fixed charges of opposite sign; to ensure continuity, each bead was arranged to communicate with both surfaces of the membrane. Reflection coefficients were determined as a function of salt concentration; in Figure 9.8 experimental results are compared with the Kedem–Katchalsky theory. At all concentrations, the modified theory of Weinstein et al. (1973) is followed; at high concentrations, the Kedem–Katchalsky theory (1963) became essentially equivalent.

9.6. FLOW AND MORPHOLOGY OF TWO-COMPONENT SYSTEMS

In Chapter 2, the criteria for phase separation and stability of morphological forms of assorted polymer blends have been reviewed. However, effects of flow on the morphology should be considered further, for one frequently wishes to extrude or mold a two-component polymer system, whether it be an "alloy" or blend, a block copolymer, or a conventional polymer containing a processing aid. (See, for example, Section 9.2.)

The deformation and hydrodynamic stability of Newtonian viscous droplets (in a continuous second phase) subjected to shearing have been studied extensively (Cox, 1969; Goldsmith and Mason, 1967). Two parameters are concerned: (1) λ, the ratio of the viscosity of the suspended fluid to that of the medium η_0, and (2) k, the ratio of the interfacial tension γ to the product of the local shear stress $\eta_0 G$ (G being the shear rate) and the particle radius a. On shearing, a droplet becomes spheroidal in shape, and more or less oriented in the direction of the velocity gradient. Cox (1969) has summarized the relationships required to define the deformation D and the orientation angle α between the axis of the deformed spheroid and the direction of the velocity gradient (Figure 9.9):

$$D = \frac{5(19\lambda + 16)}{4(\lambda + 1)[(20k)^2 + (19\lambda)^2]^{1/2}} \tag{9.1}$$

$$\alpha = (1/4)\pi + (1/2)\tan^{-1}(19\lambda/20k) \tag{9.2}$$

$$D = (L - B)/(L + B) \tag{9.3}$$

$$\lambda = \eta_i/\eta_0 \tag{9.4}$$

$$k = \gamma/\eta_0 Ga \tag{9.5}$$

where L and B are the length and height, respectively, of the deformed spheroid; η_i is the viscosity of the droplet fluid; η_0 is the viscosity of the medium; and G is the local shear rate. In all cases, small deformations are assumed.

When k and λ are large, the deformation will tend to be small. VanOene (1974) has estimated values of k for typical polymer melts subjected to a

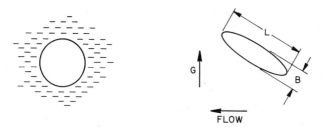

Figure 9.9. Deformation of a suspended droplet by shear. The undeformed droplet is shown on the left; the deformed droplet, undergoing Poiseuille-type flow, is shown on the right.

typical shear stress (10^5–10^6 dyn/cm^2) and having values of λ in the range 0.2–5:

$$a < 1 \ \mu\text{m} \qquad k \gg 1$$

$$a \sim 1 \ \mu\text{m} \qquad k \sim 1$$

$$a > 1 \ \mu\text{m} \qquad k \ll 1$$

The importance of the radius of the droplet is clear; the smaller the particle, the greater is k and the greater is the resistance to deformation. Indeed, large particles will tend to break up into micrometer- or submicrometer-sized ones. For any given size, of course, either λ or k may be dominant. Thus, if interfacial effects are strong (high γ and hence k), droplets will be aligned at an angle of 45° with respect to the flow direction; if, on the other hand, viscosity is most important (high λ), the droplets will be aligned in the flow direction.

Although most polymers are in fact viscoelastic rather than purely viscous in the melt, little attention has been given to stability criteria for viscoelastic suspension. However, in an interesting theoretical and experimental paper, VanOene (1972) has extended the treatment discussed to a realistic viscoelastic problem. Assuming that effects of viscosity and viscosity gradient per se on the shape of a droplet will still be valid in viscoelastic systems, VanOene considered effects of elasticity in the fluid droplet, and derived an expression to account for the difference in the free energy of deformation in flow of the two phases. Results of his analysis of droplet stability are summarized in the following expressions for the interfacial tension between two phases α and β undergoing flow:

$$\gamma_{ij} = \gamma_{\alpha\beta} = \gamma_{\alpha\beta}^0 + \tfrac{1}{6}a_\alpha[(\hat{\sigma}_2)_\alpha - (\hat{\sigma}_2)_\beta] \tag{9.6}$$

$$\gamma_{ij} = \gamma_{\beta\alpha} = \gamma_{\alpha\beta}^0 - \tfrac{1}{6}a_\beta[(\hat{\sigma}_2)_\alpha - (\hat{\sigma}_2)_\beta] \tag{9.7}$$

These two expressions correspond to a suspension of α in β and a suspension of β in α, respectively. In each case, γ_{ij} is the interfacial tension of a droplet of fluid i in the medium j; γ^0 is the interfacial tension in the *absence* of flow; a is the droplet radius; and $\hat{\sigma}_2$ is the second normal stress function of the fluid (Figure 9.10).

The second normal stress function ($\hat{\sigma}_2$) relates the stress tensor $T\langle ZZ \rangle$ in the direction of flow and the stress tensor $T\langle \theta\theta \rangle$ at an angle normal to the plane defined by the direction of flow and the velocity gradient:

$$(\hat{\sigma}_2) = T\langle ZZ \rangle - T\langle \theta\theta \rangle \tag{9.8}$$

where the caret indicates a dependence on shear stress. The value of $(\hat{\sigma}_2)$ also depends on the molecular weight and molecular weight distribution, but, for a given morphology, not on shear rate or temperature.

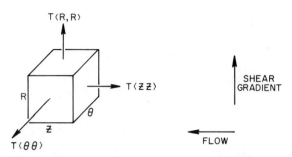

Figure 9.10. Normal stress tensors T in a fluid subjected to shear.

It is now of interest to predict under what conditions one will find spheres, ribbons, or fibers in a flowing composite melt. Several predictions can be made, based on equations (9.6) and (9.7), and on the requirement that $\gamma_{ij} \geq 0$ for a droplet to be thermodynamically stable. First, if $(\hat{\sigma}_2)_\alpha > (\hat{\sigma}_2)_\beta$, phase α will *always* form droplets (or fibers)* in phase β. On the other hand, the converse is not true; phase β can form droplets in phase α only if

$$a_\beta \leq 6\gamma_{\alpha\beta}^0/[(\hat{\sigma}_2)_\alpha - (\hat{\sigma}_2)_\beta] \qquad (9.9)$$

Assuming typical values for $[(\hat{\sigma}_2)_\alpha - (\hat{\sigma}_2)_\beta]$ (10^5–10^6 dyn/cm²) and for $\gamma_{\alpha\beta}^0$ (~5 dyn/cm) (interfacial surface tensions are treated in Section 13.4.4), the critical size for stable droplets may be calculated to fall in the range of 0.1–1 μm for polymer melts. Thus if the dimension of the dispersed phase is much larger than 1 μm (as may occur with incomplete blending), the dispersed phase must form macroscopic layers rather than droplets or fibers. If the phase dimension is small enough for stable droplets to form, another morphological variant is predicted: Since α will always form droplets in β, droplets of β formed by satisfying the drop-size criterion will always contain smaller droplets of α.

In general, VanOene's experimental studies support the conclusions of his theory. As shown in Figure 9.11, extrusion of blends of an ionomer with polypropylene gave "complementary" morphologies. When the blend (cursorily mixed) was rich in ionomer (90:10), the polypropylene stratified (phase β unstable in phase α); on the other hand, when polypropylene was the major component (10:90), the ionomer formed droplets [$(\hat{\sigma}_2)_\alpha > (\hat{\sigma}_2)_\beta$, phase α always stable in β]. It was shown that the morphology was not much affected by shear rate, residence time, or temperature.

* Topologically, with respect to hydrodynamic flow, droplets and fibers are essentially equivalent, each type presenting a circular cross section regardless of the shear stress (VanOene, 1972).

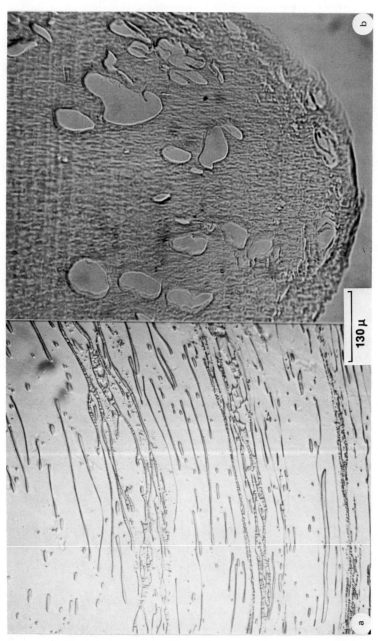

Figure 9.11. Complementary modes of dispersion for extruded blends of incompatible polymers: (a) 10/90 blend of Marlex polypropylene with ionomer, extruded at 190°C at a rate of 0.2 in./min; (b) 90/10 blend of the same polymers extruded under similar conditions. Both sections perpendicular to flow direction. (VanOene, 1972.)

Molecular weight and distribution are important. Assuming that $(\hat{\sigma}_2)$ is proportional to the steady-state shear compliance J_e and that J_e is given by

$$J_e = \frac{0.4}{RT} \frac{1}{\rho} \left[\frac{M_z M_{z+1}}{M_w^2} \right] M_w \tag{9.10}$$

we have

$$\hat{\sigma}_2 = J_e \times \tau^2 \tag{9.11}$$

where τ represents the shear stress. VanOene replaces $[(\hat{\sigma}_2)_\alpha - (\hat{\sigma}_2)_\beta]$ by

$$\frac{0.4}{RT} \left[\frac{1}{\rho_\alpha} \left(\frac{M_z M_{z+1}}{M_w^2} \right)_\alpha M_{w,\alpha} - \frac{1}{\rho_\beta} \left(\frac{M_z M_{z+1}}{M_w^2} \right)_\beta \right] M_{w,\beta} \tag{9.12}$$

where ρ is the density; M_w, M_z, and M_{z+1} are the weight-, Z-, and $(Z + 1)$-average molecular weights, respectively; R is the gas constant; and T is the absolute temperature. Thus, stability of α in β requires that the sign of σ be positive, as would be the case if α has the higher molecular weight. Experiments with poly(methyl methacrylates) (having different molecular weights) in a polystyrene matrix confirmed the prediction. When the molecular weight of the dispersed phase was higher (difference in second normal stress function positive) than that of the matrix, coarse droplets were observed; when the molecular weight was lower, layers or ribbons were more typical.

The effect of particle size was examined in several systems. When very thorough mixing was achieved using a preblended composition of polypropylene and ionomer, the ultimate mode of dispersion was droplets or fibers of the more elastic phase [higher $(\hat{\sigma})$] in a matrix of the less elastic phase (Figure 9.12). Curiously, in all systems, droplets were observed to change to fibers as the temperature was increased. Experiments with finely dispersed (preblended) mixtures of *low-molecular-weight* PMMA and PS revealed that droplet sizes tended to be less than 1 μm, and the elastic effect was no longer important, so that the dispersions were always droplets of the minor component in a matrix of the major component. As predicted above, composite droplets were noted in the PS-rich compositions (Figure 9.13).

Thus, in general, for high-molecular-weight, viscoelastic, two-component systems, the morphology is dominated by elastic contributions to the deformation. The VanOene theory should have considerable generality, and should be applicable to many real systems, including melt blending (see Section 9.2), mixing, polymer–plasticizer systems, and the extrusion of composites.

In a recent study by Work (1972*b*) on blends of poly(vinyl chloride) with chlorinated polyethylene, some of the predictions discussed above have been

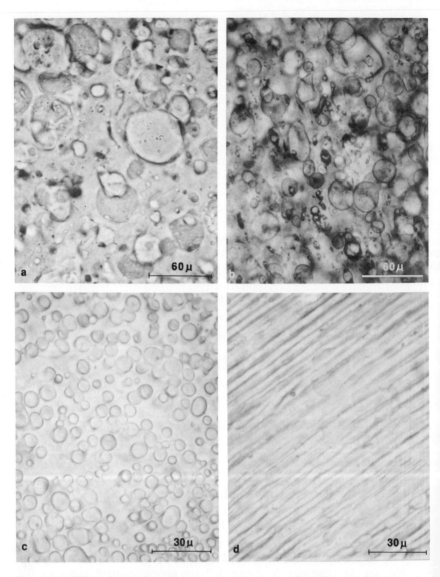

Figure 9.12. Effects of dispersion and temperature in extrusion of 70/30 blends of Marlex poly-propylene and ionomer: (a) and (b), sections perpendicular and parallel, respectively, to flow direction, extrusion at 190°C; (c) and (d), sections perpendicular and parallel, respectively, to flow direction, extrusion at 225°C. (VanOene, 1972.)

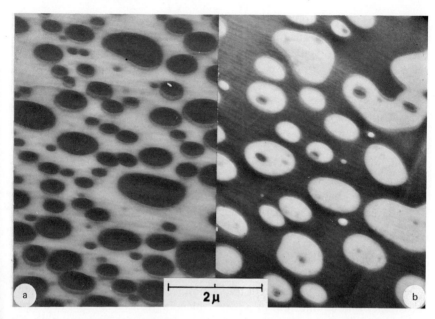

Figure 9.13. Formation of composite droplets in extruded blends of poly(methyl methacrylate) and polystyrene. (a) 70% PMMA/30% PST; (b) 70% PST/30% PMMA. Extrusion rate, 0.5 in./min; $T = 225°C$. (VanOene, 1972.)

confirmed. Thus, for all midrange compositions, the more viscous component always formed the dispersed phase. (For a discussion of the rheological behavior of block copolymers, see Sections 4.11 and 6.1.7.)

9.7. COMPATIBLE POLYMER BLENDS

As mentioned in Chapter 4, the criteria for compatibility include mechanical integrity, optical transparency, a single glass transition temperature, and homogeneity on a submicroscopic level.

MacKnight *et al.* (1971) have pointed out that, for a pair of polymers to exhibit thermodynamic compatibility as the term is presently understood, the free energy of mixing per unit mass ΔG_m must be negative for some value of s, a parameter defined to characterize the degree of mixing in terms of the average normalized size of the segmental clusters in the system. The quantity s times the degree of polymerization equals the size of the average cluster. If s corresponds to the polymer molecule itself, $s = 1$; if alternately mixing occurs at the segmental level, then $s \cong 0$. The condition $s \gg 1$ corresponds to phase separation. The quantity ΔG_m will clearly be negative for all values of s if the heat of mixing ΔH_m is negative, a condition rarely achieved. However, the size of the average cluster $s(\text{DP})$ also depends on the relative contribution of the entropy of mixing per segment ΔS_m. The entropy factor determines whether miscibility occurs in the more prevalent situation of a positive ΔH_m. Importantly, in the latter case, ΔG_m can still be negative with a minimum for $0 < s < 1$. MacKnight *et al.* (1971) point out that as s decreases, ΔS_m at first increases as a result of the greater number of segment configurations available. However, the quantity ΔS_m eventually declines because of the increasing restrictions in the number of possible configurations imposed by mixing at the segmental level.

Several types of compatible blends are now discussed.

9.7.1. Compatible Blends of Amorphous Polymers

One of the most significant implications of the above discussion is the possibility that equilibrium in polymer blending can be reached at a stage intermediate between the mixing of whole polymer molecules ($s = 1$) and the mixing of individual segments, i.e., for the case $0 < s < 1$. The blend system investigated by MacKnight *et al.* (1971) and Stoelting *et al.* (1970), poly(2,6-dimethylphenylene oxide)/polystyrene (PPO/PS), apparently falls into this interesting category. Using a dynamic elastoviscosimeter at a frequency of 110 Hz at temperatures between -180 and $240°C$, MacKnight and co-workers showed that values of the loss modulus E'' displayed broadened maxima at temperatures intermediate between the T_g's for the two homopolymers (Figure 9.14). This result, most apparent for the 50/50 composition, suggests extensive but incomplete mixing. While dielectric loss measurements yielded similar results, differential scanning calorimetry (DSC) studies revealed a single glass transition at a temperature corresponding approximately to the weighted mean of the T_g's of the constituents, 208°C for PPO and 90°C for PS. Results of the DSC study suggest that complete mixing has occurred at the segmental level, and are supported by a more recent thermooptical and DSC study by Shultz and Gendron (1972). However, it seems likely that DSC is less sensitive than mechanical spectroscopy to the presence of very small phases (MacKnight *et al.*, 1971; Stoelting *et al.*, 1970).

Studying other systems, Noland *et al.* (1971) presented data based on differential thermal analysis (DTA), thermomechanical analysis, and dilatometry to show that mechanical blends of poly(methyl methacrylate) and poly(ethyl acrylate) are compatible with poly(vinylidene fluoride)

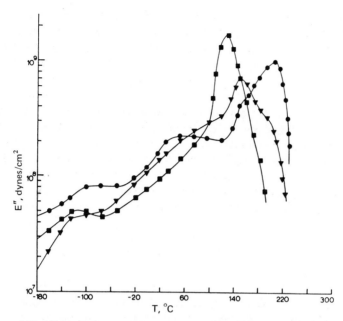

Figure 9.14. Dynamic mechanical loss at 110 Hz for three PPO–PS mixtures as a function of temperature. (■) 25 PPO/75 PS; (▼) 50 PPO/50 PS; (●) 75 PPO/25 PS. Note broadening of the loss modulus, especially for the 50/50 blend. (MacKnight *et al.*, 1971.)

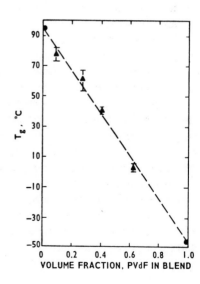

Figure 9.15. Glass transition temperatures of quenched PVdF blends. (Noland *et al.*, 1971.)

(PVdF). This study was mainly concerned with PMMA as the major component and PVdF as a permanent and light-stable plasticizer. As shown in Figure 9.15, the glass temperatures T_g of the blends follow a simple random-copolymer relationship:

$$T_g = W_1 T_{g_1} + W_2 T_{g_2} \qquad (9.13)$$

where W_1 and W_2 are the weight fractions of the two components and T_{g_1} and T_{g_2} are the T_g's of the two components. It was noted that at high PVdF concentrations, PVdF tended to separate out as a crystalline phase, especially after annealing. Generally similar results illustrating the validity of equation (9.13) for blends of compatible polymers have been observed by others, for example, by Koleske and Lundberg (1969) for blends of poly(vinyl chloride) and poly(ε-caprolactone). Other semicompatible materials are discussed in Sections 2.52 and 3.1.4.

9.7.2. Isomorphic Polymer Pairs

Interesting blends of compatible polymers can be prepared by the cocrystallization of isomorphic polymer pairs (Allegra and Bassi, 1969). When two different types of crystallizable polymers contain monomer units of approximately the same shape and volume, and their chains are able to adopt a similar chain conformation, isomorphism is possible. In such a case, each mer can fit equally well in the crystal lattice, so that a mixed crystal forms. The isomorphous mers can exist in the same molecule, as in a copolymer of vinyl fluoride and vinylidene fluoride, or in different molecules (the case of interest here).

A trivial case of macromolecular isomorphism involves the mixing of species differing only in an isotope, for example, as isotactic polypropylene and isotactic polydeuteropropylene (Natta *et al.*, 1958). More interesting examples can be realized by melting together such polymers as poly(vinyl fluoride) and poly(vinylidene fluoride) (Natta *et al.*, 1971) or poly(isopropyl vinyl ether) and poly(sec-butyl vinyl ether) (Allegra and Bassi, 1969) that form isomorphic pairs at all relative compositions.

The mechanical behavior of isomorphic macromolecular systems would be expected to be quite different from the behavior observed in bicomponent or biconstituent systems. Indeed, isomorphic systems would be expected to behave in many respects like crystalline homopolymers, except that such properties as T_m and lattice spacings may be dependent on composition. Because of the single-phase situation, the glass–rubber transition and related properties may be expected to behave as if a random copolymer were being analyzed. Unfortunately, little information on the bulk or mechanical properties of polymeric isomorphic pairs appears to be available (Allegra, 1970). In the light of the behavior of compatible polymer pairs considered in the preceding pages, a study of the melt morphology of the isomorphic pairs should be interesting. One might speculate that any material that is sufficiently compatible in the crystalline state to exhibit isomorphism in the lattice must be also highly compatible in the melt.

9.8. WOOD AS A POLYBLEND–COMPOSITE

In many ways, wood, one of man's oldest and most widely used materials, resembles the polyblends or composites discussed elsewhere (Hearmon, 1953). Since wood is a mixture of two polymers, it is a polyblend. Its behavior, however, more closely resembles that of oriented fiber composites. Since many synthetic materials are competitive with wood in commerce and resemble the structure of wood, it is of interest to understand the original natural product.

Wood contains two continuous phases: a cellular fibrous structure, largely composed of cellulose, and a network of lignin, which serves as a bonding agent. In addition, depending on the degree of drying, either water or air is contained as a third phase within the lumen structure of the cells. The phases are arranged anisotropically, so that properties must be described in terms of longitudinal, radial, and tangential axes (Figure 9.16).

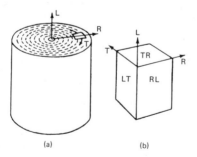

Figure 9.16. Principal axes and planes in wood. L, longitudinal; R, radial; and T, tangential (Hearmon, 1953). The rings, which arise due to the occurrence of different growth rates in spring and summer, indicate the age of the tree.

9.8.1. Chemistry and Morphology

The cell walls of most woody plants contain about 50% cellulose, 25% lignin, 24% hemicelluloses (most five carbon polysaccharides) and extractable materials, and about 1% of mineral matter (Brown *et al.*, 1949). Cellulose is a linear, highly crystalline polymer having a polyglucoside structure as shown in Table 1.1. Lignin is a highly crosslinked amorphous polymer based on *p*-coumaryl alcohol, coniferyl alcohol, and sinapyl alcohol, and their phenolic glucosides (Freudenberg, 1966). The structure of this highly irregular material is represented in Figure 9.17, in which possible crosslinking modes are illustrated (Kollman and Cote, 1968; Adler, 1961).

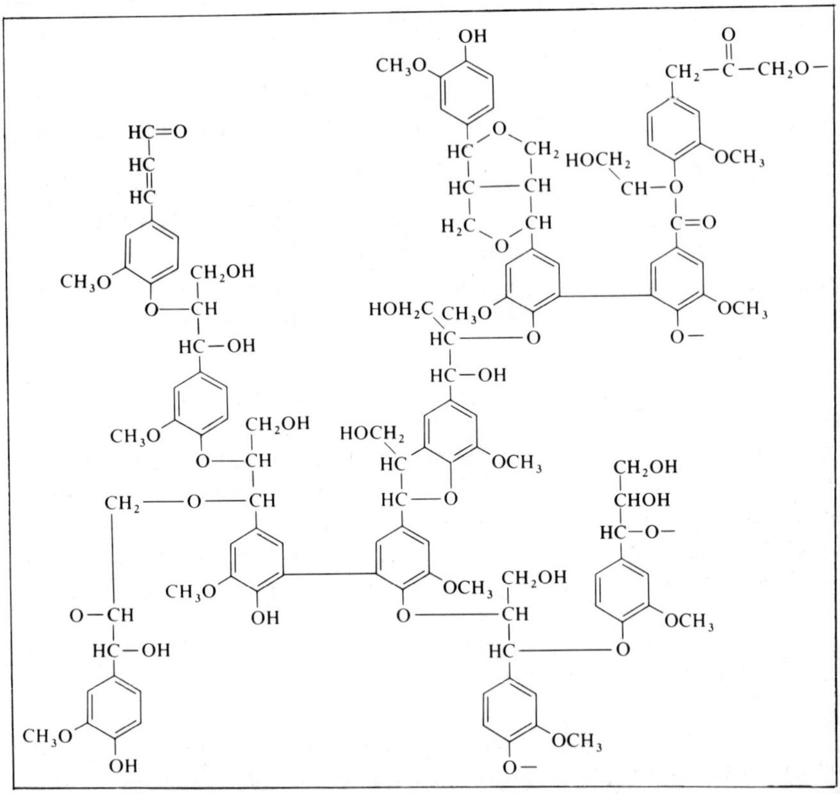

Figure 9.17. Summary of lignin structure. This natural thermoset plastic has a very irregular structure with various ether groups acting as crosslinks in many cases. (Kollman and Cote, 1968, p. 69; Adler, 1961).

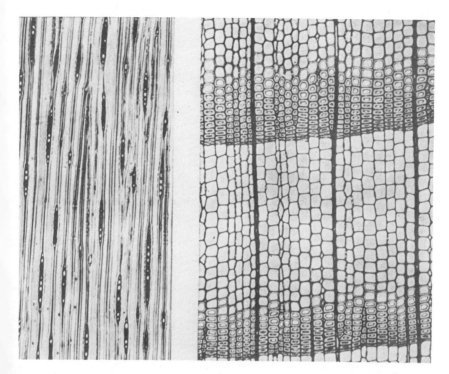

Figure 9.18. Longitudinal and transverse sections in eastern hemlock. (Brown *et al.*, 1949, Vol. I, p. 477.)

As shown in Figure 9.18, the cellular structure can be observed easily at $75 \times$ magnification (Brown *et al.*, 1949, Vol. I, p. 477). It may be seen that the cells touch at many points, thus allowing the passage of water up the stem; the dark material surrounding the cells is lignin. Within the cell walls, the cellulose molecules form bundles or microfibrils (Tsoumis, 1968, p. 69), which form three distinct layers in the cell wall (Figure 9.19). These cells (tracheids) are commonly 1–3 mm in length and about 60 μm in diameter.

9.8.2. Mechanical Behavior

Wood may then be pictured as an assembly of oriented fibers in an amorphous matrix.* The fibers are hollow, lowering the density and altering the mechanical behavior in a manner analogous to high-density foams.

* In this respect, wood is similar to glass-fiber-reinforced epoxy resins.

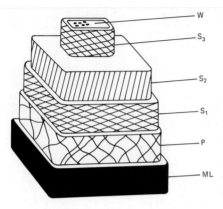

Figure 9.19. Schematic representation of cell-wall layers in a tracheid or fiber, with respective orientation of microfibrils illustrated by lines. **ML**, middle lamella, lignin; **P**, primary wall; S_1, S_2, and S_3, layers of the secondary wall; **W**, warty layer. (Tsoumis, 1968, p. 69.)

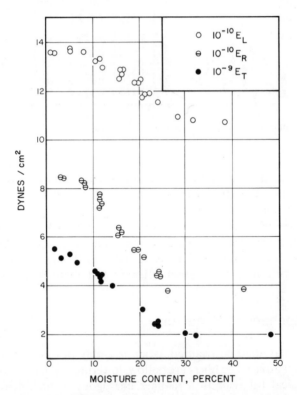

Figure 9.20. Effect of moisture content on Young's modulus for Sitka spruce. (See Figure 9.16 for the meaning of the subscripts L, R, and T.) (Hearmon, 1953.)

Table 9.1
Processes for Wood Modification[a]

Process	Description	Commercial use
Acetylation Hydroxyl groups reacted to form esters	Acetic anhydride with pyridine catalyst and swelling agent in gas phase; capillaries empty; ASE about 70% (acetyl 25%)	Thin veneers to $2 \times 6 \times 48$ in. stock
Ammonia Vapor and liquid	Wood pumped down, exposed to anhydrous ammonia vapor or liquid at 150 psi; treatment time depends upon thickness and moisture content, ranges from minutes to hours; loblolly pine requires longer exposure than hardwoods; bends in $\frac{1}{2}$-in. stock up to 90°; treated wood darker in color and sometimes streaked	Experimental products made: snowshoes, picture frames, chair backs, lamps, room dividers
Compreg Compressed wood– phenolic–formaldehyde composite	Wood soaked in water or alcohol solution of P–F, dried, heated, and compressed during polymerization; cell structure collapsed, density 1.3–1.4, ASE about 95% resistant to decay; strength, hardness, and abrasion increased	Cutlery handles, electrical insulators, thin veneers
Crosslinking with formaldehyde	Catalyst , 2% zinc chloride in wood, then exposed to *para*-formaldehyde heated to 120°C for 20 min; ASE 85% (4% weight increase); drastic loss in toughness and abrasion resistance	None
Cyanoethylation Reaction with acrylonitrile (ACN)	Southern yellow pine 5% sodium hydroxide in wood by empty cell process; full cell loading with ACN, then heat to 70–100°C; EMC 4.5% at 30% ACN; not attacked by fungi; impact strength loss	None
Ethylene oxide (EO) High-pressure gas treatment	Trimethyl amine catalyst; no deposit in capillaries; ASE to 65% (11% weight increase)	None
Impreg Wood–phenol-formal- dehyde composite (P–F)	Wood soaked in water or alcohol solution of P–F, dried, heated to polymerize; swells cell wall, capillaries filled; loading about 35%, ASE about 75%	Die models, thin veneers
Irradiated wood Beta and gamma radiation	Exposures to 10^6 rad give slight increase in mechanical properties and reduced hydroscopicity; above this level cellulose is degraded and mechanical properties decrease rapidly; lignin most resistant; wood soluble above 3×10^8 rad; low exposure used to temporarily inhibit growth of fungi	None

continued

Table 9.1—continued

Process	Description	Commercial use
Ozone 　Gas phase	Treatment degrades both cellulose and lignin; pulping action	None
Polyethylene glycol (PEG) 　Capillary structure 　filled with PEG	PEG soluble in water, low vapor pressure; wet or dry wood is soaked in PEG, replaces water in cell walls and fills capillaries; moist feel to bulked wood, hydroscopic; urethane finish used; ASE about 98 %	Art carvings, preservation of old wood
β-Propiolactone (β-P) 　Full and empty cell 　treatment	Southern yellow pine, β-P diluted with acetone , wood loaded and heated; grafted polyester side chains on swollen cell wall cellulose backbone; β-P self-polymerizing; compressive strength increase; decrease in fungi attack; carbonyl end groups reacted with copper or zinc to further decrease fungi attack	None
Staybwood 　Heat-stabilized wood	Wood heated to 150–300°C; air oven or boiling water	None
Staypak 　Heat-stabilized 　compressed wood	Wood heated to 320°C, then compressed; 400–4000 psi	Handles and desk legs

[a] Meyer and Loos (1969).

The combination is very tough: Izod impact values range from 3–18 ft-lb/in. of notch, depending on species and moisture content (Kollman and Cote, 1968).

This point will now be briefly explored, since moisture content is the single biggest variable in wood technology. Water behaves as a plasticizer, softening primarily the cellulosic portion of the wood by disrupting its hydrogen-bonded structure. While all wood of commerce contains some water (ca. 12%), wood in the living state contains much more. When the water content exceeds ca. 30%, the solid phases become saturated, and free water forms in the cell lumens. This effect is pictured in Figure 9.20, where the break at 30% moisture indicates the limit to the plasticizing action (Hearmon, 1953, Chap. 2). It should be observed that the modulus is quite different in the three directions. The impregnation of wood by polymers will be considered in Section 11.2.

9.8.3. Wood Modification

The general phenomenon of wood modification is too broad to be covered in this monograph. However, Table 9.1 (Meyer and Loos, 1969) summarizes some of the more important processes; the reader is directed to the original literature for further details. Also, for polymer-impregnated wood, see Section 11.2.

9.9. ASSOCIATION COMPLEXES

All of the materials containing two polymers can be broadly classified as either "polymer blends" or "graft copolymers" on the basis of the absence or presence of covalent chemical bonds between the two species. For special purposes, it has also been of interest to mix solutions of anionic and cationic polymers. Such materials coacervate, and form association complexes with ionic bonds. Because of the impermanent nature of such bonds, we choose to classify such materials with polymer blends. Examples of these materials are given by Leitz and Shorr (1972) and Eisenberg and King (1972). These materials are sometimes of interest because of their salt transport properties (Leitz and Shorr, 1972). See Section 9.5.

10

Reinforcement of Elastomers

Elastomers can be reinforced, or made tougher, by the addition of very small particles, typically finely dispersed carbon blacks or silicas. Similar types of reinforcement in elastomers can be brought about by the addition of a plastic polymeric phase (Sections 4.4 and 8.4) or the inclusion of poly-styrene latexes.

Elastomer reinforcement means different things to different people. Increases in modulus, tensile strength, and swelling resistance are all measures of reinforcement, but abrasion resistance is the most important index, especially to tire manufacturers. Unfortunately, abrasion resistance is a most difficult quantity to measure meaningfully.

Reinforcement, however defined, depends upon the size, surface chemistry, state of aggregation, and quantity of filler. The influence that these characteristics have on physical and mechanical characteristics will be explored. Current theories of reinforcement, both thermodynamic and viscoelastic, will be developed, in order to summarize the state of the art in this area.

10.1. HISTORICAL ASPECTS

Reinforced elastomers are one of the oldest and most important classes of composite materials (Ruffell, 1952; Sellers and Toonder, 1965; Stern, 1968, p. 278). When the automobile first became popular, the need to toughen tire rubber, especially against abrasion, became obvious. Although zinc oxide had already attained widespread use as a rubber colorant, in 1905 Ditmar realized the true importance of this material as a reinforcing agent for rubber. Many industry veterans can still remember when tires had white treads. However, such tire treads usually lasted less than 5000 miles, and the need for further improvements was imperative.

In 1904 Mote had already discovered the reinforcing value of very fine carbon blacks. Carbon black proved much superior to zinc oxide for rubber reinforcement, and replaced the latter in tires between 1910 and 1915. At first, only small amounts of carbon black were used. In 1915, for example, tire tread compounds contained only about 22 % carbon black. At the present time this has increased to 50 % or more, and tire treads lasting up to 40,000 miles are now common.

The reinforcing effect of carbon black on stress–strain behavior of natural rubber is depicted in Figure 10.1. The reinforced material has a higher modulus (is stiffer) and is less extensible.

In the period between 1939 and 1949 several types of silica were introduced as rubber-reinforcing agents. Hydrated silica, manufactured by the so-called wet process, was introduced first, followed by products of the pyrogenic and the aerogel processes. Finely divided silicas give degrees of reinforcement approaching those of the carbon blacks, and have found widespread use in rubber shoe heels and soles, in food packaging, and in many other applications where carbon blacks are objectionable.

A word should be added about the present use of zinc oxide in rubber manufacture. The discovery early in this century that many organic vulcanization accelerators are not effective without zinc oxide led to continued usage of this material for purposes other than reinforcement. By far the major use of zinc oxide in the rubber industry today is for activation of organic accelerators, and not for reinforcement.

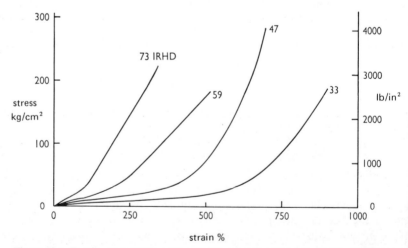

Figure 10.1. Tensile stress–strain curves for four natural rubber compounds of different hardnesses: 73 IRHD contains 50 parts of a reinforcing black, and different vulcanizing systems account for the different curves of the two gum compounds (47 and 33 IRHD). (Lindley, 1964.)

Table 10.1
Relative Reinforcement by Carbon Blacks Judged by Tire Tests in Natural
and SBR Rubbers[a]

	Relative reinforcement (HAF = 100)	Typical specific surface area, m^2/g
MT (medium thermal)	21	8
FT (fine thermal)	38	17
SRF (semireinforcing furnace)	46	25
HMF (high modulus furnace)	63	30
FEF (fast extruding furnace)	75	45
HAF (high abrasion furnace)	100	80
ISAF (intermediate super abrasion furnace)	116	115
SAF (super abrasion furnace)	125	140
EPC (easy processing channel)	85	115
HPC (hard processing channel)	88	150
Acetylene black	61	66

[a] Studebaker (1965).

10.2. TYPES OF FILLERS

Carbon blacks and silicas are both manufactured by means of several different routes, and the total number of variations on each process is large (Kraus, 1965a,b, 1971a; Kraus et al., 1970; Lindley, 1964; Peterson and Kwei, 1961; Rivin, 1971; Rowland et al., 1965; Wagner and Sellers, 1959). This section will delineate the major types of carbon blacks and silicas.

Table 10.1 gives the abbreviations, identifications, specific surface areas, and relative reinforcement values of the more important kinds of carbon black. The reader will observe that there is a good, but not perfect, correlation between reinforcement and specific surface area.

With respect to silica systems the four major types are based on the so-called pyrogenic (PP), thermal (TP), and wet (WP) processes, as well as on the natural product (N). A second set of symbols represents the degree of reinforcement: low reinforcing (LR), medium reinforcing (MR), high reinforcing (HR), super reinforcing (SR), and extremely reinforcing (ER). For example, the commercial product "Aerosil" would be classified as PP-SiO$_2$-ER. A selective list of commercial reinforcing silicas is given in Table 10.2, together with results of specific surface area measurements. There appears to be only the most general correlation between surface area and reinforcement in silicas, provided the available area is above $\sim 50\ m^2/g$.

Table 10.2
Silica Surface Areas[a]

Trade name	Surface area, m^2/g
Aerosil	175
Cab-O-Sil M-5	200
HiSil 233	150
Silene D	40

[a] Wagner and Sellers (1959).

10.3. SIZE AND SURFACE CHARACTERISTICS OF ELASTOMER REINFORCERS

Most technologists would agree that a prime requisite of any particulate reinforcing filler is that it be finely divided. Filler size dimensions are usually given in terms of surface area per gram (see Tables 10.1 and 10.2) because the first widely used modern technique for measuring small particles was the BET nitrogen adsorption method. Significant elastomer reinforcement begins when particles have greater than $\sim 50 \, m^2/g$ of surface area. Assuming simple spheres, this corresponds to particles approximately 500 Å in diameter, which is the same order of magnitude as the distance between crosslink sites in the matrix elastomer, as illustrated in Figure 10.2. Apparently, when the particles become larger than the average end-to-end distance between crosslinks, reinforcement declines due to rubber–particle adhesion failure when the attached chains become highly extended during deformation. The cases of large and small particles are compared in Figure 10.3. However, it is relatively simple to prepare both silicas and carbon blacks in the 150–200 m^2/g range, where the above aspect is not a serious problem.

A great deal of attention has been paid to reactive sites on the filler and types of bonds formed. Although most investigators think that good

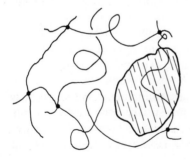

Figure 10.2. Filler particle size in relation to crosslink density.

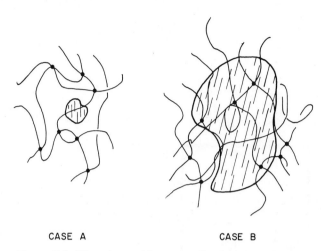

CASE A CASE B

Figure 10.3. When the particles are smaller than the end-to-end distance between crosslinks, as in case A, surface adhesion failure on extension is less likely to occur than in case B.

bonding is essential to reinforcement, two major schools of thought have appeared (Kraus, 1965a). The first school holds that primary chemical bonds are essential to reinforcement, while the second school holds that secondary physical forces are sufficient. [An interesting recent paper by Kraus *et al.* (1970), however, showed that neither is required. This will be discussed further below.]

An important point relating the two schools involves the concept of multiple physical bonds by chains lying for some distance on the filler particle surface. Assuming hydrogen bonding of 5 kcal, it will obviously take about ten or more of these to attain the strength of a single primary bond. However, statistical calculations show that the probability, under equilibrium conditions, of all ten of these bonds "letting go" simultaneously is very small. Hence, it is easier than often recognized for low-strength physical forces to attach chains to filler surfaces quite firmly (Peterson and Kwei, 1961; Rowland *et al.*, 1965).

10.3.1. Filler Surface Chemistry

The literature on this topic is too vast to cover in detail. A few essential points will be mentioned, the reader being directed to books such as those by Kraus (1965b) and Stern (1967) for more comprehensive reviews. It will be assumed below that if the polymer wets the filler, extensive physical

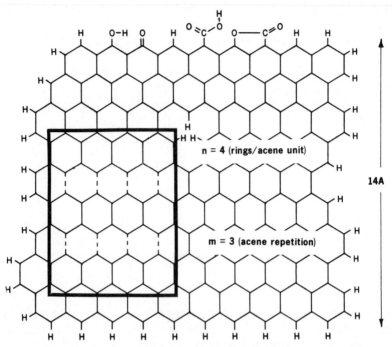

Figure 10.4. A typical carbon black "molecule." Many structures common to aromatic chemistry appear. (Rivin, 1971.)

bonding exists by means of van der Waals interactions, dispersion forces, etc. Examples of relatively nonwetting fillers include the solid propellant composites (see Chapter 12) and such substances as chalk or sand-filled materials. Except where specified, good wetting will be assumed.

Rubber-grade carbon blacks contain appreciable quantities of chemically combined hydrogen (0.1–4 %) and oxygen (0.2–1 %) and sometimes sulfur (up to 1 %). The highest oxygen and hydrogen contents are usually found in channel blacks. Some of this hydrogen and oxygen has been shown to be in the form of acid carboxyl groups, —COOH. Another type of active hydrogen results from phenolic hydroxyl groups. The quinone oxygen has been shown to be especially important in reactions involving free radicals. Also important are aldehydes, hydroperoxide, and ketones. A pictorial representation of a typical carbon black "molecule" is given in Figure 10.4.

Many of the chemical groups mentioned above have counterparts in reinforcing silicas. However, the most important groups are believed to be the silanol group, —Si—OH, and the siloxane group, —Si—O—Si—. The former is sometimes deliberately masked by reaction with methyl or butyl alcohols, since this aids in filler dispersion.

10.3.2. Morphology and Microstructure of Carbon Blacks

Within the last few years electron micrographs have become available which show a characteristic layer orientation within the primary carbon black particles (Rivin, 1971). Electron micrographs of furnace blacks, for example, demonstrate a continuous, oriented network with no identifiable crystallites, as shown in Figure 10.5. Heat treatment causes a marked improvement in layer alignment and orientation, attaining the highly ordered capsular structure of graphitized black shown in Figure 10.6. Surprisingly, while the relatively disordered structure of Figure 10.5 yields a highly reinforcing black, the structure illustrated in Figure 10.6 is essentially nonreinforcing.

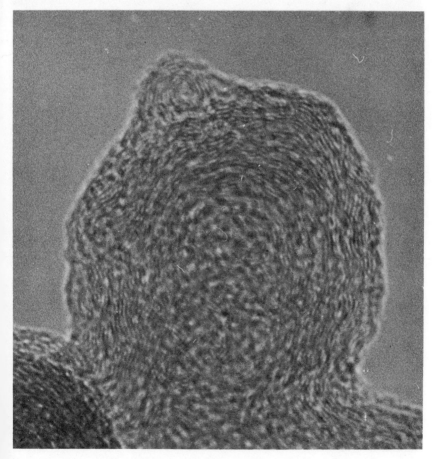

Figure 10.5. High-resolution electron micrograph of a furnace black particle which received no heat treatment. (Rivin, 1971.)

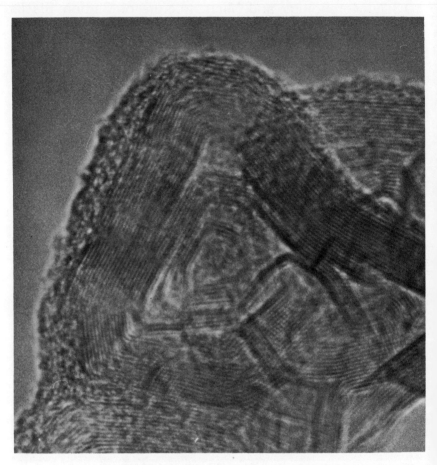

Figure 10.6. After extensive heat treatment (2700°C) the now graphitized furnace black shows improved layer alignment and orientation, but reduced reinforcing ability. (Rivin, 1971.)

10.4. AGGREGATION AND AGGLOMERATION

Up to this point only the surface chemistry and size of the primary filler particles have been examined. However, important evidence exists that these particles are not randomly distributed throughout the elastomeric matrix. Indeed, much evidence shows that a complex state of aggregation is important for reinforcement in both carbon black and silica fillers. Two levels of structure have been identified in reinforcing fillers beyond the primary particles: aggregates of primary particles, which are bonded together rather strongly, and weakly bonded collections of these aggregates, which are sometimes called agglomerates.

10.4.1. Structure in Carbon Blacks

Many types of carbon black exhibit a characteristic chainlike structure, aggregation, as shown in Figures 10.7 and 10.8 (Kraus, 1971a; Medalia, 1970; Sambrook, 1970, 1971). These structures are carried over into the rubber compound (Sweitzer, 1961). Carbon blacks exhibiting a high degree of chainlike aggregation produce high degrees of shear during mixing, which aids in proper dispersion. This type of aggregation also results in a higher modulus and in higher levels of abrasion resistance. The optimum degree of aggregation depends on the chemical nature of the elastomer, butyl compounds requiring less structure than styrene–butadiene–rubber (Ford et al., 1963).

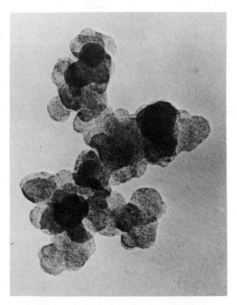

Figure 10.7. HAF carbon black exhibits a high degree of aggregation as seen by electron microscopy. (Medalia, 1970; Sambrook, 1970, 1971.)

Figure 10.8. Three-dimensional model of the particle shown in Figure 10.7, showing an 82-particle aggregate. Model constructed by Medalia (1970) and Sambrook (1970, 1971).

10.4.2. Dispersion of Carbon Black within the Elastomer*

Dispersed within the rubber as a reinforcing filler, carbon black retains much of the aggregation displayed by the pure phase. In addition, a more complex state of agglomeration develops within the rubber, as shown in Figure 10.9. These structures often appear like elongated strings or chains of interconnected carbon black particles. In a review paper, Kraus (1971) pointed out that secondary aggregation (agglomeration) is responsible for the large increase in the modulus of filled elastomers at low strains. The breakdown in agglomeration at high strains substantially reduces the modulus, usually permanently.

It should be emphasized that high reinforcement results because of, not in spite of, the aggregation of the filler particles. In Section 10.3 a model was developed that emphasized that the filler particles must be the same size as or smaller than the chain end-to-end distance for maximum reinforcement. This model can be extended to filler aggregates surrounded by polymer network chains. The close interlocking of filler aggregate and polymer network, both structures having similar dimensions, allows for maximum reinforcement.

* The corresponding problem of filling plastic materials with particulate matter and fibers will be treated in Chapter 12. These particles are usually much larger than the carbon blacks and silicas considered here.

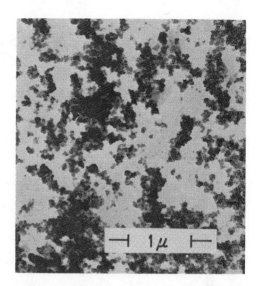

Figure 10.9. Electron micrograph of ALS–HAF in butyl vulcanizate, dry mix. The state of agglomeration can be controlled by the mixing technique employed. (Hess, 1965.)

It is common today to employ elastomer/elastomer polyblends in the manufacture of automotive tires and other products. (See Section 9.1.) This blending complicates the problem of dispersing the reinforcing carbon black properly within both phases (Boonstra, 1970; Callan *et al.*, 1971). Since the first occurrence in mixing involves penetration of the voids between carbon black particles by the elastomer, the less viscous, easier flowing phase will penetrate first. As a result, the primary carbon black aggregates contain more of this elastomer, and the amount of transfer to the other phase depends on subsequent mixing. The situation is complicated further by the tendency of the elastomers to form bound rubber (Section 10.5). Figure 10.10 illustrates typical results that can be obtained if conditions are not adequately controlled.

10.4.3. Structure in Silica Fillers

The state of aggregation and agglomeration of reinforcing silicas also appears complex. Some time ago Vold (1963) hypothesized a particle structure organized on three levels after observing the highly caducous behavior of silicas suspended in chlorobenzene. The corresponding state of agglomeration of reinforcing silica particles in organic elastomers also has been recognized for some time (Bartung *et al.*, 1964; Sellers and Toonder, 1965). Through the means of mixing dissolved silicones with reinforcing silicas followed by film formation, Chahal and St. Pierre (1969) recently showed that agglomeration also could be important in the silicone elastomers.

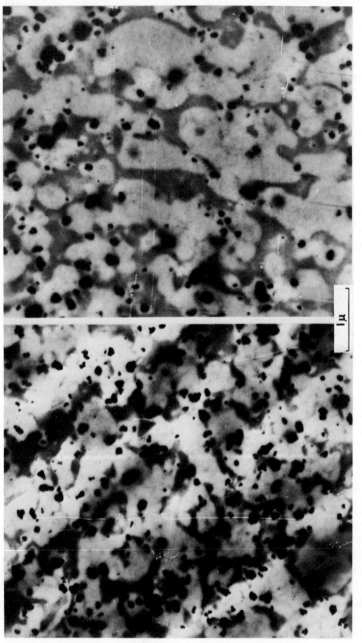

Figure 10.10. Micrograph of natural rubber masterbatch containing 80 phr of HAF black after dilution with polybutadiene and SBR, respectively. Natural rubber is the lighter phase. In the NR/BR blend (left) much of the black is in the BR phase, whereas with NR/SBR (right) all the black seems to be in the NR. (Callan *et al.*, 1971.)

In 1970, Galanti and Sperling (1970a) reported on the agglomeration characteristics of several silicas milled into silicone elastomers under conditions approaching those employed industrially. Figure 10.11 shows the state of dispersion of HiSil in silicone elastomer as a function of filler level. The smallest particles visible average $\sim 5\ \mu m$ in diameter, and appear roughly spherical. Some of these, in turn, appear clustered together further to form agglomerates of 20–30 μm in diameter. It should be observed here that the actual size of the individual particles is $\sim 200\ \text{Å}$, as seen in the electron microscope. Thus this study strongly suggests that the three-level structural organization previously observed by Vold (1963) may be carried over into elastomer dispersion. The three levels of dispersion are : (1) individual primary particles of about 200 Å diameter, (2) aggregates of $\sim 5\ \mu m$ in diameter, and (3) agglomerates of ~ 20–30 μm in diameter.

10.5. BOUND RUBBER

The phenomenon of bound rubber has played a crucial role in the chemistry of elastomer reinforcement. Very simply, if vulcanized rubber is masterbatched with a reinforcing filler, a certain fraction of the rubber is found to become insoluble, and to remain as a gel. This gel, a swollen mass of rubber and dispersed carbon particles, is called bound rubber. The percentage of bound rubber depends, of course, on the quantity of filler employed, and also on mixing conditions and choice of solvent (typically benzene). Working under standardized conditions, the percentage of bound rubber was found to correlate with the specific surface area of the filler as shown in Figure 10.12.

There are several reactions postulated as being important in bound rubber besides physical adsorption. For example, mastication under shear causes degradation of organic polymer molecules with the formation of free radicals (Casale and Porter, 1971 ; Watson, 1955):

$$\text{P (polymer)} \rightarrow 2\text{R}\cdot \text{ (initiation by shear)}$$

$$2\text{R}\cdot + \text{F} = \text{RFR (crosslinking by filler F)}$$

The percentage of bound rubber increases with increasing unsaturation of the polymer.

Attention is called to the following: (1) The size of the carbon black particles is comparable to the size of the dispersed phase in the block-copolymer thermoplastic elastomers, and (2) the physical bonding/grafting combination between the rubber and the filler functionally resembles the

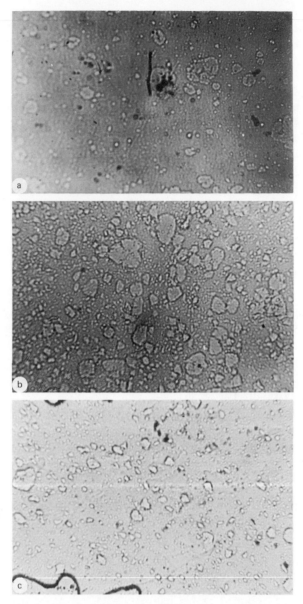

Figure 10.11. HiSil-filled silicone rubber at (a) 2.4 vol %; (b) 7.5 vol %; (c) 15.5 vol %. Magnification 100 ×. A broad particle-size range can be seen. This is clearest in (b), where the smaller particles retain their identity in the second level of agglomeration. (Galanti and Sperling, 1970a.)

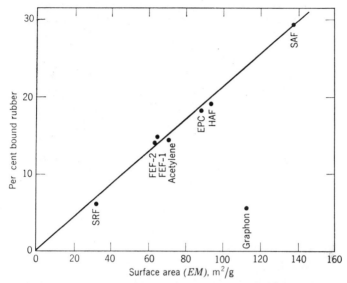

Figure 10.12. Correlation between bound rubber and specific surface area. The data are for 50 parts of black per 100 parts of SBR 1500. Graphon, a graphitized carbon, does not follow the correlation. (Kraus, 1965c.)

bonding developed between the phases in the block copolymers (see Section 4.4). This analogy will be treated further in Section 10.13.

10.6. THE MULLINS SOFTENING EFFECT

Mullins softening may be defined as the reduction of modulus caused by successive stretching (Bachmann et al., 1959; Boonstra, 1965; Bueche, F., 1965; Dannenberg, 1966; Harwood et al., 1967; Mullins, 1969; Mullins and Tobin, 1956; Sellers and Toonder, 1965). This phenomenon is characteristic of truly reinforcing fillers, but occurs to a lesser extent in gums and non-reinforced systems (Kraus, 1971).

Considering Figure 10.13, the stress–strain behavior of a virgin elastomer may be assumed as curve ABCD, with failure at D. A particular strip of this material is first stretched only to point C, and then released. On the second stretch, curve AEC will be followed, with further extension along line CD. The area encompassed by ABCEA may be replotted vs. elongation to yield characteristic Mullins softening curves, as shown in Figure 10.14. The pattern shown by type A, with a relatively sharp maximum in the softening

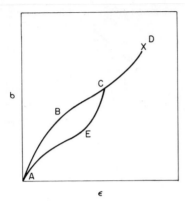

Figure 10.13. The Mullins effect on stress–strain curves in reinforced elastomers. The modulus on second and subsequent curves is lower than obtained on first extensions.

plot at higher extensions, is characteristic of highly reinforcing materials. Pattern type B is characteristic of clay, calcium carbonate, and similar nonreinforcing fillers.

Explanations of the Mullins effect have included failure of weak linkages, failure of network chains extending between adjacent filler particles, and polymer–filler bond failure, a form of dewetting. Brennan *et al.* (1969) and

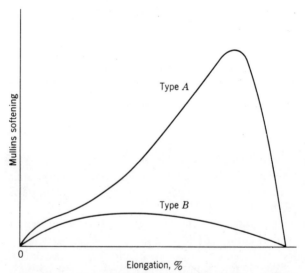

Figure 10.14. Characteristic types of Mullins softening in silica-filled rubber vulcanizates. Highly reinforcing fillers give more stress-softening than slightly reinforcing or nonreinforcing materials. (Sellers and Toonder, 1965.)

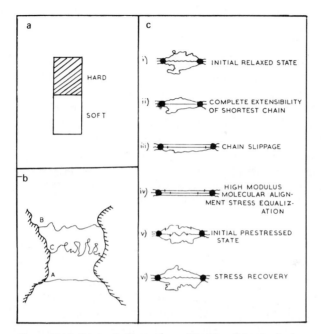

Figure 10.15. Schematic illustrations of theories of stress softening. (a) Mullins and Tobin (1956) considered the filled rubber as a heterogeneous system comprised of hard and soft phases. Deformation breaks down the hard phase, but the degree of breakdown depends on the maximum extension of the sample. (b) F. Bueche (1965) attributed stress-softening to the breakage of network chains attached to adjacent filler particles (A molecule breaks first). (c) Dannenberg (1966) and Boonstra (1965) suggested that reinforcement can be understood through chain slippage mechanisms. The slippage is shown by the chain marks. (Smith and Rinde, 1969.)

Boonstra (1970) have suggested that stress-softening is in part a rate process. Upon the initial straining of the macroscopic sample, molecular slippage mechanisms operate at the carbon–rubber interface to redistribute the stress. By so doing, the level of stress on the most highly stressed chains is reduced (Mullins, 1969). Figure 10.15 illustrates this hypothesis for several models (Harwood *et al.*, 1967). The degree of softening is related to pigment particle size, particle size distribution, and the nature of the pigment surface. The Mullins effect should be distinguished from a low-elongation softening that sometimes takes place at elongations less than about 20 % (Bachmann *et al.*, 1959). This latter is thought to be the result of a breakdown of filler agglomerate structure. (See Section 10.4.) Both phenomena are essentially irreversible.

10.7. VISCOELASTIC RUPTURE OF REINFORCED ELASTOMERS

While it is true enough that all failures start at a weak point, time- and temperature-dependent viscoelastic flow at the growing tip of the flaw controls the all-important propagation of the crack (Scott, 1967; Smith and Rinde, 1969). (See Section 1.6.) With some modification, Halpin and Bueche (1964) have applied the concept of viscoelastic flow to the rupture of reinforced elastomers. As shown in Figure 10.16, the failure envelopes obtained on a carbon black-loaded SBR rubber depend on the level of reinforcement. One finding of particular importance is that tensile strengths decrease rapidly in the high-temperature portion (lower left) of Figure 10.16. This finding was recently confirmed by Shuttleworth (1968, 1969). The failure envelope in Figure 10.16 is seen to depend on the extent of filler loading. In general, more highly reinforced elastomers are less extensible, but have a higher stress to break than less reinforced elastomers. See Section 1.6.3.

Another viscoelastic aspect of failure that Halpin and Bueche investigated involved time-to-break experiments. When dead loads are applied to polymeric materials, failure does not occur either immediately or never, but after finite lengths of time. Figure 10.17 compares a gum SBR with HAF-reinforced material. Each curve exhibits a portion in which failure

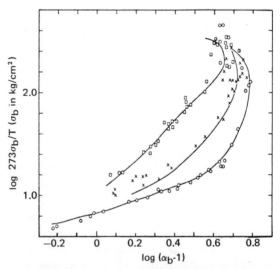

Figure 10.16. Comparison of failure envelopes for an SBR rubber as a function of reinforcement concentration. (□) 30 HAF; (×) 15 HAF; (○) 0 HAF. (Halpin and Bueche, 1964.)

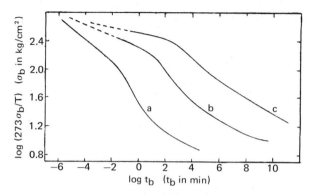

Figure 10.17. Comparison of master time-to-break curves at
$-10°C$. SBR vulcanizate containing (a) 0, (b) 15, (c) 30 parts
HAF black. (Halpin and Bueche, 1964.)

takes place in the glassy state, followed by a transition zone and then a rubbery zone, with decreasing load and increasing time. (Increasing times, of course, are equivalent to an increasing temperature.) The more highly reinforced materials are seen in Figure 10.17 to take longer times to break, or to withstand greater stress at any specified time.

These workers also showed that the apparent energy of activation of the failure process could be calculated assuming an Arrhenius mechanism. As illustrated in Table 10.3, addition of reinforcing filler raises the apparent activation energy of the viscoelastic failure processes. Halpin and Bueche ascribe the enhanced reinforcement to those processes that spread the viscoelastic motions of the filler–rubber complex over a much wider time scale, and concluded that the lower strength observed at elevated temperatures was due to the increased rate at which viscoelastic response to deformation

Table 10.3
Apparent Activation Energies in Reinforced
Systems (HAF-Filled SBR Rubbers)[a]

Percentage filler	E, kcal/mol
0	26
15	30
30	41

[a] Halpin and Bueche (1964).

occurred. One of the important consequences of the Halpin–Bueche investigation was to show that the time–temperature superposition principle was applicable to failure phenomena in reinforced elastomers.

10.8. RESTRICTION OF MOLECULAR MOBILITY BY FILLER SURFACES

If an elastomer is really tightly bound or adsorbed to reinforcing filler surfaces, should not molecular mobility of the polymer chains be restricted accordingly? In particular, should not the glass transition temperature, thermal expansion coefficient, and free volume depend upon reinforcement? Surprisingly, these properties turn out to be only modest functions of filler level.

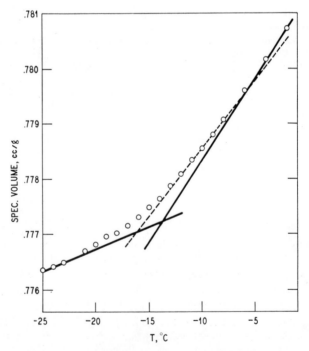

Figure 10.18. The transition region of unvulcanized styrene–butadiene polymer filled with 100 parts of HAF black. Details of transition region. (Kraus and Gruver, 1970.)

For instance, Kraus and Gruver (1970) found that the T_g of a styrene–butadiene copolymer increased only 0.2°C for every ten parts per hundred by weight of reinforcing carbon black added, and that the coefficient of thermal expansion of the polymer component in the rubbery region was substantially unaffected by the presence of filler. While Yim and St. Pierre (1969) found that the T_g of silicone rubber increased up to 8°C with the addition of 40 parts per hundred by weight of reinforcing silica, this effect is still rather modest.

A detailed diagram of dilatometric behavior of carbon black-reinforced elastomer in the vicinity of T_g shows clearly the exact effects of the filler on the transition (Figure 10.18). The solid lines in Figure 10.18 represent the least square fits of the data far from the transition, and the dashed line represents the best straight line that can be drawn through T_g of the unfilled rubber $(-16°C)$ and the experimental points in the broadened portion of the transition region. Broadening occurs only on the high-temperature portion of the transition, showing that the effect is to raise T_g. The intersection of the rubbery line at $-6°C$ presumably represents the T_g of the surface layer. Kraus and Gruver calculated from the relative changes in slopes that about one-fourth of the polymer is surface rubber, corresponding to an affected depth of about 30 Å from filler surfaces. We see here a picture of unaffected bulk rubber, and variably, but only slightly, affected rubber in the vicinity of filler surfaces. This result strengthens the hypothesis that only wetting and slight adsorption of polymer on filler surface is required for reinforcing action by filler. Further discussion on filler–matrix interactions is given in Section 12.3.

10.9. THERMODYNAMIC ASPECTS OF REINFORCEMENT

In the preceding sections the chemical, morphological, and viscoelastic aspects of reinforcement have been emphasized. While it is true enough that bonding to the polymer surface, agglomeration, and macromolecular slippage stand forth as vital aspects of reinforcement, what about the thermodynamic considerations? (Galanti and Sperling, 1970b,c.) As an elastomer becomes reinforced, its modulus, tensile strength, and swelling resistance all increase. More work is required to attain a given extension. Is this extra work stored in entropic or energetic modes?

All changes in free energy at constant volume are the sum of two terms:

$$\Delta G = \Delta E - T \Delta S \tag{10.1}$$

The basic equation of state of rubber elasticity, from Chapter 1,

$$f = nRT\left(\alpha - \frac{1}{\alpha^2}\right) \tag{10.2}$$

was derived on the major assumption that the change in internal energy ΔE is small compared to the entropic change $T\,\Delta S$. Recognition that, in general, the ΔE term was nonzero led to the incorporation of a "front-factor" term (Tobolsky and Shen, 1966; Tobolsky et al., 1961) X/X_0, which multiplies the rhs of equation (10.2):

$$f = nRT\left(\frac{X}{X_0}\right)\left(\alpha - \frac{1}{\alpha^2}\right) \tag{10.3}$$

For gum elastomers, the quantity X/X_0 relates chain dimension characteristics to *trans–gauche* energy differences and volume changes, and is usually fairly close to unity (Flory et al., 1960; Shen, 1969, 1970; Sperling and Tobolsky, 1966; Tobolsky and Shen, 1966; Tobolsky et al., 1961). When reinforcing filler is added, internal energy changes become more pronounced (Bueche, F., 1965; Flory and Rehner, 1943; Galanti and Sperling, 1970b,c) and the definition of X/X_0 must be broadened to include the sum of the internal energy changes of the filler–polymer complex. In fully reinforced elastomers, this last term is believed to dominate the X/X_0 term. It is convenient experimentally to measure the internal energy component of the retractive force f_e as the ratio f_e/f. It can be shown (Galanti and Sperling, 1970c) that

$$X/X_0 = T^{f_e/f} \tag{10.4}$$

Substitution of equation (10.4) into equation (10.3) and normalizing yields for the reduced Young's modulus

$$\log \frac{E}{E_0} = \log \frac{n}{n_0} + \left(\frac{f_e}{f} - \frac{f_{e0}}{f_0}\right) \log T \tag{10.5}$$

where the subscript zero represents the conditions existing in the non-reinforced gum elastomer. Equation (10.5) relates the modulus increase on reinforcement to changes in crosslink density, internal energy, and temperature, with no adjustable parameters.

10.9.1. Modulus Increases

Values of f_e/f vs. E/E_0 for a series of silica-reinforced silicone elastomers are shown in Figure 10.19. The most surprising feature is the relatively large value of f_e/f, which approaches 0.6 for fully reinforced materials. This means

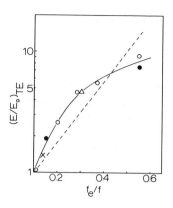

Figure 10.19. Logarithmic plot of reduced Young's modulus E/E_0 vs. f_e/f for the several loadings of HiSil, Quso, TK, and Celite. Values of E/E_0 determined from the thermoelastic data at 70°C and $\alpha = 1.55$. Dashed line represents equation (10.5); (○) HiSil; (●) Quso; (△) TK; (×) Celite; (⊙) gum. (Galanti and Sperling, 1970c.)

that over half of the retractive force is now energetic in origin. (The total retractive force f increases nearly tenfold, as suggested by the ordinate axis.)

The dashed line in Figure 10.19 was obtained from equation (10.5) assuming $n = n_0$. Inspection of Figure 10.19 shows that n/n_0, the ratio of crosslink densities of filled and unfilled materials, may have values as high as 1.5 below $E/E_0 = 4$, but then decreases to slightly below unity for the highest values of E/E_0. This behavior was taken to indicate that n is relatively constant, independent of reinforcement level—a conclusion in contrast with earlier treatments, which assumed that n increases with reinforcement, X/X_0 implicitly remaining constant (Bueche, F., 1965). If n is taken as roughly constant, it becomes obvious that equation (10.5) is incomplete, lacking as yet unidentified terms.

10.9.2. The Flory–Rehner Equation

Special insight into the question of rubber reinforcement can be gained by examining an equation developed by Flory and Rehner, which shows how swelling in networks is controlled by crosslink density. Before considering possible uses of the Flory–Rehner equation in a rubber reinforcement context, however, the basic premises of the original theory will be reviewed. The equilibrium swelling theory of Flory and Rehner (Flory, 1950; Flory and Rehner, 1943) treats simple polymer networks in the presence of small molecules. The theory considers forces arising from three sources:

1. The entropy change caused by mixing polymer and swelling agent. The entropy change from this source is positive and favors swelling.

2. The entropy change caused by reduction in numbers of possible chain conformations on swelling. The entropy change from this source is negative and opposes swelling.

3. The heat of mixing of polymer and swelling agent, which may be positive, negative or zero. The Flory–Rehner equation may be written

$$-[\ln(1 - v_2) + v_2 + \chi_1 v_2^2] = \hat{v}_1 n(v_2^{1/3} - \tfrac{1}{2}v_2) \qquad (10.6)$$

where v_2 is the volume fraction of polymer in the swollen mass, $\hat{v}_1$ is the molar volume of the solvent, and χ_1 is the Flory solvent–polymer interaction term.

The important point is that the original Flory–Rehner equation contains no thermoelastic "front factor" term, yet simple analogy with the mechanical equations shows that one must exist, even for unfilled elastomers.* The need becomes more obvious on considering reinforced elastomers, where equilibrium swelling is greatly reduced. In an empirical manner, equation (10.6) may be modified to

$$-[\ln(1 - v_2) + v_2 + \chi_1 v_2^2] = \hat{v}_1 n(X/X_0)(v_2^{1/3} - \tfrac{1}{2}v_2) \qquad (10.7)$$

This concept leads to an equation having a form similar to equation (10.5). The relationship between f_e/f and v_2 is shown in Figure 10.20 (Galanti and Sperling, 1970c). Equation (10.7) suggests that at least part of the swelling restriction is due to internal energetic processes, and not to increased crosslink density.

10.9.3. Ultimate Tensile Strength

Landel and Fedors (1965) pointed out that the ultimate tensile strength σ of an elastomer may be written under certain conditions as follows:

$$\sigma = E(t)f(\varepsilon) \qquad (10.8)$$

where $E(t)$ is the time-dependent, strain-independent modulus, and $f(\varepsilon)$ is a nonlinear function of the strain and is independent of time. If $f(\varepsilon)$ remains constant on reinforcement, the reduced tensile strength can be expressed as

$$\log \frac{\sigma}{\sigma_0} = \log \frac{n}{n_0} + \left(\frac{f_e}{f} - \frac{f_{e0}}{f_0}\right) \log T \qquad (10.9)$$

A plot of $\log(\sigma/\sigma_0)$ vs. f_e/f is predicted to be linear at constant T if n/n_0 is also constant. The result is shown in Figure 10.21. The agreement between theory and experiment for the reduced Young's modulus (Figure 10.19) was significantly better than the agreement for the reduced tensile data

---

* The analogy here is to simple extension. In both cases (stretching, swelling) a change in *trans–gauche* energy state populations is observed on deformation. If the *trans–gauche* energy states are different, an energetic term must exist.

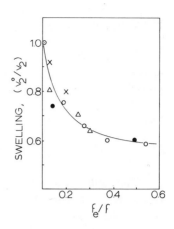

Figure 10.20. Variation of degree of swelling restriction with f_e/f. Values of f_e/f were computed from the thermoelastic data. The quantity v_2^0 stands for the volume fraction of elastomer in the swollen gum, and v_2 stands for the volume fraction of elastomer in the swollen filled polymer after correcting for the volume of the filler. ($\bigcirc$) HiSil; ($\bullet$) Quso; ($\triangle$) TK; ($\times$) Celite; ($\odot$) gum. (Galanti and Sperling, 1970c.)

(Figure 10.21). For arguments similar to the modulus case, it may be concluded that the discrepancy is not due to large variations in n. However, it should be noted that there are no undetermined constants in these equations. The temperature prediction of equation (10.9) needs to be examined further. Experimental observations show that f and hence E decrease with increasing temperature at constant time of measurement, due to kinetic effects [see theory of Halpin and Bueche (1964)] which mask the thermodynamic aspects. For the case of equilibrium swelling, there is some indication that the counterpart of equation (10.9) does hold, swelling increasing with $\log T$ (Sperling, unpublished).

One aspect that may tend to clarify the near constancy of n involves the fact that for normally crosslinked elastomers, values of n are largely controlled

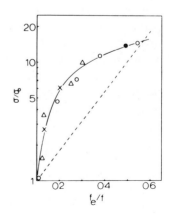

Figure 10.21. Logarithmic plot of the ratio of ultimate tensile strength of filled to unfilled rubber as a function of f_e/f. Each value of the ultimate tensile strength is an average of several such points. Dashed line represents equation (10.9); ($\bigcirc$) HiSil; ($\bullet$) Quso; ($\triangle$) TK; ($\times$) Celite; ($\odot$) gum. (Galanti and Sperling, 1970c.)

by the physical entanglements (Bueche, A. M., 1956; Meissner, 1967). True chemical crosslinks prevent molecular flow, but are usually far outnumbered by the entanglements. Thus the introduction of reinforcing filler, which may increase the number of chemical crosslinks in the system significantly, probably has only a minor effect on the total number of effective chains in the system.*

10.9.4. Mechanisms and Mechanics of Reinforcement

The above discussion teaches that reinforcement serves to augment greatly the energetic portion of the free energy change on deformation. Let us reexamine the carbon and silica agglomerate structures previously discussed and search for deformation mechanisms potentially rich in energetic shifts. These complex structures contain many filler–filler interactions involving both primary and secondary chemical bonds. These agglomerates may be bent, stretched, or have internal parts temporarily dissociated during deformation. Likewise, many polymer–filler interactions are active, and short chain segments joining two filler particles are especially subject to nonaffine deformations (Tobolsky, 1960, pp. 92, 304). The modulus inside the filler agglomerates and nearby chain structures is generally accepted as variable and higher than that of the surrounding material. It is thought that only relatively slight agglomerate deformations would be required to produce large f_e/f ratios by means of the several above mechanisms.

The theoretical treatment shows that the experimentally determined modulus and swelling changes are direct consequences of increased f_e/f values. In the case of tensile strength the situation is more complicated. In Chapter 3 on impact polyblends, it was pointed out that the inclusion of a rubbery phase increases impact resistance because the presence of rubber particles favored crazing and crack branching, bringing about a large dissipation of strain energy without causing failure. In reinforced rubber, a similar phenomenon may be happening, but with the dispersed phase stiffer than the continuous phase. The presence of agglomerate or chain filler particles in a variable-modulus medium may "confuse" a crack, causing it to divide into smaller, relatively harmless cracks. This would relieve large stress concentrations and prevent catastrophic failure. (See also Chapter 12.)

---

* This effect can be observed experimentally by measuring the shear modulus by $G = nRT$, and comparing the value of n with the actual numbers of chemical crosslinks introduced.

10.10. SWELLING ANOMALIES IN FILLED ELASTOMERS

Above, we presented a thermodynamic theory of swelling restriction in filled, reinforced elastomers. An interesting and different approach to the structure of filled elastomers involves observations of swollen systems. According to Sternstein (1972) and Kotani and Sternstein (1971), a swelling anomaly arises due, in part, to the inhomogeneous stress field generated by the restriction of swelling at the particle interface. Assuming that the rubber remains firmly bound to the particle during swelling, a radial-type strain will occur as a result of the swelling.

Besides the direct measurement of swelling restriction, light scattering studies with polarized light on the swollen elastomers yield quantitative information about the affected regions (Picot et al., 1972). The Stein and Wilkes (1969) theory predicts an enhanced H_v scattering component, where a vertically polarized incident beam is observed (after scattering) through a horizontal polarizer. In a recent examination involving both model glass beads and HiSil 233 reinforcing silica fillers, Stein and Wilkes (1969) found characteristic cloverleaf small-angle light scattering patterns, which were used to confirm the theoretical finding of inhomogeneous swelling over distances greatly exceeding the diameters of the particles.

10.11. GUTH–SMALLWOOD RELATIONSHIP

In the mid-1940's Guth (1945) and Smallwood (1944) developed a widely employed (Cohan, 1947; Kraus, 1965c, Chapter 4) equation expressing rubber reinforcement directly in terms of filler concentration. An important form of the equation can be written as

$$E = E_0(1 + 2.5v_f + 14.1v_f^2) \tag{10.10}$$

where E and E_0 are the moduli of the filled and unfilled rubber, respectively, and v_f is the volume fraction of filler in the rubber. Equation (10.10) is based on the famous viscosity equation of Einstein (1906, 1911), who considered simple dispersions of hard spheres in Newtonian liquids. Equation (10.10) correctly predicts the relatively modest increases in modulus developed by the addition of inactive or nonreinforcing fillers such as calcium carbonate, but has been found relatively unsatisfactory for highly reinforced systems, where large positive deviations occur. The equation has been employed as a correction factor in other theories (Galanti and Sperling, 1970c).

Many other equations have been applied to rubber reinforcement. Perhaps the most important include the rule of mixtures laws, the Mullins equation, and the upper- and lower-bound relations (Broutman and Krock, 1967). Since these relationships often apply more satisfactorily to plastic rather than elastomeric composites, they will be broadly developed in Section 12.1.

10.12. CRYSTALLIZING ELASTOMERS

Many of the elastomers discussed in this chapter, such as SBR, contain sufficient chain irregularity to prevent crystallization. Such materials are always amorphous, whether in the glassy or rubbery state. However, some elastomers, particularly natural rubber and *cis*-polybutadiene, crystallize on extension to great advantage. In the cured and filled state, crystallization is suppressed sufficiently that they are usually 100 % amorphous when relaxed. On extension, however, the aligned chains crystallize, stiffening the material. The crystallites also offer additional modes of strain energy dissipation, which serve to prevent crack growth. Such crystallites are, of course, well bonded to the remaining elastomeric portion. The result is that at critical stress levels, an important type of internal reinforcement is developed. It is this ability to crystallize that has caused natural rubber to remain the rubber of choice for many applications, and caused the recent large growth in the manufacture of *cis*-polybutadiene.

While the melting point of undeformed natural rubber is just over room temperature, the melting point of crystallites in stretched samples is much higher. At the highest elongations possible, natural rubber melts at about 100°C (Greensmith *et al.*, 1963, esp. p. 264), placing a temperature limitation on this mode of reinforcement.

When noncrystallizing elastomers are reinforced, modulus, tensile strength, and abrasion resistance are all greatly increased. In fact, useful properties cannot be attained without reinforcement. However, reinforcement of crystallizing elastomers produces increased initial modulus and higher abrasion resistance, but only slightly higher tensile strength and high-elongation modulus. In other words, reinforcement is greatest under conditions when the elastomer remains amorphous.

10.13. MODEL ELASTOMER–FILLER SYSTEMS

In Section 4.4 we saw that the thermoplastic elastomers behave as self-reinforcing systems. Earlier in this chapter, we examined silica- and carbon black-reinforced elastomers. In each of these materials the elastomeric matrix was firmly bound to the filler by direct chemical bonds and/or extensive physical bonds. It was also suggested that considerable reinforcement might be obtained regardless of bonding if only the particles were small enough.

Morton *et al.* (1968, 1969) and Kraus *et al.* (1970) tested this proposition employing polystyrene latexes embedded in a continuous styrene–butadiene elastomer. This system exhibits a high degree of incompatibility (see Sections 2.5 and 3.2.1), the two phases having only London forces holding them together. Each studied effects of varying filler loading, and in addition Morton *et al.* studied effects of filler size and bonding to matrix. Most of the important results were obtained using polystyrene latexes in the 400–500-Å range. Kraus *et al.* investigated the loss and storage

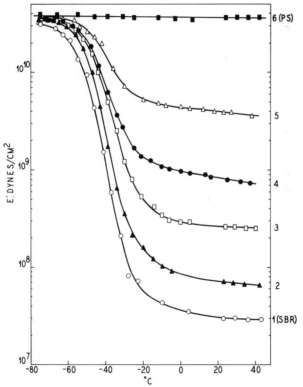

Figure 10.22. Storage moduli of polystyrene-latex-reinforced SBR. Polystyrene contents: 1, 0%; 2, 11.7%; 3, 23.7%; 4, 34.0%; 5, 45.9%; 6, 100%. (Kraus *et al.*, 1970.)

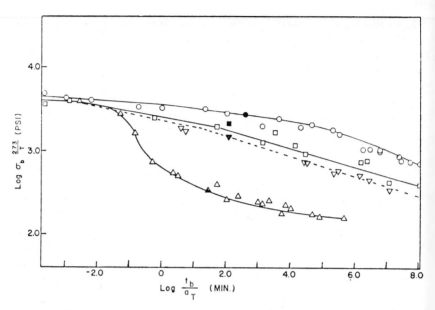

Figure 10.23. Effect of model fillers on tensile strength at 0°C. (△) SBR; (□) 15% PS; (○) 25% PS; (▽) 15% SB 10. Filled points refer to 25°C and 20 in./min. (Morton *et al.*, 1969.)

moduli of their materials, while Morton *et al.* concentrated on ultimate failure behavior.*

The storage modulus vs. temperature behavior of these model systems resembles that of the ordinary reinforced elastomers, as shown in Figure 10.22. While the T_g of the composite increases only slightly with as much as 46% polystyrene, the rubbery plateau modulus increases by two orders of magnitude! The loss modulus curves were correspondingly broadened, but the temperature of maximum loss was unaffected. Kraus *et al.* concluded that effects of reinforcing filler need not be caused by restriction of segmental motion in the rubber resulting from elastomer–filler interactions, as previously thought by some workers. The mere presence of high-modulus particles having diameters smaller than the chain end-to-end distances seems to be significant, however.

As with simple vulcanizates, the tensile strength of reinforced materials

---

* These materials also serve as models of the block-copolymer thermoplastic elastomers. In this case, post-curing of the continuous elastomer phase was required to attain normal rubber behavior. However, the polybutadiene portion is not covalently bonded to the polystyrene latex spheres.

depends upon the strain rate and temperature, and is subject to the application of the time–temperature superposition principle. Tensile strength σ_b vs. time to break t_b for the gum vulcanizates and several reinforcing compositions is shown in Figure 10.23. While no reinforcement occurs on the leftmost portion of the figure (due to the fact that both phases behave as glasses in this region), tensile strengths in the rubber region can be increased by a factor of ten by adding 25% of polystyrene. Returning to Figure 10.22, we find the increase in tensile strength matches the increase in rubbery plateau modulus for the corresponding composition almost exactly, support-ing the thermodynamic theory of reinforcement in the theoretical section, since similar X/X_0 values can be calculated. Using the failure envelope theory of Smith (1958), Morton *et al.* found that the curves of the several vulcanizates could be superimposed as shown in Figure 10.24. The failure envelope of Figure 10.24 does not show the effect of reinforcement in Figure 10.23. Morton *et al.* also found that a filler of smaller particle size shows a greater relative tensile strength at any given temperature and strain rate. The major conclusion that Morton *et al.* reached was that reinforcing fillers act by increasing matrix viscosity. (See also composite theory in Chapter 12.)

It is of interest to compare the mechanical behavior of these model reinforced elastomers with the thermoplastic elastomers discussed in the

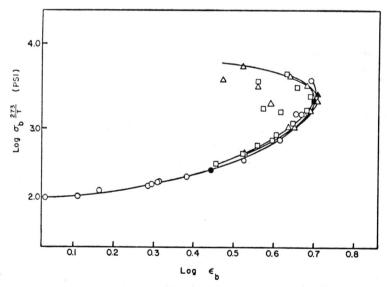

Figure 10.24. Effect of polystyrene latex filler content on failure envelope: (○) SBR; (□) 15% PS; (△) 25% PS. Filled points refer to 25°C and 20 in./min. (Morton *et al.*, 1969.)

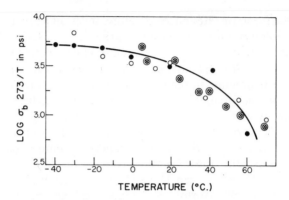

Figure 10.25. Tensile strength data for Kraton 101 (SBS triblock polymer) (●) compared with those for an SBR vulcanizate containing 25% by volume of 350-Å polystyrene latex spheres (○) (Morton and Healy, 1968) and SBR vulcanizate reinforced with 30% by weight HAF carbon black filler (⊗). (Morton and Healy, 1968; Smith, 1970.)

chapter on block copolymers. In the present context, the model reinforced elastomers may be considered as block copolymers with two changes in fine structure: (1) All chemical bonds between the blocks have been cut, and (2) the elastomeric portion has been vulcanized. The latter, of course, is required to prevent gross flow.

Figure 10.25 illustrates the similarity of tensile behavior of ABA blocks and the model reinforced elastomers (Morton and Healy, 1968; Smith, 1970). For comparison, the tensile strength of SBR containing 30% by weight HAF carbon black is also included (Bueche, F., 1965, p. 17). Within experimental error, no significant difference exists. This emphasizes again that chemical bonds between phases are not required to obtain high tensile strengths.

The major conclusion obtained in both of these investigations is that neither primary chemical bonding nor strong secondary physical forces are required for simple reinforcement. It is expected that these results will discomfit those workers whose main theme centers on strong filler interactions, although the latter obviously do play an important role.

10.14. EFFECT OF PRESSURE ON REINFORCEMENT

In a unique experiment, Tschoegl (Tschoegl, 1971a,b; Lim and Tschoegl, 1969) recently showed that in model reinforced systems where little or no wetting occurred (glass beads, for example), vacuole formation could be reduced and reinforcement increased by application of external hydrostatic pressures. At the level of several thousand pounds per square inch, even normally unbonded materials became highly reinforcing. The model system studied by Tschoegl is of interest to the solid propellant industry (Oberth, 1967; Schwarzl, 1967). During combustion, the propellant is subjected to several thousand pounds per square inch hydrostatic pressure in compression and the normally nonreinforcing filler becomes effective (see Figure 10.26). The interesting note is that the hydrostatic pressure must prevent the formation of vacuoles, which serve as weak spots and thus encourage failure. As long as the polymer and filler remain in intimate contact, proper transmission of stress is retained and the material remains reinforced. This point will be further developed in Section 12.1.

Tschoegl's result is especially interesting in the light of a recent proposal by Shuttleworth (1968, 1969) that equilibrium polymer–filler debonding is responsible for decreased tensile strength at elevated temperatures. This is contrary to the viscoelastic mechanism of high-temperature failure of Halpin and Bueche (1964), which was developed in an earlier section. A possible resolution of the relative importance of the two proposed mechanisms could lie in the application of Tschoegl's experiment to carbon black- or silica-reinforced materials.

10.15. RELATIVE IMPORTANCE OF REINFORCEMENT MODES

In the foregoing, several different modes of elastomer "reinforcement" were discussed and developed separately. A brief critical analysis of the relative importance of the several modes will now be given.

1. Polymer–filler bonds. Since copious amounts of surface area are required for maximum bonding, the filler particles must be small. Both primary (chemical) and physical (hydrogen, van der Waals) bonds are important. While scientists are divided in their views, at least some chemical bonding appears necessary for maximum toughness. However, several weak bonds joining the same chain to a particle yields the capability of controlled debonding under stress, allowing otherwise destructive amounts of mechanical energy to be absorbed chemically. While some debonding is irreversible (the Mullins effect), other modes are reversible.

2. Structure in the filler. Both silica and carbon black form various but different!) states of aggregation and agglomeration, with interparticle bonding. A controlled amount of nonrandom dispersal appears to impart

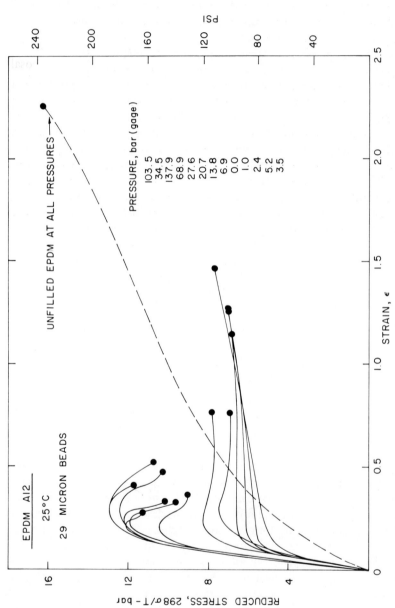

Figure 10.26. Stress–strain curves under different hydrostatic pressures; $v_f = 0.430$; the dots at the ends of the curves are break points. (Lim and Tschoegl, 1969.)

a range of moduli on the microscopic scale, causing any crack to grow by a nonuniform mechanism. Probably more important, however, the filler–filler bonds can absorb vast amounts of deformational energy, toughening the material.

3. The very presence of the high-modulus filler imparts a certain stiffness to the system, as given by the lower-bound-type Guth–Smallwood relationship. In normally reinforced systems, however, the thermodynamic reinforcement mode, by way of internal energy increases, is much more important. Thermodynamic reinforcement acts in addition to the Guth–Smallwood mode. While the presence of the filler tends to increase the crosslink density slightly, this is apparently not a major cause of elastomer modulus increases.

It must be emphasized that reinforced elastomers are tough materials because of the simultaneous operation of multiple reinforcement mechanisms. The several mechanisms interact during stress so that when one is "exhausted," another comes to the fore. Such a multiplicity also ensures that no matter what the type of stress imposed, one or more mechanism is always available to absorb the energy and thus resist failure.

Filled Porous Systems

11.1. INTRODUCTION

In most of the two-component systems discussed so far, such as rubber-reinforced polyblends, the density or specific volume is approximately an average of the values for each component. There may be some exceptions to this generalization; for example, if the smaller component fills free volume available in the major phase (Chander, 1971; Harmer, 1962; Huang and Kanitz, 1969). A similar but more important phenomenon exists when the volume available for filling by a monomer comprises not only "free" volume elements, but also gross pores, which may range in size from tens of angstroms to the order of micrometers or more. Examples of matrices may include partially sintered polymers, ceramics or metals, cement, concrete, minerals and rocks, paper, and wood (American Chemical Society, 1973). Clearly such systems tend to be complex; even the matrix itself is often a multiphase material.

The properties of filled porous systems must, for a given matrix type, depend on several factors: (1) the degree of porosity, (2) the nature of the porosity (size, degree of pore connectivity, shape, and distribution of sizes), (3) the properties of the polymeric filler, including its state, and (4) the nature of the filler–matrix interface.

Rationalization and prediction of behavior will, of course, depend on the classification that applies, e.g., interpenetrating networks or discontinuous phases in a matrix (American Concrete Institute, 1973a; Krock, 1966). The classification itself must be general, and not specific to polymers, metals, or ceramics. When porosity is negligible, models such as those discussed in Chapter 12 may be appropriate; otherwise the effects of porosity and of changes in porosity must be taken into account.

In general, because of the strong interest in the practical behavior of this class of two-component materials, technology tends to have advanced more rapidly than fundamental understanding. The discussion that follows summarizes our state of knowledge about the preparation and behavior of filled porous composites, and indicates possible directions for future research. Major emphasis in this chapter will be placed on polymer-impregnated concrete and polymer-impregnated wood, which have been the systems most extensively studied.

11.2. POLYMER-IMPREGNATED WOOD

For many years, modified woods have been prepared by impregnation with prepolymers such as phenolic resins, followed by curing under heat and pressure. By this means considerable improvement in dimensional stability and mechanical properties may be obtained (Tarkow et al., 1970), and products of this type are often encountered in articles of commerce, such as knife handles. In general, densely crosslinked thermosetting resins have been used in these applications. The structure of wood itself is briefly considered in Section 9.9.

In the 1950's, as part of a broad program to develop applications for nuclear irradiation, the Division of Isotopes Development, U.S. Atomic Energy Commission, sponsored a number of projects on the radiation polymerization of vinyl monomers. One of the projects was concerned with the impregnation of wood with monomers such as methyl methacrylate, and polymerization in situ using gamma irradiation (Anon, 1962). Preliminary results of a generally similar study by a Russian group had been published previously (Karpov et al., 1960). As with the phenolic impregnations, mechanical properties and dimensional stability were found to be much improved in comparison with the parent wood; indeed, the combination of properties was, for many applications, apparently superior to that of the earlier impregnated woods.

Following the initial reports, research and development was intensified, especially at West Virginia University (Boyle et al., 1971; Kent et al., 1968) and Syracuse University (Duran and Meyer, 1972; Langwig et al., 1968, 1969; Meyer, 1965; Siau and Meyer, 1966; Siau et al., 1965a, 1965b, 1968), where major programs have been conducted on the impregnation process, polymerization kinetics, effects of dose rate and other factors on conversion, and the physical and mechanical properties of the wood–plastic composites [sometimes abbreviated as "WPC's" (Boyle et al., 1971; Kent et al., 1968)]. Other programs were developed elsewhere, both in the U.S. (Feibush, 1963) and in wood-producing countries such as Canada, Finland, Germany, Austria, Czechoslovakia, and Hungary. A voluminous literature has developed, for which reference may be made to conferences of the International Atomic Energy Agency (1963, 1968) (also Kinell and Aagaard, 1968). As a result of the initial studies, commercial processes were developed for the manufacture of polymer-impregnated wood (e.g., "Novawood"), using both gamma and electron irradiation—the latter being especially adaptable to the polymerization of relatively thin layers of imbibed monomer, as in flooring (Taylor *et al.*, 1972).

Polymerization may also be effected by conventional free-radical initiation (Boyle *et al.*, 1971; Duran and Meyer, 1972; Feibush, 1963; Karpov *et al.*, 1960; Kent *et al.*, 1968; Langwig *et al.*, 1968, 1969; Meyer, 1965; Siau and Meyer, 1966; Siau *et al.*, 1965a, 1965b, 1968), but, as with polymer–concrete combinations discussed below, the use of irradiation rather than chemical initiators to generate free radicals for polymerization gives a somewhat better combination of properties in the composite (Siau and Meyer, 1966). Irradiation may be more effective than, say, peroxide initiation in the induction of grafting of the growing polymeric radicals to the matrix; the lower temperature used for irradiation will also minimize the loss of monomer by evaporation. Since most research has been concerned with irradiation, much of the discussion following will be concerned with irradiated systems. However, the principles should be generally similar for systems initiated by other means.

11.2.1. Experimental Conditions

Although a typical wood contains about 50–70% voids, the voids are interconnected in a very tortuous manner so that permeation of a monomer, even a highly fluid one, is difficult (Stamm, 1964, p. 363), and only part of the water initially contained in the pores can be replaced. To improve the permeation, the wood is evacuated (at a pressure of say, 5 mm Hg), excess monomer is added, pressure is applied using an inert gas, and the wood is soaked in the monomer for several hours. The monomer-impregnated specimen is then subjected to irradiation until polymerization of the monomer is essentially complete (90–100% conversion); the specimen is then dried in air, or in a kiln.

Many types of hardwoods and softwoods have been used, including oak, pine, poplar, birch, elm, and sycamore (Boyle *et al.*, 1971; Hills *et al.*, 1969). Not surprisingly, the response to impregnation is quite dependent on the nature of the wood selected. Thus, for a series of woods, polymer loadings of from 5 to 110% were obtained (Hills *et al.*, 1969), depending on the wood and on the monomer. Methyl methacrylate, styrene, and styrene–acrylonitrile combinations are among some of the monomers used successfully. If a monomer mixture is used, the composition of the copolymer will, of course, be heterogeneous unless the monomers have appropriate reactivity ratios (Hills *et al.*, 1969). Additives such as methanol may also be used to enhance the imbibition of monomer or the polymerization rate (Kenga *et al.*, 1962; Ramalingam *et al.*, 1963).

11.2.2. Polymerization Characteristics

Whether initiated by radiation or by the thermal decomposition of free radical initiators, and whether in the bulk or in the imbibed state, the mechanisms of free radical polymerizations of monomers in wood should be essentially similar. As with any free radical polymerization, three basic steps must be involved: initiation, propagation, and termination; also, chain transfer reactions may occur, depending on the monomer, additives, and on the mode of initiation (Chapiro, 1962; Siau *et al.*, 1965a; see also Appendix A, Chapter 1). In such cases, the rate of polymerization should depend on the square root of the concentration of initiating radicals, which, in turn, should depend on the dose or on the concentration of free radical initiator:

$$R_p = K[\text{R}\cdot]^{1/2}$$
$$R_p = K'[\text{DR}]^{1/2} = K''[\text{I}]^{1/2}$$

(11.1)

where R_p is the rate of polymerization, K, K', and K'' are constants, $[\text{R}\cdot]$ is the concentration of initiating radicals, and $[\text{DR}]$ and $[\text{I}]$ are the dose rate and initiator concentration, respectively. Expressions of this type may be expected to hold only at low rates and at low conversions, i.e., under conditions that minimize complications such as autoacceleration (see below) or adiabatic temperature rises.

Kent *et al.* (1968) have, in fact, observed the square-root dependence at low dose rates and low conversions. However, as would be expected, at high dose rates, which raise the concentration of radicals and favor parasitic termination steps, the exponent is observed to drop below 0.5. At high conversions, which favor diffusion control of the reaction and a lowering of the rate of termination with respect to propagation, the rate and the exponent increase sharply. Typical results of further experiments (Kent *et al.*, 1968) on dose rate and total dose are shown in Figures 11.1 and 11.2. At a moderate dose rate, 0.1 Mrad/hr, an effect of the wood matrix may be noted. Although white pine and yellow poplar exhibit similarly shaped curves, the pine is a less effective substrate than the poplar—a phenomenon attributed to the presence of resinous inhibitors in the pine. For another typical system, birch impregnated with methyl methacrylate, increasing the *dose rate* was found to

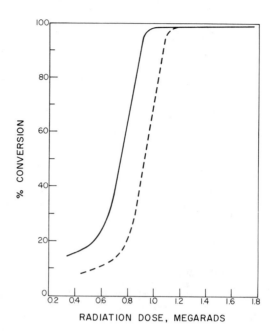

Figure 11.1. Radiation polymerization of methyl methacrylate in white pine and yellow poplar at 0.11 Mrad/hr; (—) poplar; (--) pine. [Adapted from Kent *et al.* (1968).]

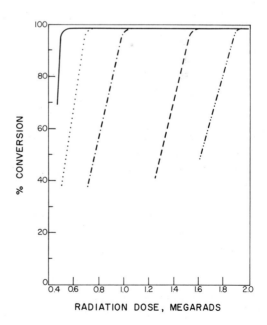

Figure 11.2. Radiation polymerization of methyl methacrylate in birch at several dose rates in Mrad/hr: (—) 0.01; (···) 0.04; (- ·) 0.11; (– –) 0.43; (– ··) 0.82. [Adapted from Kent *et al.* (1968).]

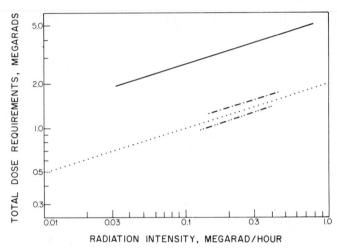

Figure 11.3. Effect of radiation intensity on the total dose requirements for complete polymerization of various systems: (– · – ·) 80/20 ethyl acrylate/acrylonitrile; (—) 60/40 styrene/acrylonitrile; (· · ·) methyl methacrylate; (– · · – · ·) 12/88 Phosgard/methyl methacrylate. [Adapted from Kent *et al.* (1968).]

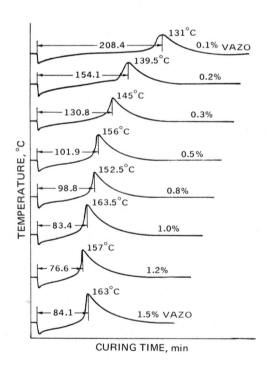

Figure 11.4. Temperature–time curves. Effects of varying concentrations of Vazo catalyst on the polymerization exotherm of basswood–methyl methacrylate composite. (Duran and Meyer, 1972.)

reduce the residence time for polymerization (curve not shown), but to increase the *total dose* required, because increasing the free radical concentration increases the rate of termination (Chapiro, 1962). The total dose required for this system varied from about 1 to 2 Mrad, as the dose rate was varied from 0.1 Mrad/hr to 1 Mrad/hr; in comparison, the total dose required for a birch–styrene–acrylonitrile system was about 5 Mrad at a dose rate of 1 Mrad/hr. In each case, the total dose required varied with the dose rate to the 0.3 power; deviations were observed, however, for white pine. Also, at a given dose rate the styrene-containing systems were consistently less responsive to radiation, as may be expected in view of the generally sluggish response of styrene itself to irradiation (Chapiro, 1962).

Polymerization of vinyl or methacrylic monomers (especially in conjunction with crosslinking monomers) within the wood often results in an autoacceleration during the latter phase of the polymerization; this phenomenon is known as the Trommsdorff or "gel" effect in homopolymerization reactions (Duran and Meyer, 1972; Trommsdorff *et al.*, 1948). The gel effect arises from a decrease in the termination rate of the free radical polymerization, caused in turn by the effect of the local viscosity on the diffusion rates of the growing polymer chains. Since the heat of polymerization cannot be removed rapidly enough to maintain isothermal conditions, autoacceleration is characterized by a strong exotherm; the intensity of the exotherm depends on the catalyst level, as illustrated in Figures 11.4 and 11.5 (Siau *et al.*, 1968).

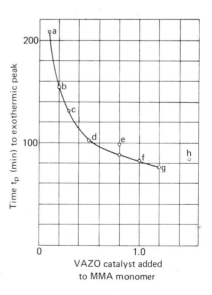

Figure 11.5. Time to exothermic peak t_p vs. percent Vazo catalyst added to basswood–methyl methacrylate (MMA) composite. Exothermic peak temperatures and percent monomer conversions are: (a) 131°C, 85.4%; (b) 139.5°C, 85.8%; (c) 145°C, 87.8%; (d) 156°C, 86.6%; (e) 152.5°C, 85.9%; (f) 163°C, 83.5%; (g) 157°C, 78.7%; (h) 163°C, 78.0%. (Duran and Meyer, 1972.)

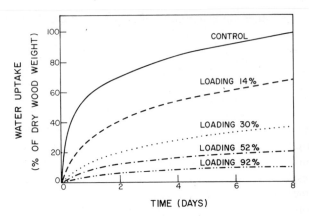

Figure 11.6. Water uptake with increasing time for sycamore impregnated with 60:40 styrene/acrylonitrile. (Adapted from Hills *et al.*, 1969.)

11.2.3. Properties: Dimensional Stability

As mentioned above, one of the major advantages of a wood–plastic composite is its improved dimensional stability (see also Section 11.3). When saturated with water, dry wood swells by about 45%, based on volume. Swelling or deswelling of this magnitude implies either a corresponding dimensional change, if the specimen is unrestrained, or severe splitting or warping, if restrained by external forces. In general, as shown by Karpov *et al.* (1960) and Kinell and Aagaard (1968), impregnated wood absorbs only about 25% as much water as the corresponding untreated wood; even lower percentages may be observed in individual cases (see Figure 11.6). As a result, radial and longitudinal changes in dimensions are reduced by about 75% in impregnated woods. The subject of dimensional stabilization has been reviewed by Siau *et al.* (1965*a*).

Improvements of this magnitude in dimensional stability may lead to a diversity of applications, e.g., cutlery handles, industrial patterns, flooring, veneers, sports equipment, and even musical instruments.* It should be noted, however, that even partial rehydration may in some cases lead to problems with polymers, such as poly(methyl methacrylate), that tend to bloom and

---

* Hills *et al.* (1969) refer to two curious applications especially suited to the damp climate of England. One is to golf club heads, which are often stored under moist conditions with minimum ventilation. The second is to organ consoles in old cathedrals and churches where the atmosphere has become more variable since the recent introduction of central heating. In fact, one wood–plastic composite console is said to have been installed in an English cathedral.

whiten. It seems that the phenomena of dehydration and hydration may be worth further study because of their importance to appearance and presumably other properties.

11.2.4. Properties: Mechanical

As shown in Table 11.1, mechanical properties of a wood–plastic composite are generally improved in comparison with those of an untreated wood. The major improvements are in compressive strength (increased by a factor of up to $1.5 \times$), shear strength ($2 \times$), hardness ($11 \times$), and abrasion resistance ($8 \times$) (Hills et al., 1969; Langwig et al., 1968, 1969; Meyer, 1965; Siau and Meyer, 1966; Siau et al., 1965a, 1965b, 1968).

One property of considerable importance in applications such as patterns and flooring is hardness. Table 11.2 gives results of an investigation into the effects of various loadings of several monomers on resistance to indentation, using a standardized test (Hills et al., 1969). Even a relatively small polymer loading (14–16 wt %; dose, 0.5 Mrad) gave a significant increase in the resistance to penetration. At maximum loading (73–90 wt %; dose 2 Mrad), values of the depth of penetration by a standard indentor for a given time and load were decreased by a factor of three or four, depending on the monomer system used, with a styrene–acrylonitrile system showing the greatest effect. It may be noted that specimens of wood–plastic composites may be machined and polished to a high gloss, and that wear simply exposes a fresh surface of the composite rather than a bare substrate as in the case of a worn protective coating. In order to minimize costs, partial impregnation may be used.

Table 11.1
Properties of Wood–Plastic Composites[a]

Property	Improvement, %
Compression strength	70–140
Static bonding strength	40–70
Elastic modulus	12–33
Tensile strength	20–65
Shear strength: radial	25–95
tangential	30–110
Hardness: radial	500–1100
tangential	500–1100
longitudinal	400–600
Abrasion resistance	100–800

[a] Hills et al. (1969).

Table 11.2
Effect of Dose and Impregnant in Sycamore on Resistance to
Indentation[a]

Impregnant	Dose, Mrad	Average penetration[b]
Control: a	—	637
b	—	611
Styrene/acrylonitrile 60:40	0.50	541
	0.75	360
	1.00	257
	1.25	241
	1.50	227
	2.00	166
Styrene/acrylonitrile 70:30	0.50	348
	0.75	358
	1.00	317
	1.25	315
	1.50	146
	2.00	147
Methyl methacrylate	0.50	524
	0.75	400
	1.00	300
	1.25	275
	1.50	259
	2.00	155

[a] Hills et al. (1969).
[b] Diameter of indentor, 1/16 in.; load, 300 g for 30 sec.

Although relatively few stress–strain curves for WPC's have been published, several have been given recently in papers by Meyer and co-workers (Langwig et al., 1968, 1969; Siau et al., 1968). As shown in Figure 11.7, tests of basswood impregnated with poly(methyl methacrylate) indicate a slightly higher elastic modulus and significantly higher tensile strength, as reported by others (Hills et al., 1969). However, the toughness, as indicated by the area under the stress–strain curve, was also increased significantly in compressive and bending tests (see Section 11.3.3.1 for a discussion of analogous curves for polymer-impregnated concrete). Some effects of monomer structure are also described in Figure 11.8 (Langwig et al., 1969), the greatest improvement in properties being noted for a crosslinked epoxy resin and for a linear polymer of t-butylstyrene, which is low in volatility, less prone to evaporate, and less prone to shrinkage during polymerization than methyl methacrylate.

Siau et al. (1968) have suggested that an upper-bound equation (Section

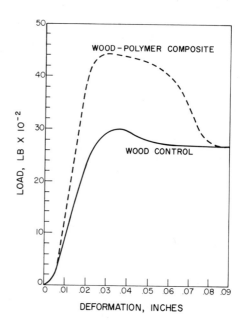

Figure 11.7. Example of typical compression test data for basswood impregnated with poly(methyl methacrylate). (Adapted from Langwig et al., 1968.)

12.1.1.1) should hold for polymer-impregnated wood:

$$E_c = E_w v_w + E_p v_p \tag{11.2}$$

where E and v refer to elastic moduli and volume fractions, respectively, and the subscripts c, w, and p refer to the composite, cell wall material in the wood, and polymer. The composite was assumed to consist of parallel, hollow columns of wood fibers and polymer, perfectly bonded to each other

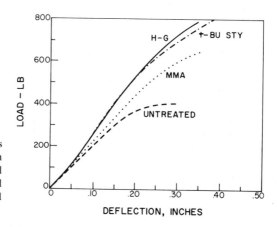

Figure 11.8. Load–deflection curves for basswood impregnated with an epoxy resin (H-G), poly(t-butyl styrene) (t-BuSty), and poly(methyl methacrylate) (MMA). (Adapted from Langwig et al., 1969.)

along their lengths and subject to equal strains along their axes. Assuming that the value of E_w is given by

$$E_w = E_g/v_w \qquad (11.3)$$

where E_g is the modulus of elasticity of gross untreated wood, good agreement between prediction and experiment was found.*

Theoretical equations were also derived for dielectric constant and ac resistivity using the same model, and again reasonable agreement was found. Thus a major effect of polymer impregnation appears to be due to the filling of void space. As predicted by Equation (11.2) for Young's modulus, the properties of the polymer must also play a significant role (see also Section 11.3.3.5). As shown in Figure 11.8, an epoxy resin is more effective than t-butylstyrene in raising the tensile strength at a lower concentration. However, quantitative comparisons are not possible at this time.

11.2.5. General Comments

Even a cursory examination of the literature reveals that the correlation of physical properties with polymerization and other process conditions has received a great deal of attention. Thus effects of dose and dose rate have been extensively studied in relation to properties. With a few exceptions [for example, Siau et al. (1965a) and Timmons et al. (1971)], less attention appears to have been given to the structure and morphology in a fundamental way. Studies of such topics as hydration and dehydration behavior, heterogeneity of loading [consider, for example, variations in local densities noted by Hills et al. (1969)], detailed microstructure, and models for mechanical behavior should be worthwhile in seeking to guide research that is otherwise necessarily rather empirical.

11.3. POLYMER-IMPREGNATED CEMENTS, MORTARS, AND CONCRETES

It has been known since the 1950's that the incorporation of various polymers into portland cement mortars and concretes can considerably improve many physical and chemical properties (American Chemical Society, 1973; American Concrete Institute, 1972, 1973a, 1975; Auskern and Horn, 1971; Coughlin et al., 1970; Dikeou et al., 1969, 1971, 1972; Dow Chemical

---

* This model is analogous to the porosity model for polymer-impregnated concrete discussed in Section 11.3.3.5.

Co., n.d.; Gebauer and Coughlin, 1971; Steinberg, 1973; Steinberg *et al.*, 1968, 1969; Wagner, 1965, 1966, 1967). For example, both tensile and compressive strengths can be increased by factors of from three to five; in addition, the resistance to freeze–thaw cycling, to water permeation, and to degradation by inorganic acids, sulfate ions, or water, is considerably increased. This section is concerned with polymer-modified cement mortars (i.e., cement plus sand) or concretes (i.e., cement, sand, and aggregate), not with so-called "resin" or "polymer" concretes, in which resins are used to bind together materials such as silica. For discussions of the latter composites, often referred to as "PC's," see Dikeou *et al.* (1971, 1972), Solomotov (1967), Steinberg (1973), and Steinberg *et al.* (1968, 1969). Such materials are essentially particulate composites (Chapter 12) with the polymer as the minor constituent.

Polymers may be incorporated basically in three ways: (a) by adding a polymerizable monomer to a concrete or mortar mix, and then curing both concrete and polymer; (b) by adding a latex or an aqueous solution of a polymer to a mortar, or concrete mix, and then curing the composition in the presence of the polymer; and (c) by impregnating a cured mortar or concrete with a monomer, and then polymerizing the monomer using thermal or radiation catalysis.

Types (a) and (b) are commonly referred to as "PCC," and type (c) as "PIC."

Differences do exist among these three types of material, depending on the method of incorporation. The curing of a cementitious material involves the hydration of cement constituents; the presence of an additional phase may alter the hydration behavior and lead to properties different from those that one would get by impregnating an already cured material. Thus impregnation polymerization tends to increase the elastic and flexural moduli, while the use of a polymer latex in a mortar mix tends to increase the elastic modulus in some cases but decrease it in others. In general, maximum levels of properties tend to be obtained in materials impregnated after curing.

These favorable results have led to several major research and development programs, mainly in the U.S.A., Europe, and Japan. Latexes based on copolymers of vinylidene chloride, and of styrene with butadiene, are available commercially, and the Atomic Energy Commission and the Office of Saline Water have supported a great deal of research and development on impregnated concretes at the Brookhaven National Laboratory and at the laboratory of the Bureau of Reclamation. The use of modified cements for specialized purposes such as the surfacing of bridge decks and for the repair of masonry is well established, and major research and development programs on highway applications have been initiated by the Federal Highways Administration and state highways organizations. In the following sections, typical preparation methods and resulting properties will be discussed, and models that have been proposed to describe observed behavior will be presented.

11.3.1. Monomer-Impregnated Cement Mixes

In principle, the addition of monomers or prepolymers to fresh concrete mixes is an attractive route to a polymer-modified concrete. Monomers studied have included water-soluble monomers such as acrylamide and substituted acrylamides, and water-insoluble monomers such as acrylonitrile, vinyl acetate, acrylates, epoxies, and polyester–styrene mixtures (DePuy et al., 1973; Dikeou et al., 1971, 1972; Gebauer and Manson, 1971; Steinberg et al., 1968, 1969). Results, however, have generally been not promising. In most cases, strength properties were not improved significantly,* though a reduction in water absorption was sometimes noted. In the authors' laboratories, a slight increase in strength ($\sim 20\%$) was observed on using a substituted acrylamide (Gebauer and Manson, 1971), but the water absorption was not decreased to a useful level.

Possible reasons for this behavior are not difficult to envisage (Dikeou et al., 1971, 1972). Homogeneous mixing is difficult to achieve, though it may be accomplished in some cases by grinding the monomer and clinker together during the preparation of the cement powder (Gebauer and Manson, 1971). Monomer may interfere with the hydration reactions; differential thermal analysis has confirmed this in the case of methyl methacrylate (Coughlin et al., 1970; Gebauer and Coughlin, 1971). Even if no specific interference occurs, as with styrene, the styrene still acts to dilute the water and slow the hydration (Coughlin et al., 1970). Degradation of monomers also is a problem in some cases due to the strong alkalinity characteristic of a cement mix. Thus a successful monomer must not interfere with hydration, must be chemically stable in an alkaline medium, and must itself be polymerizable in the presence of water and base. Though epoxy resins do enhance compressive strengths (Dikeou et al., 1972; Nawy et al., 1975) and styrene does improve corrosion resistance (Gebauer and Coughlin, 1971), the ideal combination has not yet been found.

11.3.2. Latex–Concrete Mixtures

The incorporation of a latex (an aqueous emulsion of a polymer) into a cement mix makes it possible to achieve both an intimate blend of the

---

* This observation conflicts with claims in the voluminous patent literature, which, however, tends not to provide quantitative data for comparison.

Table 11.3
Variation of Compressive and Tensile Strength of Portland Cement Mortars
with Polymer Latex Modification[a]

Polymer latex type	Compressive strength, psi	Tensile strength, psi
I None	4930	250
II Poly(vinylidene chloride-co-vinyl-chloride)	8800	970
III Poly(acrylic ester)	3665	565
IV Poly(styrene-co-butadiene)	5610	760
V Polystyrene	4275	345
VI Poly(vinyl chloride)	4800	405
VII Poly(vinyl acetate)	3040	600

[a] After Wagner (1965), with permission from *Ind. Eng. Chem., Prod. Res. and Dev.* **4**, 191 (1965). Copyright by The American Chemical Society.

polymer particles with the cement gel and an acceptable degree of hydration (Pierzchala, 1969; Wagner, 1965, 1966, 1967). Parasitic reactions between a monomer and cement constituents are thus avoided. Many different kinds of polymers have been used: latexes of such polymers as vinylidene chloride copolymers (Dow Chemical Co., n.d.), styrene–butadiene copolymers (Dow Chemical Co., n.d.), acrylates (Wagner, 1965, 1966, 1967), and epoxy resins (Sussman, 1970); dispersions of a melamine resin (Aignesberger *et al.*, 1969), have also been studied. Typical mechanical properties are given in Table 11.3 (Wagner, 1965, 1966, 1967).

Table 11.4
Adhesion Values of Cement Paste to Silica[a]

Composition,[b] parts by weight	Mean value of force to rupture, kg
100 PC, 32 W	6.1
100 PC, 16 poly(vinylidene chloride-co-vinyl-chloride),[c] 16 W	42
100 PC, 16 Neoprene, 16 W	15
100 PC, 16 poly(acrylic ester), 16 W	25
100 PC, 16 poly(vinylidene chloride-co-vinyl-chloride),[d] 16 W	16
100 PC, 16 polyethylene, 16 W	11
100 PC, 16 poly(styrene-co-butadiene), 16 W	20

[a] After Wagner (1967), with permission from *Ind. Eng. Chem., Prod. Res. and Dev.* **6**, 225 (1967). Copyright by The American Chemical Society.
[b] PC, portland cement. W, water.
[c] Film-forming.
[d] Non-film-forming.

The improvements in properties noted may be linked to several factors: the reduction made possible in the water/cement ratio, a reduction in gross porosity, the entrainment of air as a fine dispersion, and specific interactions or bonds with the cement "gel." Using electron microscopy, Wagner (1965, 1966, 1967) and Aignesberger et al. (1969) have shown that melamine and vinyl resins form networks that interpenetrate the network formed by the cement gel. Also, using similar techniques, Pierzchala (1969) concluded that poly(vinyl acetate) effectively becomes a constituent of the cement gel, and makes it chemically resistant, even to hydrofluoric acid.

A series of comprehensive studies by Wagner (1965, 1966, 1967) casts additional light on the question of mechanism. Assuming that adequate hydration is achieved (e.g., by preventing water evaporation or imbibition by a dry substrate), Wagner (1965, 1966, 1967) showed that the strength of the adhesive bond between sand and matrix is significantly increased by incorporation of a polymer in a mortar mix. (See Tables 11.3–11.6.) Using ingeniously designed tests, Wagner was able to determine bond rupture strengths for the adhesion of matrix to silica (silica gel being used to simulate sand) and for the adhesion of latex films to hydrated cement particles.

Several observations are worth noting. The incorporation of a non-film-forming polymer, i.e., an emulsion that does not coalesce on drying to form a continuous film, resulted in an increase in the mean rupture force of the cement-to-silica bond by a factor of from two to three. However, better results were obtained by the use of a film-forming emulsion, which led to forces up to seven times the values for the ordinary cement. This result is consistent with the conclusions of Aignesberger et al. (1967), Pierzchala (1969), and others. On the other hand, much smaller variations were seen in the values found for the strength of the *polymer*–cement bonds (Table 11.5).

Table 11.5
Adhesion Values of Polymer to Hydrated Cement[a]

Latex	Mean force to rupture, kg
Poly(vinylidene chloride-co-vinyl-chloride), film-forming	13
Neoprene	6.5
Polyacrylic[b]	12
Poly(vinylidene chloride-co-vinyl-chloride), non-film-forming	7.5
Polyethylene	5
Poly(styrene-co-butadiene)	9.5

[a] After Wagner (1967), with permission from *Ind. Eng. Chem.* **6**, 225 (1967). Copyright by The American Chemical Society.
[b] Only partial coverage of surface by polymer film noted.

Table 11.6
Relative Modulus of Rupture[a] Following Sealed Hardening, as Related to Matrix–Silica Adhesion[b]

Polymer type	Matrix–silica adhesion value, kg	Modulus of rupture, at polymer level		
		High	Intermediate	Low
Poly(vinylidene chloride-co-vinyl-chloride), film-forming	42	29	10	8
Polyacrylic	25	29	20	15
Poly(styrene-co-butadiene)	20	11	14	10
Poly(vinylidene chloride-co-vinyl-chloride), non-film-forming	16	23	—	17.5
Neoprene	15	17	9	8
Polyethylene	11	10	12.5	12.5
Unmodified	6	14	14	14

[a] $R = 3Ll/2ba^2$, where L, l, b, and a are the breaking load, length, width, and thickness, respectively.
[b] After Wagner (1967), with permission from *Ind. Eng. Chem.* **6**, 225 (1967). Copyright by The American Chemical Society.

It may also be noted that the rupture force tends to be greater, the greater the inherent strength of the polymer itself.

A clear correlation was seen between the degree of matrix–silica adhesion and the relative modulus of rupture. (The modulus of rupture is defined as $R = 3Ll/2ba^2$, where L, l, b, and a are the breaking load, length, width, and thickness of the test specimen; for a given specimen, $R \propto L$). As shown in Table 11.6, the modulus of rupture (which should also reflect the tensile strength) tends to correlate with the strength of the matrix–silica interface. Examination of fracture surfaces revealed that rupture occurred at, or near, the interface between sand grains and the cement–polymer matrix. The sand grains were less shiny in the polymer-containing specimens, indicating either some etching of the surface or, more likely, the presence of a thin, presumably strongly bonded, layer of matrix. Similar observations for impregnated cement have been noted (Gebauer and Manson, 1971).

Further confirmation of these conclusions has been provided by Isenburg *et al.* (1971) and Isenburg and Vanderhoff (1973), who examined the microstructure of a latex-modified mortar using scanning electron microscopy. The interpenetrating network of polymer appears to encapsulate the sand grains with a thin layer of polymer and to resist the propagation of microcracks when a load is applied; such cracks are always present in

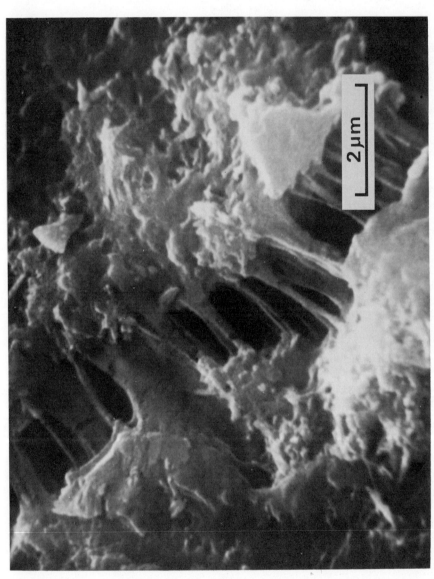

Figure 11.9 SM-100 styrene–butadiene copolymer latex-modified mortar overlay on concrete substrate fractured at 90°

cement due to shrinkage on curing. Indeed, as shown in Figure 11.9, although the polymer may be deformed by a growing crack to the point of forming fibrils, it effectively holds the crack together. Since the fracture energy of the polymer is higher than that of the mortar, even a little polymer can contribute significantly to the toughness of the composite (see also Section 11.3.3.5). Evidence of strong chemical interaction between the polymer and matrix also exists (Eash and Shafer, 1975).

Thus the microstructure and the bonding between phases are profoundly affected by the presence of a polymer, especially a film-forming polymer. Further studies of the adhesive and morphological characteristics should make it possible to improve the efficiency of the polymer still further. A somewhat analogous dispersion of rubber latex particles in a plastic matrix is discussed in Section 3.2.2; the related interpenetrating polymer networks discussed in Chapter 8 should also be mentioned.

11.3.3. Polymer-Impregnated Cements and Concretes

In this type of system, a matrix with a more or less continuous pore system is filled with a second component. The pore system is, however, fixed during the formation of the matrix itself. Thus one might expect some differences between polymer-impregnated concretes and the latex-modified concretes discussed above; for example, the latter are inherently more porous because porosity will always develop during curing, regardless of the presence of polymer. In both cases, however, the cement and polymer phases appear to form continuous and intertwined networks.

A major study in this area has been the joint program conducted by Brookhaven National Laboratory and the Bureau of Reclamation (DePuy *et al.*, 1973; Dikeou *et al.*, 1971, 1972; Steinberg, 1973; Steinberg *et al.*, 1968, 1969); additional studies have been conducted elsewhere in a rapidly increasing number of laboratories (American Concrete Institute, 1972, 1973*a*, 1973*b*; Coughlin *et al.*, 1970; Dahl-Jorgensen and Chen, 1973, 1974; Dahl-Jorgensen *et al.*, 1974; Fowler *et al.*, 1973; Gebauer and Coughlin, 1971; Gebauer and Manson, 1971; Heins, 1973; Lenschow *et al.*, 1971; Manning and Hope, 1971; 1973; Manson *et al.*, 1975; Singer *et al.*, 1971; Tazawa and Kobayashi, 1973; Whiting *et al.*, 1973; Vanderhoff *et al.*, 1973; Whiting *et al.*, 1974). Although methyl methacrylate has been the monomer studied to the greatest extent, studies have also been made of other monomers, such as styrene, acrylonitrile, divinyl and trivinyl monomers, vinyl chloride, and substituted styrenes. The polymer is usually present at levels in the range 5–10 wt %, with the cement porosity (to water) of concretes usually in the range from 10 to 30 vol. %.

As is the case with wood, impregnation is usually accomplished by drying the specimen, evacuating the matrix cement or concrete, filling with a liquid monomer (usually under pressure), and polymerizing using a free-radical initiator or gamma irradiation (dose ~ 6 Mrad) (Steinberg et al., 1968). Gamma irradiation tends to give somewhat higher strengths than thermal free-radical initiation, e.g., a 35% higher strength in the case of styrene. Again, as with wood substrates, it seems possible that irradiation results in an increased interaction between polymer and substrate, possibly in the form of grafting. Thus the radiation may induce active sites in the inorganic phase, which can either initiate polymerization or serve to enhance adhesion. It is also possible that losses of monomer by evaporation are lower with radiation catalysis, due to the lower temperatures used.

Penetration appears to take place by a capillary rise mechanism; in such a case, the height of rise h of a fluid of viscosity η ascending a capillary pore of radius r and contact angle θ is given by a relationship such as the Rideal–Washburn equation (Rideal, 1922; Washburn, 1921):

$$h = [(\gamma/2\eta)(r \cos \theta)t]^{1/2} \tag{11.4}$$

where γ is the surface tension, and t represents time. Dependence of the rate of penetration on $t^{1/2}$ has in fact been noted in studies by Heins (1973) and Vanderhoff et al. (1973) for a variety of monomers, and confirmed by Godard et al. (1974), who also proposed a diffusion model, and by Weyers et al. (1975). Complications may arise, however, if the monomer can react with the substrate, as is the case with, for example, acrylic acid. In any case, pressure may be applied in order to enhance the rate of penetration (Mehta et al., 1975c).

11.3.3.1. Mechanical Properties

Typical mechanical properties for polymer-impregnated concrete are given in Table 11.7 (Auskern and Horn, 1971). As may be seen, the incorporation of 7 wt % poly(methyl methacrylate) results in an increase in both compressive and tensile strengths by factors in the range 3–4. Indeed, fracture of impregnated specimens tends to occur through the aggregate (stone) rather than the cement matrix, or the matrix-aggregate interface, as is usual in normal concrete. Even larger factors may be obtained fortuitously from time to time—an observation consistent with the difficulties of optimizing properties in a very heterogeneous system. Young's modulus, modulus of rupture (which is, strictly speaking, a strength), flexural modulus, and hardness are also increased by lesser factors—in the range 1.5–2. It is interesting that the flexural modulus in latex-modified concrete is, in contrast, often reduced by the presence of the polymer (Dow Chemical Co., n.d.). In general the value of a mechanical property varies directly with the percentage of polymer incorporated.

Table 11.7
Summary of Properties of Concrete–Polymer Material (PIC)[a]

Property	Concrete control specimen[b]	Polymer-impregnated concrete[c]
Compressive strength, psi	5267	20,255
Tensile strength, psi	416	1,627
Modulus of elasticity, psi	3.5×10^6	6.3×10^6
Modulus of rupture, psi	739	2,637
Flexural modulus of elasticity, psi	4.3×10^6	6.2×10^6
Water absorption, %	5.3	0.29
Freeze–thaw durability		
Number of cycles	590	2,420
Weight loss, %	26.5	0.5
Corrosion by 15% HCl on 84-day exposure, % wt loss	10.4	3.6
Corrosion by sulfates on 300-day exposure, % expansion	0.144	0
Corrosion by distilled H_2O	Severe attack	No attack

[a] Auskern and Horn (1971).
[b] Concrete mix design: type II cement, water/cement ratio = 0.51; cement:sand:aggregate ratio = 1:2:5:3.4 by maximum aggregate size $\frac{3}{4}$ in. Fog-cured 28 days.
[c] Concrete with up to 6.7 wt % loading of poly(methyl methacrylate) polymerized with ^{60}Co γ-radiation.

Surprisingly, relatively few data have been published on stress–strain, creep, or fatigue behavior. Auskern and Horn (1971) and Manning and Hope (1971) have reported stress–strain relationships for samples of concrete containing up to 12 vol. % poly(methyl methacrylate); see Figure 11.10 (Auskern and Horn, 1971). More comprehensive stress–strain data have been

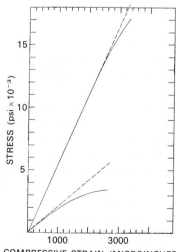

Figure 11.10. Stress–strain curves for concrete and concrete–polymer (CP-type concrete, cylinders 3 in. in diameter and 6 in. high). The upper (solid) curve is for CP concrete (PMMA, loading 5.4 wt %, $E = 5.5 \times 10^6$ psi) and the lower (solid) curve is for plain concrete (unloaded, $E = 1.8 \times 10^6$ psi by the U. S. Bureau of Reclamation method, $E = 1.3 \times 10^6$ psi by secant method). The upper ends of the curves correspond to fracture. (Auskern and Horn, 1971.)

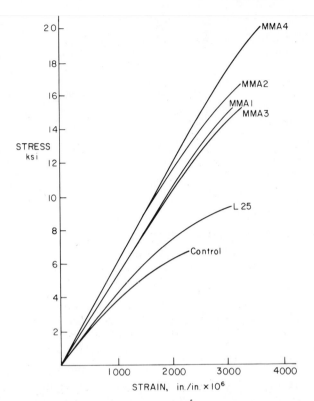

Figure 11.11. Compressive stress–strain curves for concrete impregnated with poly(methyl methacrylate). Polymer loading: MMA1 > MMA2 > MMA3. Specimen L25 is a latex-modified concrete. (Dahl-Jorgensen and Chen, 1973.)

reported by the group at Lehigh (Dahl-Jorgensen *et al.*, 1975; Manson *et al.*, 1973) for mortar and cement impregnated with several monomer systems; see Figures 11.11–11.14.

  It is clear not only that the stress (or load) at rupture is increased by the presence of a rigid polymer such as poly(methyl methacrylate), but also that the nature of the variation of stress with strain is significantly affected (Figures 11.10–11.14). The control specimen, i.e., plain concrete, behaves in a nonlinear fashion from almost the very beginning of the compression, the stress increasing at a slower rate than the strain, so that, at rupture, the curve is almost horizontal. In contrast, the rigid-polymer-impregnated concrete behaves in a linear (Hookean) manner until 80% of the ultimate load is applied. Only then does a deviation from linearity occur, the deviation being only slight even at rupture. In other words, the impregnated concrete

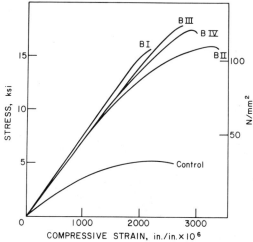

Figure 11.12. Compressive stress–strain curves for concrete impregnated with various combinations of methyl methacrylate (MMA), n-butyl acrylate (BA), and a crosslinking agent, trimethylolpropane trimethacrylate (TMPTMA). Specimen BI, 90 MMA/10 TMPTMA; BII, 70 MMA/20 BA/10 TMPTMA; BIII, 60 MMA/30 BA/10 TMPTMA; BIV, 50 MMA/40 BA/10 TMPTMA. (Dahl-Jorgensen et al., 1975.)

behaves much more like a classical elastic solid than the unimpregnated concrete. Auskern and Horn (1971) concluded that the polymer improves the bond between the elastic cement matrix and the elastic aggregate particles. On the other hand, if polymer is not present, the bond fails progressively as the strain is increased. These conclusions are in agreement with those discussed above for the case of latex-modified concrete.

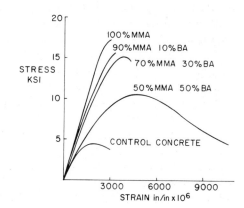

Figure 11.13. Polymer-impregnated concrete: compressive stress–strain curves as a function of polymer composition (hydraulic tester); KSI = psi × 10^{-3}. MMA and BA refer to methyl methacrylate and n-butyl acrylate, respectively. (Manson et al., 1973.)

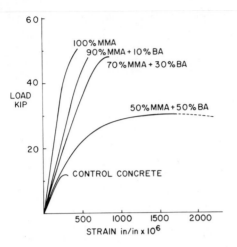

Figure 11.14. Polymer-impregnated concrete: tensile load–strain curves as a function of polymer composition (hydraulic tester). MMA and BA refer to methyl methacrylate and n-butyl acrylate, respectively. Unit KIP = lb load $\times 10^{-3}$. (Manson *et al.*, 1973.)

While the observations discussed are generally valid for impregnation with a rigid polymer, different behavior may be seen in other cases. Thus, as seen in Figures 11.12–11.14, the stress–strain behavior is dependent on the state of the polymer, and also on the testing conditions (Dahl-Jorgensen *et al.*, 1973, 1974, 1975; Manson *et al.*, 1973, 1975a, 1975b). As the polymer is changed from a glassy to a rubbery one, in this case poly(n-butyl acrylate), the increase in modulus and tensile strength observed on impregnation with the glassy polymer is essentially lost when the impregnant is rubbery. Similar effects have been noted when the state of a polymer is changed by heating or cooling. Thus, when a rigid-polymer PIC is heated, the strength (Gebauer *et al.*, 1972a) and modulus (Manson *et al.*, 1975a, 1975b) decrease as the glass temperature T_g of the polymer is exceeded; when a rubbery PIC is cooled below the T_g of the polymer, the modulus (and presumably strength) is increased to a level typical of rigid-polymer systems. At the same time, ductility is higher, the greater the rubberiness of the polymer, as shown by the development of, or by the onset of, significant yielding and increased elongation prior to failure (Figures 11.13, 11.14). Such results suggest that the properties of the polymer must play a major role in determining the ability of the cement–aggregate bond to withstand stress, and are consistent with many of the observations on latex-modified systems*

---

* It should be recalled, however, that the greatest improvement of properties in latex systems is seen with film-forming polymers, whose T_g's are near room temperature. This is probably so because a continuous network is desirable. A hard polymer would still be desirable from the standpoint of strength, as in the case of PIC; the strength of PIC containing a rigid polymer is greater than that of a latex-modified material (Figure 11.11).

(Wagner, 1965, 1966, 1967). The ability to control the shape of the stress–strain curve in this way is of considerable potential importance in the design of structures.

Other mechanical properties are, of course, important. Although some studies of creep (so far inconclusive) have been described (Dikeou *et al.*, 1971), fatigue behavior, though important, seems scarcely to have been studied at all. Preliminary measurements using mortar specimens in a bending mode have been given by Gebauer and Manson (1971) (Table 11.8). A significant effect of impregnation was observed, even though experimental scatter is high. The presence of about 9 wt % poly(methyl methacrylate) permits cycling at loads five times higher than is possible with control specimens. Even at the higher loads, the numbers of cycles to failure is increased at least severalfold; about half of the specimens had not broken at 10^6 cycles. It seems likely that the increased resistance to fatigue is related to the conclusions by Wagner (1965, 1966, 1967) that polymer strengthens the sand–matrix bond, and to the similar conclusions of Auskern and Horn (1971) for concrete.

Thus all the observations for both latex-modified and polymer-impregnated mortar or concrete seem to be consistent in several important respects, even though the latex modification involves factors not concerned in impregnation (Dow Chemical Co., n.d.). In each case interpenetrating networks of polymer and cement are evidently formed (Aignesberger *et al.*, 1969; Auskern, 1973) see Figure 11.15. The presence of polymer may effectively "strengthen" or toughen the bonds between sand or aggregate and cement, so that progressive failure is minimized as the specimen is

Table 11.8
Fatigue Behavior of Polymer-Impregnated Cement (PIC) Mortar[a,b]

Property	PIC	Control
Ultimate flexural strength (UFS), psi	5,300	900
Upper stress level (75% of UFS), psi	3,960	682
Lower stress level (25% of UFS), psi	1,320	228
Speed of cycling, cpm	2,400	2,400
Number of cycles to failure:		
specimen 1	236,100	116,000
specimen 2	190,700	90,100
specimen 3	370,000	77,700
specimen 4	900,000[c]	254,300
specimen 5	900,000[c]	188,600

[a] Gebauer and Manson (1971).
[b] Containing 8.6 wt % poly(methyl methacrylate).
[c] Fatigue failure was not obtained after 900,000 cycles.

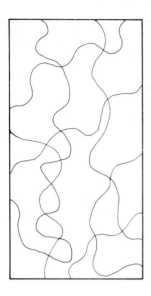

Figure 11.15. Schematic showing polymer network in polymer-impregnated cement. For evidence based on electron microscopy, see the original micrographs of Auskern (1973).

strained. At the same time, porosity is reduced, especially in the case of impregnation, and microcracks resulting from shrinkage during curing are presumably filled. In any case, some combination of such factors must play an important role with respect to strength, modulus, and general toughness (see Section 11.3.3.5).

11.3.3.2. Water Absorption and Related Behavior

As shown in Table 11.7, water absorption in PIC is decreased considerably by filling pores with polymer—a reduction of about an order of magnitude being characteristically observed (Auskern and Horn, 1971; Mehta *et al.*, 1975c). Permeability is generally reduced as well (Steinberg *et al.*, 1968) by a factor of 70–80%, probably reflecting in large part the reduction in water absorption. In view of the coupling of a reduced water sensitivity with greater mechanical strength, one might expect the resistance to cyclic freezing and thawing to be increased by the presence of polymer. Indeed, this is the case; see Table 11.7. While untreated concrete exhibited a 25% weight loss after 6×10^2 test cycles, a poly(methyl methacrylate)–concrete specimen showed only a 0.5% weight loss after 24×10^2 cycles. Significant improvements have also been reported by Fowler *et al.* (1973) and Mehta *et al.* (1975c). The permeability to salts, an important factor in the corrosion of steel in concrete (Section 11.3.3.4), is reduced by over an order of magnitude in PIC mortars containing poly(methyl methacrylate) (Dahl-Jorgensen *et al.*, 1975; Liu and Manson, 1975).

11.3.3.3. Thermal Properties

One factor which is somewhat adversely affected by the presence of polymer is the coefficient of expansion, which has been shown to be increased about 25% by the presence of 6 wt % poly(methyl methacrylate) or polystyrene in concrete (Steinberg *et al.*, 1968). This phenomenon is a consequence of the polymer exhibiting a larger coefficient of expansion than the cement. Slight ($\sim 5\%$) increases in thermal diffusivity and slight decreases in thermal conductivity were also noted (Steinberg *et al.*, 1968).

11.3.3.4. Environmental Resistance

As indicated in Table 11.7, resistance to distilled water, dilute aqueous hydrochloric acid, and sulfates is significantly improved by polymer impregnation (DePuy *et al.*, 1973; Dikeou *et al.*, 1971, 1972; Gebauer and Coughlin, 1971; Mehta *et al.*, 1975c; Steinberg *et al.*, 1968, 1969). Ordinary concrete is quite susceptible to exposure to these three media. Distilled water leaches away components of the cement, hydrochloric acid attacks the basic constituents, and sulfate ions undergo exchange with carbonates and expand the crystal structure, thus causing crumbling and ultimate failure. On the

Figure 11.16. Effect of dilute hydrochloric acid etching on fracture cross sections of a core from a concrete slab impregnated with poly(methyl methacrylate). Impregnation (in this case from the bottom end, through a thin sand layer) was confined to the light area, in which the stone aggregate was etched rather than the matrix. The untreated region (top) was severely attacked. (Mehta *et al.*, 1975.)

other hand, the presence of the polymer appears to serve as an internal protective coating, hindering access of the aggressive medium to the cement.

Thus the onset of failure in dilute hydrochloric acid was delayed from about 100 days to 800 days for typical specimens containing polystyrene, poly(methyl methacrylate), or other polymers. A typical example of increased resistance to acid etching is given in Figure 11.16; note that the limestone aggregate and the impregnated portion of the matrix are attacked, but not the unimpregnated matrix. Whereas untreated specimens showed a significant degree of expansion when exposed to sulfates in a standard test, the degree of expansion was reduced by a factor of two by polymer impregnation.

Figure 11.17. Scanning electron photomicrographs of near-end sections of steel reinforcing rods in salt-contaminated concrete slabs after freeze-thaw testing: (A) polymer-impregnated core; (B) unimpregnated core. (Mehta *et al.*, 1975.)

Resistance to sulfuric acid, which combines both acidity and sulfates, and to distilled water, is also improved by impregnation with polymers.

The ability of polymer impregnation to reduce the susceptibility of concrete to corrosion by salts finds particular applications in such structures as highway bridge decks (Blankenhorn *et al.*, 1975; Dikeou, 1973; Dikeou *et al.*, 1972; Fowler *et al.*, 1973; Kuckacka *et al.*, 1972; Manson *et al.*, 1973, 1975a, 1975b; Mehta *et al.*, 1975a, 1975b, 1975c; Vanderhoff *et al.*, 1973). While the improved mechanical properties may or may not be needed in this and related applications, restriction of the diffusion of salts to the steel reinforcement in the concrete is very important if adequate service life is to be obtained (see Section 11.3.3.2). Confirmation of the ability of polymer impregnation to reduce corrosion of reinforcing steel in even heavily salt-contaminated concrete has been shown by Mehta *et al.* (1975c); as shown in Figure 11.17, no corrosion of steel was detected after freeze-thaw cycling of PIC containing salt, while severe pitting was observed in control specimens.

Resistance to heat and flame will also be important in many applications in structures. Although little has been reported on flammability, preliminary tests indicate that PIC's tend to be either self-extinguishing or nonburning, at least under the conditions of the tests used (DePuy *et al.*, 1973). Deterioration of properties at elevated temperatures may be a more serious problem than flammability per se (DePuy *et al.*, 1973; Lenschow *et al.*, 1971).

11.3.3.5. Models for Mechanical Behavior

The development of models for the mechanical behavior of porous bodies in general must take into account porosity and the sizes, shapes, and distributions of the pores. (The case of impregnated wood has been discussed in Section 11.2.4.) Porosity acts to reduce the amount of load-bearing material in any given plane subjected to stress; see Figure 11.18.

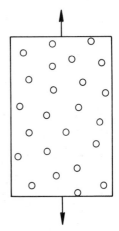

Figure 11.18. Model for a porous body subjected to tensile loading.

This increases the effective shear in the system; slippage occurs at a lower stress than in a corresponding fully dense body. Pores also serve to concentrate stresses—a generally deleterious effect which depends on the nature of the pores. At the same time, pores may serve to interact with cracks in the matrix, and in some cases to effectively obstruct further growth of cracks.

Due to this complexity, it is not surprising that empirical measurements of, for example, strength as a function of porosity often follow a complex functional relationship. Theoretical interpretation is hindered by the fact that the precise relationship often depends on the microstructure and that it may not be possible to prepare a completely nonporous standard of reference.

While considerable progress has been made with ceramics, cement and concrete offer special difficulties with respect to porosity problems. Cement and concrete contain several components, each of which has its own characteristic porosity, which, in turn, is also sensitive to the method of preparation and the composition. Also, the difficulty of preparing cement specimens with a porosity lower than about 10% makes it hazardous to extrapolate exponential strength–porosity functions.*

In spite of these fundamental difficulties, it should be possible to at least develop simple semiempirical or empirical models that will have some utility in the correlation and prediction of behavior. Such an approach has been taken by Auskern (1970, 1973) and Auskern and Horn (1971) and by Manning and Hope (1971), who have attempted to combine effects of porosity with effects due to a filler, which in this case is a polymer. Each factor is considered to decrease the strength from the value for a hypothetical fully dense cement.

The approach of Auskern and Horn starts with a typical mixing law†

$$S_c = S_1 v_1 + f(S_2 v_2) \qquad (11.5)$$

where S_c is the composite strength; v is the volume fraction; and subscripts 1 and 2 refer to the cement and polymer, respectively. In the ideal case, $f(S_2 v_2) = S_2 v_2$. It is assumed further that the cement strength is related to porosity by an expression such as the following:

$$S_1 = S_{1,0} f(P) \qquad (11.6)$$

where $S_{1,0}$ is the strength of the ideal pore-free matrix cement, P is the porosity, and $f(P)$ is some function (usually exponential) of the porosity.

---

* New developments indicating that still lower porosities can be obtained by special techniques should aid in resolution of this question.

† This type of mixing law represents an upper bound for the modulus of mixtures, and corresponds to the assumption that strains are equal in both components (Krock and Broutman, 1967); see also Section 12.1.1.

Often $f(P)$ is taken as $f(P) = (1 - P)^n$, where n varies depending on the nature of the material. In a typical calculation for cement, Auskern and Horn took n to be 6, a choice which, it must be admitted, was somewhat arbitrary. Thus

$$S_1 = S_{1,0}(1 - P)^6 \qquad (11.7)$$

A value of 120,000 psi was estimated for S_{10}, based on measurements of the strength and porosity of hardened cement paste specimens. (It must be remembered that values of $S_{1,0}$ are subject to considerable uncertainty.) Since $S_1 \gg S_2$, in the simple case

$$S = S_1 v_1 = 120{,}000 v_1 \qquad (11.8)$$

Porosity is now admitted by combining equations (11.7) and (11.8):

$$S = 120{,}000 v_1 (1 - P)^6 \qquad (11.9)$$

The separate effects of filler content [equation (11.8)] and porosity alone are shown in Figure 11.19. A similar approach was taken by Manning and Hope (1971) in order to predict moduli, using a value for n of 3.

In this view, the apparent increase in the strength or modulus of concrete obtained by adding polymer evidently comes about by enabling the specimen

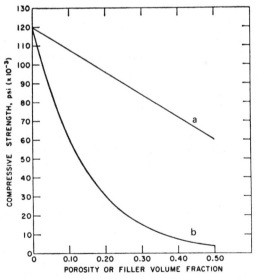

Figure 11.19. Effect of filler and porosity on compressive strength of cement. (a) Filler effect, $S = 120{,}000\, v_1$; (b) porosity effect, $S = 120{,}000(1 - P)^6$. (Auskern and Horn, 1971.)

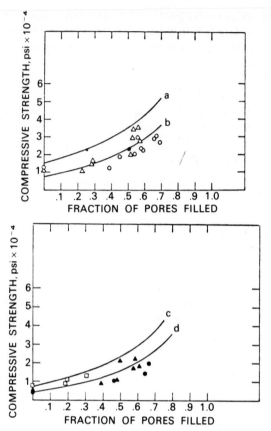

Figure 11.20. Comparison of theory and experiment for the compressive strength of cement paste as a function of polymer loading. The solid curves give calculated values with the following initial porosities: (a) 30% (b) 40%; (c) 40%; (d) 50%. ($\bigcirc$) $W/C = 0.30$, $P = 30\%$; ($\triangle$) $W/C = 0.35$, $P = 33\%$; ($\square$) $W/C = 0.43$, $P = 41\%$; ($\blacktriangle$) $W/C = 0.50$, $P = 47\%$; ($\bullet$) $W/C = 0.58$, $P = 50\%$. (Auskern and Horn, 1971.)

to approach its theoretical limit more closely, not by causing a nonadditive effect. While not included in the above derivation, polymer properties and pore shape may, however, be important (Hasselman *et al.*, 1972; Manning and Hope, 1971; Manson *et al.*, 1973a, 1975).

Although the uncertainties are great, it is interesting to compare experimental data with predicted values based on equation (11.9); see

Figures 11.19 and 11.20. Except that values observed are less than those predicted (a fact possibly due to error in S_1), plausible agreement is obtained.

Moving to a typical concrete, which contains still another phase, the aggregate, Auskern and Horn (1971) modified equation (11.5) by taking $f(S_2 v_2) = 0.3BS_d v_d$, where B is a term characteristic of the effectiveness of the matrix–aggregate bond, and the subscript d refers to the aggregate. Thus one obtains

$$S_c = S_m v_m + 0.3BS_d v_d \qquad (11.10)$$

where S_c is the strength of the concrete and the subscript m refers to the matrix.

For comparison with experiment (Figure 11.21) the following values were used: $v_m = 0.2$, $v_d = 0.8$, S_m given by equation (11.6), $S_d = 30,000$ psi, and $B = 0.5$ ($B = 1$ for perfect adhesion). Good agreement was found. While B is an empirical constant, it does rise with polymer content. Thus this type of expression, though empirical, takes account of the interfacial bonding, which has been shown above to play a significant role in determining the strength and related properties of cement and concrete. Other relation-

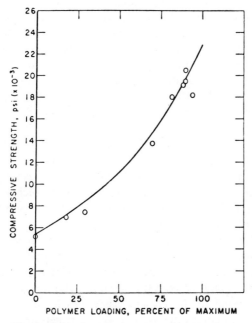

Figure 11.21. Compressive strength vs. polymer loading for PIC specimens compared to strength model (solid line). (Auskern and Horn, 1971.)

ships have been examined to a limited extent; for example, equations proposed by Counto (1964) and Hobbs (1969) have been used by Manning and Hope (1971) and Auskern (De Puy *et al.*, 1973), respectively. When porosity was allowed for, Manning and Hope (1971) found reasonable agreement between experiment and theory for a variety of concretes; Auskern (DePuy *et al.*, 1973) also reported fairly good agreement for mortars and some concretes. A porosity model was also used successfully with impregnated wood (Section 11.2.4).

Nevertheless, while the filling of pores is certainly important in the stiffening and strengthening of concrete by impregnation, filling cannot in itself account for all observed results. For example, as mentioned above, the state of the polymer plays a major role in determining stress–strain behavior. The effect of a given polymer in a porous ceramic is greater when the polymer is glassy than when it is in the rubbery state (Gebauer *et al.*, 1971); similar results were obtained by Manson *et al.* (1975*a*, 1956*b*) and Dahl-Jorgensen *et al.* (1975) by changing the composition from a glassy to a rubbery state; see Figures 11.12–11.14. Effects of this kind are, of course, not accounted for in composite equations such as equations (11.9) and (11.10), though more empirical equations may do so (Auskern, 1970, 1973; Counto, 1964). Consideration of strengthening from the viewpoint of fracture mechanics may also be useful (Sections 1.6 and 12.1.2.4). Tazawa and Kobayashi (1973) have suggested that impregnation increases the modulus, decreases the critical flaw size, and increases the fracture energy. Each of these effects would tend to increase strength [equation (1.22)]; preliminary calculations suggested that the increase in fracture energy (which is much higher for the polymers than for the cement phase) may be dominant. Auskern (1974) has used a similar approach, though with a different physical interpretation of the flaw concerned. Also, the ability of the polymer to transfer stress from the matrix will be greater the higher its modulus, and the greater the flatness of the pore (Hasselman and Gebauer, 1973; Hasselman *et al.*, 1972, 1973; Manson *et al.*, 1973). If, as seems likely, the pores are, in fact, fairly flat on the average, composite equations such as (11.10) require adaptation; a modified composite equation has been proposed (Hasselman *et al.*, 1972) to take account of pore shape.

Manning and Hope (1971) have summarized the possible mechanisms for strengthening by polymer impregnation as being due to the ability of the polymer (i) to act as a continuous, randomly oriented, reinforcing network, (ii) to increase the bond between the aggregate and the cement paste, (iii) to repair microcracking in the cement paste, (iv) to absorb energy during deformation of the composite system, (v) to penetrate and reinforce the micropores of the cement paste, and (vi) to bond with the hydrated or unhydrated cement. These mechanisms are not, of course, mutually exclusive. In any case, as of this writing, a fully satisfactory quantitative and general treatment of strengthening in polymer-impregnated concrete (or other porous systems) is yet to be developed.

11.3.3.6. Related Systems

A number of interesting extensions of polymer impregnation may be noted, for example, to fiber-containing cements and mortar (Dikeou *et al.*, 1971; Flajsman *et al.*, 1971; Steinberg *et al.*, 1968), to a variety of light concretes (Auskern, 1970), to rocks (U.S. Bureau of Mines, 1971), to porous ceramics (Gebauer *et al.*, 1972a,b; Hasselman and Gebauer, 1973; Hasselman and Penty, 1973; Hasselman *et al.*, 1972), and to marble and other kinds of statuary (Gauri, 1970). Uses as a binder for such materials as solid wastes (Steinberg, 1973) are receiving increasing attention.

The reinforcement of concrete or mortar by the use of glass or steel fibers results in an improvement in properties such as ultimate flexural strength and toughness, at the expense of compressive strength (Flajsman *et al.*, 1971). However, corrosion of glass or steel by the basic mortar environment or by gases that permeate from the outside creates a problem. In principle, the addition of a polymer to such a reinforced system might be

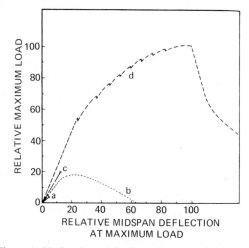

Figure 11.22. Load vs. deflection curves for plain and fiber-reinforced mortars, both natural and polymer-impregnated. (a) Plain mortar; (b) plain, steel-fiber-reinforced mortar; (c) polymer-impregnated mortar; (d) polymer-impregnated, steel-fiber-reinforced mortar. (Flajsman *et al.*, 1971.)

Table 11.9
Mechanical Properties of Fiber- and Polymer-Reinforced Mortars[a]

Mortar	Fibers, vol %	Polymer, wt %	Ultimate flexural strength, kgf/cm²	Increase in strength, %	Toughness, kg-cm	Compressive strength, kgf/cm²
Plain	0	0	12	100	0.06	109
Plain (moist-cured)	0	0	40	330		371
Polymer-impregnated	0	7.4	61	510	1.05	413
Glass-fiber-reinforced	2.0	0	54	450	2.72	39
Glass-fiber- + polymer-reinforced	2.0	10.7	136	1130	3.16	513
Steel-fiber-reinforced	2.0	0	54	450	2.09	100
Steel-fiber- + polymer-reinforced	2.0	9.1	338	2820	61.50	894

[a] Flajsman et al. (1971).

expected to have a dual effect: improvement of the fiber–matrix adhesion, and reduction of the permeation of deleterious substances. Indeed, Steinberg et al. (1968) and Dikeou et al. (1971) found a synergistic effect when concrete was both reinforced (with 2 vol % steel fibers) and impregnated with poly-(methyl methacrylate). For example, flexural strengths measured were about 200 kg/cm², compared with about 120, 80, or 40 kg/cm² for concrete that had been impregnated, reinforced, or untreated, respectively. Using a different concrete, which had been modified in porosity to permit easy impregnation by capillary absorption, Flajsman et al. (1971) confirmed the synergism noted by Steinberg et al. (Figure 11.22; Table 11.9). It should be noted that the relative improvement is greatest in the case of the most porous plain mortar, and less in the case of the steam-cured mortar, which showed a lower porosity. This observation is consistent with the model discussed in the previous section; the more porous the matrix, the greater the relative enhancement of properties by impregnation (Dikeou et al., 1971).

As with plastic-impregnated concrete itself, the shape of the stress–strain curve is of interest, or, in this case, the load vs. deflection curves (Figure 11.22). In the fiber-reinforced mortar, failure apparently occurred by pulling out of the fibers, as indicated by the drop in load as deflection is increased. However, with polymer impregnation (to a level of about 11 %), failure occurred by progressive failure of the fibers, as shown by the discontinuities in the load–elongation curves, and by microscopic examination. Thus, improved bonding between the fibers and the matrix evidently permits the inherently stronger fibers to carry a greater proportion of the applied load.

The impregnation of lightweight concretes, i.e., foamed cements, or concretes in which the aggregate consists of a porous material such as perlite or foamed glass, has been studied by Auskern (1970) and Dikeou *et al.* (1971). Compressive strengths as high as 7100 psi were found for a perlite sample that contained 95 wt % polymer, in comparison to 100 psi for the unimpregnated specimen. (It may be preferable to consider such a system as a highly filled polymer.) Attempts were made to correlate the data using an approach similar to that used previously (a combination of porosity with a rule of mixing), with limited success.

Gebauer *et al.* (1972*b*) have demonstrated a significant increase in the strength of a porous ceramic on impregnation with poly(methyl methacrylate) or polystyrene. Again, the greatest relative improvement is noted with the more porous bodies. In a related study, Gebauer *et al.* (1972*a*) showed that the state of the polymer, not just its proportion, plays a role in determining the strength. In the case of impregnation with polychlorostyrene and poly(*t*-butylstyrene), strengths were found to decrease at a temperature around the glass transition temperature. The general strengthening behavior observed has been interpreted by Hasselman *et al.* in terms of a mechanical reinforcement (Hasselman and Gebauer, 1973; Hasselman and Penty, 1973; Hasselman *et al.*, 1972). (See also Section 11.3.3.5.)

As mentioned above, the impregnation of other inorganic materials, such as rock (U.S. Bureau of Mines, 1971) and statuary (Gauri, 1970) has received some attention. In these cases, stabilization is the desired goal, whether to prevent the falling or rocks in a mine, or to prevent atmospheric corrosion of irreplaceable works of art.

11.3.4. General Comments

The impregnation or polymer modification of cementitious materials certainly leads to many improvements in properties. In a practical sense, one can foresee many applications in highways, bridges, marine structures, housing, pipes, etc. Indeed, the composite should be treated as a material in its own right, for its properties make it possible to design structures without the limitations of conventional concrete. For example, a polymer-impregnated concrete bridge deck or floor in a tall building can be integrated into the structure as a load-bearing component rather than as dead weight.

Many fundamental questions remain. Further confirmation of the role of the polymer and further studies of chemical reactivity, corrosion resistance, and the effects of the state of the polymer will all be of great interest.

12

Particle- and Fiber-Reinforced Plastics

As noted previously, the modulus of a two-phase system is some kind of average of the moduli for the individual components. Thus when a rigid, high-modulus phase such as silica, glass, or steel is incorporated into a lower modulus matrix such as a polymer, the modulus of the composite is increased, in some proportion to the volume fraction of filler added. Other properties may be affected also; for example, toughness, thermal expansion characteristics, and permeability (Sections 12.1.2 and 12.1.3).

In general, the behavior of the composite depends not only on the individual properties of the two components and on their relative proportions, but also on the size, shape, and state of agglomeration of the minor component, and on the degree of adhesion between the filler and the matrix. For the sake of convenience, one may consider fillers in two main groups: particulate and fibrous. Particulate phases are usually called fillers, or if the interphase adhesion is high, reinforcing fillers. Fibrous phases are usually referred to as reinforcing, for the fibers themselves bear an important fraction of any load applied. This chapter deals primarily with matrices that are more or less rigid rather than elastic at normal use temperatures; reinforced elastomers are discussed in detail in Chapter 10. Particulate fillers have received less attention than the fibrous reinforcements of interest in high-performance composites and will therefore be emphasized; however, for the sake of completeness, fibrous composites will be discussed briefly.

12.1. PARTICULATE PHASES

In this section, effects of a particulate filler* on the mechanical and other physical properties of a polymeric composite will be considered.

---

* It should be noted that the size of the filler may indirectly play an extremely important role in affecting the relaxation behavior of the polymeric matrix. Very small particles may not only stiffen a matrix by purely mechanical means, but may raise the glass temperature to a significant degree (see Section 10.8). Effects of large particles on relaxation behavior, on the other hand, tend to be less marked, though they, too, can exhibit both stiffening and relaxation effects (see Section 12.3.1).

Major aspects of mechanical and relaxation behavior have been reviewed by several authors, including Nielsen (1962; 1967a; 1974, Vol. 2), Broutman and Krock (1967, 1974), Ferry (1970), Ashton et al. (1969), and Corten (1971). However, since a considerable body of recent work of interest to the chemically oriented reader has not been included within the scope of these useful reference works, an updating and a broadening of scope are worthwhile.

12.1.1. Mechanical Behavior at Small Strains

12.1.1.1. Static and Dynamic Modulus

A polymeric matrix is strengthened or stiffened by a particulate second phase in a very complex manner. The particles appear to restrict the mobility and deformability of the matrix by introducing a mechanical restraint, the degree of restraint depending on the particulate spacing and on the properties of the particle and matrix. In the simplest possible case, two bounds have been predicted for the composite elastic modulus E_c (Broutman and Krock, 1967, Chapters 1 and 16; Lange, 1974; see also Section 2.6.4 of this book):

upper bound : $E_c = v_p E_p + v_f E_f$ (case of equal strains)   (12.1)

lower bound : $E_c = \dfrac{E_p E_f}{E_p v_f + E_f v_p}$ (case of equal stresses)   (12.1)

where v represents the volume fraction, E the modulus, and the subscripts p and f the polymeric matrix and particulate filler phases, respectively.* In many cases, observed values fall in between these bounds, though when matrix constraint is large, the bounds may be exceeded.

In practice, more complex equations may be useful, such as those proposed by Kerner (1956b), Eilers (1941), van der Poel (1958), Sato and Furukawa (1963), Guth (1944), Smallwood (1944), Wu (1966), Halpin and Tsai (Ashton et al., 1969, p. 77), Hasselman (1962), Mooney (1951), Hashin and Shtrikman (1963), and Nielsen (1970b); several typical equations are given in the appendix. Good examples of recent critical studies and discussions of typical equations are the articles by Nielsen (1967a, 1970b), Ziegel and Romanov (1973), Brassell and Wischmann (1974), Jenness (1972), Lange (1974), Nielsen and Lewis (1969), Lewis and Nielsen (1970), and

---

* Throughout this chapter, the subscripts c, p, and f (referred to components) will always indicate composite, polymer, and filler, respectively; v will be used throughout for volume fraction, though ϕ is often used for this purpose.

Ashton *et al.* (1969, Chapter 5). Over certain ranges of filler concentration and modulus, some of these equations become approximately equivalent on simplification, while over other ranges, considerable divergence may be noted.

For example, the equations proposed by Halpin and Tsai (Ashton *et al.*, 1969; Brassell and Wischmann, 1974) and Nielsen and Lewis (Nielsen and Lewis, 1969; Lewis and Nielsen, 1970; Hashin and Shtrikman, 1963) are essentially generalized versions of the Kerner equation (1965*b*); the Hashin–Shtrikman (1963) expression is analogous, as is the Takayanagi equation (Section 2.6.4). As will be seen below, the Kerner equation (see Figure 12.1) in one form or another is especially useful in predicting the moduli of composites of a spherical filler randomly dispersed in a glassy (not elastomeric) matrix (Lewis and Nielsen, 1970). Kerner's original expression is as follows (assuming good adhesion between the phases):

$$\frac{E_c}{E_p} = \frac{G_f v_f / [(7 - 5v)G_p + (8 - 10v)G_f] + v_p / [15(1 - v)]}{G_p v_f / [(7 - 5v)G_p + (8 - 10v)G_f] + v_p / [15(1 - v)]} \tag{12.3}$$

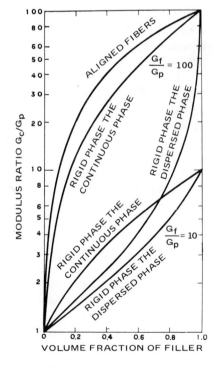

Figure 12.1. Modulus ratio as a function of filler concentration for filler-to-matrix moduli ratios of 10 and 100 according to Kerner's (1956*b*) equation. Bottom curves or loops are for cases where the filler (dispersed phase) is the more rigid phase, while the upper curves are for the inverted case, where the more rigid material is the continuous phase. The top curve is the elastic Young's modulus of a material filled with very long fibers aligned in the direction of the applied tensile load. (Nielsen, 1967*a*.)

where E_c and E_p are the Young's moduli of the composite and polymeric matrix, respectively; G_p and G_f are the shear moduli of the plastic and filler (i.e., the continuous and discontinuous phase, respectively); v is Poisson's ratio of the plastic; and v_p is the volume fraction of the plastic.

In general, particle size per se does not appear as a variable in most equations, and if the modulus of the filler is very high relative to that of the matrix, the polymer moduli themselves may disappear as explicit variables on simplification, leaving an expression depending only on the volume fraction of the filler. However, particle size, shape, and agglomeration may, of course, play roles, as may the question of whether the high-modulus filler is in fact dispersed or continuous. As predicted (Ashton *et al.*, 1969; Brassell and Wischmann, 1974; Broutman and Krock, 1967; Corten, 1971; Eilers, 1941; Ferry, 1970; Guth, 1944; Hashin and Shtrikman, 1963; Hasselman, 1962; Hirai and Kline, 1972; Jenness, 1972; Kerner, 1956*b*; Lange, 1974; Mooney, 1951; Nielsen, 1962, 1967*a*; Sato and Furukawa, 1963; Smallwood, 1944; Wu, 1966; van der Poel, 1958), the modulus of a polymer containing a rigid particulate filler is, in fact, generally increased, even if the filler does not interact strongly with the matrix. Some of the equations mentioned may be used to treat systems containing particles of nonspherical shape (Ashton *et al.*, 1969; Wu, 1966) and systems in which the reinforcement is continuous (Ashton *et al.*, 1969; Nielsen, 1967*a*). In general, lamellar or fibrous fillers raise the modulus to a greater extent than do spherical fillers (Halpin, 1975; Halpin and Kardos, 1972).

One of the first fundamental studies illustrating the effects of fillers on modulus was described by Nielsen *et al.* (1955), who showed that the shear modulus of polystyrene in the glassy state was increased by the incorporation of from 20 to 60 vol % of mica, calcium carbonate, or asbestos. The magnitude of the increase depended on the filler, about eightfold for mica, but less for asbestos. In addition, as shown by damping measurements, the apparent glass temperature was increased by the presence of the filler, by up to 15°C.

It was proposed that the modulus of a filled glassy or rigid polymer could be represented by an equation of the form

$$G_c = G_p v_p + A G_f v_f \tag{12.4}$$

where G_p, G_f, and G_c are the shear moduli of the pure polymer, filler, and composite, respectively, v_p and v_f represent volume fractions of polymer and filler, respectively, and A is an empirical term to give a measure of the filler–matrix adhesion. Under some conditions, this equation is approximately equivalent to Kerner's (1956*b*) lower-bound-type relationship. In any case, the constant A empirically allows for the fact that upper bound values for modulus are not found consistently in practice with such systems.

Although there are frequently disparities between experiment and predictions of particular equations, more recent studies have demonstrated reasonable agreement in certain cases. For example, Wambach *et al.* (1968) have shown that values of Young's modulus for poly(phenylene oxide) filled with glass beads follow approximately van der Poel's (1958) equation. At least in the range of concentration reported (up to a volume fraction of filler of 0.25), van der Poel's equation is approximately equivalent to Kerner's (Trachte and DiBenedetto, 1971). Similar agreement had been reported earlier by Schwarzl *et al.* (1965) for filled poly(propylene oxide) in the glassy state. Interestingly, it may be noted (Brassell and Wischmann, 1974; Trachte and DiBenedetto, 1971) that treatment of the glass with a silane coupling agent,* which might be expected to change the adhesion factor, does not necessarily significantly affect the modulus. It was suggested that residual compressive stresses might mask the lack of adhesion in the untreated system. In contrast, positive effects of silanes have been reported for the flexural modulus of epoxy-resin laminates containing small glass spheres (Wells, 1967) and for epoxy resins filled with glass beads or powders (Wells, 1967). Discrepancies of this type are often found in studies of filled systems, the precise characterization of which is difficult (Nielsen, 1967a).

Additional evidence for the validity (to a good approximation) of a Kerner-type equation—in this case, the unmodified Kerner equation—as applied to rigid polymeric matrices filled with rigid particles (up to a filler volume fraction of 0.5) has been given by Kenyon and Duffey (1967), Ishai and Cohen (1967), Moehlenpah *et al.* (1970, 1971), Manson and Chiu (1972) [based on Chiu (1973)], and Brassell and Wischmann (1974).

Modification of the basic equation may be useful in some cases; the modification proposed by Nielsen (1970b) has been shown by Lewis and Nielsen (1970) and Nielsen and Lewis (1969) to be effective with glass-bead-filled epoxy resins and by Narkis and Nicolais (1971) with glass-bead-filled styrene–acrylonitrile copolymers at a temperature close to the glass temperature. A comparison of the Mooney, Kerner, and modified Kerner equations is given in Figure 12.2 (Lewis and Nielsen, 1970). Thus, in order to test predictions of moduli under dynamic conditions and to study damping in particulate-filled polymers, Lewis and Nielsen (1970) studied the dynamic mechanical behavior of an epoxy resin filled with glass beads (5–90 μm in size; volume fractions up to 0.40) over a wide temperature range, from 100°C below to 50°C above the glass temperature (Figure 12.3). It was found that the relative shear modulus G_c/G_p in the glassy state was somewhat

---

* A coupling agent is a compound, frequently a silane, which improves the bonding of an inclusion or filler to the matrix, or at least improves the mechanical coupling between them (Plueddemann, 1974b; Ranney *et al.*, 1974).

higher than predicted by the Kerner (1956*b*) equation, which serves in effect as a lower bound. Better representation of the data was obtained by taking into consideration the maximum possible packing fraction ϕ_m. As shown in Figure 12.2, the following modification of Kerner's equation was shown to fit experimental data rather well (Lewis and Nielsen, 1970; Nielsen and Lewis, 1969):

$$\frac{G_c}{G_p} = \frac{1 + ABv_f}{1 - B\psi v_f} \tag{12.5}$$

where G_c and G_p are the shear moduli for the composite and matrix resin,

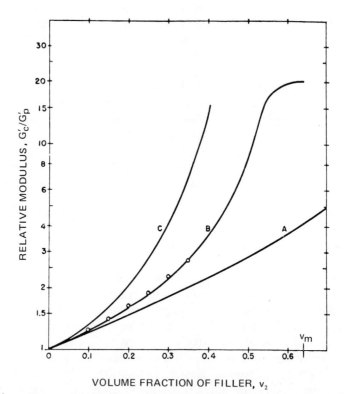

VOLUME FRACTION OF FILLER, v_2

Figure 12.2. Dependence of relative modulus (composite/polymer) on concentration for (A) the original Kerner equation; (B) the modified Kerner equation; (C) the Mooney equation. Circles correspond to experimental results. Solid curves are calculated with $v = 0.35$, G_f/G_p (relative modulus: filler/polymer) $= 25$, and $\phi_m = 0.64$.

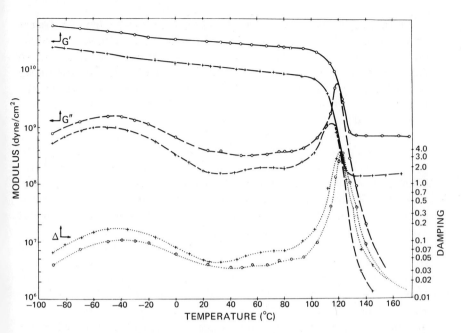

Figure 12.3. Shear modulus (—), loss modulus (--), and damping (···) vs. temperature for unfilled epoxy (+) epoxy with 0.41 volume fraction (○) spheres of diameter 10–20 μm (Lewis and Nielsen, 1970.)

respectively, v_f is the volume fraction of filler, and A, B, and ψ are given by*

$$A = \frac{7 - 5v}{8 - 10v} \tag{12.6a}$$

$$B = \frac{G_f/G_p - 1}{G_f/G_p + A} \tag{12.6b}$$

$$\psi v_f = 1 - \exp\frac{-v_f}{1 - (v_f/\phi_m)} \tag{12.6c}$$

or

$$\psi = 1 + [(1 - \phi_m)/\phi_m^2]v_f \tag{12.6d}$$

with v the Poisson ratio of the matrix and G_f the shear modulus of the filler. In effect, this equation is a more general version of the equation proposed by Halpin and Tsai (Ashton et al., 1969; Nielsen, 1970b).

---

* The two functions proposed for ψv_f both fulfill the boundary conditions imposed: however, one may give a better fit than the other in a given case.

Good agreement with equation (12.5) was also obtained by Narkis and Nicolais (1971), who studied the effect of glass bead concentration on styrene–acrylonitrile copolymers at 110°C, a temperature close to the glass temperature, the best fit being obtained with equation (12.6d) to represent ψv_f. As expected, the filler shifted the relaxation curves to longer times, in proportion to the volume fraction of filler—an observation consistent with reported increases in T_g (see below).

Instead of using equations (12.5) and (12.6a)–(12.6d) to correct for a slight underprediction of moduli noted for the original Kerner equation [equation (12.3)], Jenness (1972) found that the relatively untested Wu equation (1966) was able to account for the glassy moduli of an epoxy resin filled with a variety of particulate fillers, including glass spheres, silica, alumina, carbon, and graphite; the effect of particle shape was also reasonably accounted for by using the appropriate form of the equation. The effect of aspect ratio (length/thickness) of platelets has been studied by Lusis et al. (1973), who found that the relative modulus of a mica-reinforced resin varied strongly with aspect ratio up to a value of about 200 [see also Ashton et al. (1969), Halpin (1975), and Halpin and Kardos (1972) for discussion of particle shape].

In at least one case, a second-order effect has been observed. Lewis and Nielsen (1970) found that relative shear modulus in their glass-bead-filled epoxy-resin systems increased slightly but significantly with temperature up to the glass temperature; this effect was ascribed to a reduction in modulus of the matrix due to thermal stresses arising from the inequality of coefficients of expansion (Lewis and Nielsen, 1970; Nielsen and Lewis, 1969). In all cases, a slight apparent effect of particle size (attributed to poor distribution of the larger particles) was noted as well as an effect of stresses induced by thermal inequalities; all data for the rubbery modulus G_r were therefore extrapolated to zero filler particle size and to T_g. This point is worth noting, for it may help account for discrepancies between various sets of data and between predictions and experiment.

Although relative moduli were found to be increased by presence of the glass spheres, only a slight increase in T_g was noted, up to 3°C for the most highly filled systems. Curiously, maximum increases were noted for samples that contained no silane coupling agent to enhance adhesion. Thus the effect of filler on relaxation behavior was, as pointed out previously, relatively small, though significant. Somewhat larger increases in glass-bead-filled epoxy resins have been reported by Manson and Chiu (1972, 1973a).

In the rubbery plateau region, the modulus is also increased by the presence of a filler. For example, Maewaka et al. (1967) reported increases in both E' and E'' (storage and loss moduli, respectively) as a function of filler content in rubber–calcium carbonate systems. However, the Kerner and related equations are not suitable for quantitative predictions of rubbery moduli.

The Eilers (1941) equation has been frequently used to describe the behavior of such filled elastomeric systems, e.g., by Blatz (1956), Schwarzl *et al.* (1965), and Landel (1958) and Fedors and Landel (1967):

$$G_c/G_p = [1 + 1.25v_f/(1 - v_f/\phi_m)]^2 \tag{12.7}$$

where G_c/G_p is the modulus of the composite relative to the polymer, and v_f and ϕ_m are the volume fraction of filler and maximum packing fraction, respectively. At moderate volume fractions, the relationship is equivalent to van der Poel's (1958) formulation. With typical filled systems, the rubbery plateau region also appears to be extended to higher temperatures (or lower frequencies) than for the corresponding unfilled systems, perhaps due to the occurrence of a mechanical network which prevents the onset of viscous flow (Ferry, 1970, p. 459). Reinforcement in elastomeric systems is discussed specifically in Chapter 10.

The increase in modulus may also often be expressed in terms of the slightly different Mooney (1951) or Guth (1944)–Smallwood (1944) equations. For example, with glass-bead-filled epoxy resins, Lewis and Nielsen (1970) found agreement between predicted and observed values of modulus in the rubbery region using the Mooney (1951) equation:

$$\ln(G_c/G_p) = K_E v_f/[1 - (v_f/\phi_m)] \tag{12.8}$$

where K_E is the Einstein coefficient (Einstein, 1905, 1906, 1911) (2.5 for dispersed spheres). Manson and Chiu (1972) also found the Mooney equation to be useful for a rather similar system, though slightly better agreement at low filler contents was given by the Guth (1944)–Smallwood (1944) relationship:

$$G_c = G_p(1 + 2.5v_f + 14.1v_f^2) \tag{12.9}$$

In any case, the difference between the two equations is small at volume fractions up to, say, 0.25. Equation (12.9) was also used successfully by Landel (1958) to characterize the modulus of glass-bead–polyisobutylene composites in the long-time (rubbery) region of stress relaxation.

In addition to the expressions mentioned for predicting moduli in the elastic state, blending equations developed by Ninomoya and Maekawa (1966) have been adapted to predict frequency-dependent moduli of filler–polymer systems. Compliances were considered to be additive, and the following relations for relative moduli (D_p/D_c) were tested using a rubber–

calcium carbonate system:

$$D'_c(\omega) = f(v_f)D'_p(\omega\lambda_2)$$
$$D''_c(\omega) = f(v_f)D''_p(\omega\lambda_2)$$

(12.10)

where D' and D'' are the tensile storage and loss compliances at frequency ω, and the subscript c refers to the composite; $f(v_f)$ is a function of the volume fraction of filler v_f; and λ_2 represents the effect of the filler on the matrix relaxation time.

Good agreement was found up to filler concentrations of 50% if λ_2 was taken as equal to 1 and $f(v_f) = [1 - (v_f/0.74)]^{2.5}$, values which correspond to an equation proposed originally by Landel et al. (1965). For polymer–polymer systems, it may be necessary to use quite different values of λ_2 (Horie et al., 1967; Kambe and Kato, 1971).

Crystallites may also be considered to act as reinforcing fillers. For example, the rubbery modulus of poly(vinyl chloride) was shown by Iobst and Manson (1970, 1972, 1974) to be increased by an increase in crystallinity; calculated moduli in the rubbery state agreed well with values predicted by equation (12.9). Halpin and Kardos (1972) have recently applied Tsai–Halpin composite theory to crystalline polymers with considerable success, and Kardos et al. (1972) have used in situ crystallization of an organic filler to prepare and characterize a model composite system. More recently, the concept of so-called "molecular composites"—based on highly crystalline polymeric fibers arranged in a matrix of the same polymer—has stimulated a high level of experimental and theoretical interest (Halpin, 1975; Lindenmeyer, 1975).

In addition to the prediction of moduli as a function of filler concentration, it is of interest to consider the relationships between modulus and viscosity. For example, it may be useful to predict the modulus of the composite that may be expected to result from the solidification or curing of a given filled polymeric liquid whose viscosity can be determined. It is usually assumed that the viscosity η_c and shear modulus G_c of the composite are related as follows (Nielsen, 1967a):

$$\eta_c/\eta_p = G_c/G_p$$

(12.11)

where the subscripts c and p denote the value of viscosity or shear modulus for the filled and unfilled polymer, respectively. As pointed out by Nielsen (1968), however, the viscosity ratio is often larger than the modulus ratio. Two reasons for the discrepancy may exist: a greater disparity between the moduli of the two components in the viscous than in the solid state, and the fact that Poisson's ratio v for the solid matrix is not, as implicit in equation (12.8), equal to 0.50, but rather is close to 0.30. A third factor, the effect of a mismatch in coefficients of expansion, was later suggested by Nielsen and Lewis (1969) and Lewis and Nielsen (1970). Assuming that Kerner's (1956b) equation [equation (12.3)] gives a reasonable approximation to the shear modulus of filled plastics and that it fits viscosity data over a wide range of filler concentrations for an assumed value of Poisson's ratio of 0.5, Nielsen calculated the deviations to be expected when the matrix has a Poisson ratio significantly less than 0.5 while the liquid has a value close to 0.5. Indeed, if $v = 0.3$ for the matrix, and the filler is much more rigid than the matrix, the viscosity ratio may be about 20% higher than the modulus ratio.

12.1.1.2. Stress Relaxation and Creep

Just as with polymers themselves, the behavior of a filled viscoelastic polymer is dependent on both temperature and time (frequency, loading rate), and may often be characterized using the Williams–Landel–Ferry (Williams *et al.*, 1955) time–temperature superposition [see Section 1.13, and Ferry (1970, p. 386)]. Thus, Landel (1958), studying polyisobutylene loaded with small glass beads, showed that the temperature dependence of relaxation times could be expressed in terms of the Williams–Landel–Ferry (WLF) equation (Section 1.5.7). By using an adjustable value for T_s, the WLF reference temperature, and selected values for the WLF constants, Landel found that T_s was a nearly linear function of the volume fraction of filler; at a volume fraction of 0.37, T_s was increased by 7°C. Presumably the increase in T_s implies a corresponding increase in T_g. Other evidence for the applicability of superposition techniques to various filled polymer systems may be found in the work of Becker and Oberst (1956), Blatz (1956), Bills *et al.* (1960), and Payne (1958). [For a critical discussion of all these studies, see Ferry (1970).] There are, however, differences in the shapes of relaxation spectra. For example, Payne (1958) has shown that a filler flattens both relaxation and retardation spectra, and causes a shift of the transition region to lower frequencies (probably corresponding to a small increase in T_g).

A typical example of recent studies of time–temperature–modulus relationships may be found in papers by Moehlenpah *et al.* (1970, 1971), who examined crosslinked epoxy resins filled with glass beads, fibers, or air bubbles. The initial tangent modulus in compression was seen to increase with a decrease in strain rate; flexural and tensile moduli were reported to behave in a similar fashion. The WLF shift factor was essentially independent of the type of filler used and of the mode of loading. Kerner's equation was found to hold for the particulate composites in the glassy range.

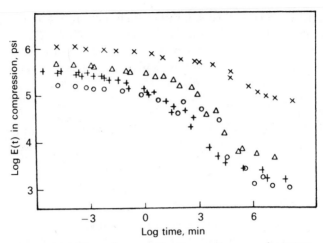

Figure 12.4. Flexural stress relaxation master curve for epoxy composites ($T_{ref} = 50°C$). ($\times$) Continuous transverse; ($\triangle$) particulate-filled; (+) unfilled; (O) foam. (Moehlenpah *et al.*, 1970, 1971.)

In the same studies, Moehlenpah *et al.* (1970, 1971) obtained master curves for the stress relaxation of their epoxy systems, at least into the glass-to-rubber transition region (Figure 12.4), and demonstrated similar behavior of both the stress relaxation modulus and the tensile modulus as a function of strain rate. As with the strain rate studies mentioned, no effect of filler type on the WLF shift factor was observed. All solid fillers increased the modulus of the system, the fibers being more effective than the spheres. The bubbles, as expected (Nielsen, 1967*a*), decreased the modulus.

The effect of fillers on creep phenomena (essentially the inverse of stress relaxation) is also of interest; a detailed study by Nielsen (1969*b*) of creep in filled polyethylene is illuminating. Kaolin and wollastonite were used as fillers, both treated with a silane coupling agent and untreated. A major aim was to discover whether the major effect of a filler is due to its effect on elastic modulus, or whether a filler also changes the viscoelastic nature of the system. As reported in previous work by others, the presence of a filler did in fact reduce the creep, the relative effect being nearly independent of the applied stress. The nature of the filler and the surface treatment were also found to be important. In experiments at a constant volume fraction (0.2), kaolin was more effective than wollastonite. Silane treatment of the filler surface tended to decrease creep, especially if the specimens had been soaked in water.

In general, the creep elongation could be estimated in terms of the shear moduli alone:

$$\frac{\varepsilon_p(t)}{\varepsilon_c(t)} \simeq \frac{G_c}{G_p} \tag{12.12}$$

where $\varepsilon(t)$ and G represent the elongation at time t and the shear modulus, respectively, and the subscripts c and p refer to the filled and unfilled polymers, respectively. However, at long times, the creep is slightly less than the value predicted by equation (12.12), the deviation being attributed to the occurrence of filler–polymer interactions. As mentioned above, such interactions were invoked by Landel (1958) to explain the ability of glass beads to increase the glass temperature of polyisobutylene.

12.1.1.3. Damping

The damping behavior of polymeric composites is frequently complex (Galperin and Kwei, 1966; Morgan, 1974; Nielsen, 1969b). Often a filler increases the relative damping, especially if the matrix is rubbery (Galperin and Kwei, 1966; Kardos et al., 1972; Landel, 1958; Lewis and Nielsen, 1970; Morgan, 1974; Nielsen, 1969b). For example, in Nielsen's (1969b) study of filled polyethylene, the damping, expressed as the logarithmic decrement Δ, was found to exceed the values expected from the following relation, which would hold if damping were due to polymer only:

$$\Delta_c \simeq \Delta_p v_p \tag{12.13}$$

where the subscripts c and p refer to the filled and unfilled resin, respectively, and v is the volume fraction of polymer. The logarithmic decrement Δ is itself defined by

$$\Delta = \ln \frac{A_1}{A_2} \tag{12.14}$$

where A_1 and A_2 are the amplitudes of successive oscillations. Thus, at higher concentrations or longer times, additional damping mechanisms must come into play. This observation was confirmed in Lewis and Nielsen's (1970) study of the transitions in bead-filled epoxy resins (see Figure 12.3), as well as in extensive studies by Hirai and Kline (1972) and Jenness (1972) using epoxy reins containing a variety of particulate fillers. A major mechanism may involve friction between particles or between particles and polymer, and provide additional means of dissipating energy. In addition to such an increase in damping due to filler–matrix shear interaction, contributions from the inhomogeneous distribution of particles and from thermal stresses may also exist. Damping was also affected significantly by silane treatment of the glass spheres, the treated systems exhibiting higher damping. On the other hand, damping was decreased by the use of aggregates of spheres, apparently because of trapping of matrix resin within the aggregates and subsequent removal of any contribution of the trapped resin to the damping. Decreases in relative damping have also been observed in other systems, for example, TiO_2-filled poly(vinyl acetate) (Galperin and Kwei, 1966) and silica- and carbon-filled SBR in the glassy state (Morgan, 1974); there is also evidence for a decrease in damping in the region of the γ transition in filled epoxy resins, perhaps due to interaction between the filler surface and unreacted epoxy groups (Hirai and Kline, 1972).

The high damping capacity of filled viscoelastic polymers makes them useful as vibration-damping materials (Thurn, 1960); damping is not only increased, but high damping exists over a wider range of temperature than in an unfilled system (see Section 13.5).

12.1.2. Mechanical Behavior at Large Strains

12.1.2.1. Large vs. Small Strains

The behavior of filled polymers is very sensitive to the magnitude of the deformation imposed; nonlinear behavior is often observed at much lower strains than in unfilled systems (Ferry, 1970, p. 460; Payne, 1961). For example, in carbon-filled elastomers the storage modulus of specimens subjected to a cyclic deformation has been shown to decrease markedly, and $\tan \delta$ to increase, as the amplitude is increased at a low total strain (Fletcher and Gent, 1953). Also, Payne (1960) has shown that the absolute Young's modulus of a filled rubber decreases with increasing strain (above 0.1 %), approaching that of the unfilled rubber. This effect, which may be to some extent reversible, is usually attributed to a dewetting of the filler particles or to a breakdown of filler agglomerations. Dewetting behavior is clearly dependent on the filler and its surface characteristics; the stronger the filler–resin interaction, the greater the number of primary chemical bonds which must be broken, and the larger the strain required for dewetting (DeVries et al., 1973; Lange, 1975) (see Section 10.6). Thus, in any attempt to characterize temperature–time behavior in filled polymers, strain must be considered as a potentially important variable, even at low levels.

12.1.2.2. Stress Relaxation

While the stress relaxation of a filled plastic as a function of temperature can be characterized in terms of a WLF-type function (Section 12.1.1.2; Williams et al., 1955), a significant effect of strain has been noted by Narkis and Nicolais (1971) for copolymers of styrene and acrylonitrile filled with glass beads and examined at temperatures above the T_g. As mentioned above, such an effect tends to occur in filled polymers at strains which are relatively low, though, to be sure, strain effects occur in unfilled polymers as well at other than very low values (Andrews et al., 1948). Whereas the modulus $E(t)$ varied by as much as 40% when the strain was varied from 0.1 to 0.3, the data could be fitted by the following empirical function:

$$\frac{[E_{c\varepsilon}(t)]_2}{[E_{c\varepsilon}(t)]_1} = \frac{1 + (\varepsilon_p)_1}{1 + (\varepsilon_p)_2} \tag{12.15}$$

where $[E_{c\varepsilon}(t)]_1$ and $[E_{c\varepsilon}(t)]_2$ are the moduli for the composite measured at time t and strain ε; ε_p is the strain in the polymer component; and the subscripts 1 and 2 refer to two different times of measurement. All strains were defined by

$$\varepsilon = \ln(L/L_0) \tag{12.16}$$

where L and L_0 are the final and initial lengths, respectively. The effect of filler volume fraction v_f is implicit in the value of ε_p, which was found to conform to the following equation giving ε_c, the strain in the composite, as a function of ε_p and v_f:

$$\varepsilon_c = \varepsilon_p + (1 - Kv_f^{1/3}) \tag{12.17}$$

for which the best fit was obtained using $K = 1$. This expression is equivalent to one for yield strain derived earlier by Smith (1959), with a value for K of 1.105; it has also been used by Nielsen (1966) (see Section 12.1.2.3).

In other words, just as was shown by Andrews et al. (1948) for unfilled polymers, the ratio of two relaxation moduli for a given system at a common time but at different strains was found to be essentially independent of time, at least within a certain range of time and strain. Thus stress relaxation curves for a filled polymer at different strains ε_c could be shifted along the modulus axis to give a master curve at a reference strain (Table 12.1). The corresponding modulus function becomes

$$E_c(t) = \left[1 + \frac{\varepsilon_c}{(1 - v_f^{1/3})}\right]E_{c\varepsilon}(t) = (1 + \varepsilon_p)E_{c\varepsilon}(t) \tag{12.18}$$

where v_f is the volume fraction of filler. In other words, a strain-independent modulus may be determined by multiplying the relaxation modulus at a given strain by the factor $(1 + \varepsilon_p)$, with ε_p being determined by equation (12.17).

12.1.2.3. Yield and Strength Phenomena

Clearly, the incorporation of a rigid particulate filler must invariably increase the modulus of a lower modulus matrix. However, in any optimization of properties, the effects of fillers on ultimate properties such as yielding,

Table 12.1

Strain-Independent Modulus Calculated from Relaxation Data
at Different Strains at 127°C[a]

Time, min	$1.248E_{0.1}(t)$	$1.496E_{0.2}(t)$	$1.744E_{0.3}(t)$
1	8.85	9.05	10.0
2	6.85	6.95	7.20
3	5.75	5.83	5.75
5	4.55	4.65	4.62
10	3.18	3.21	3.14
20	2.00	2.10	2.00
40	1.33	1.35	1.34

[a] Modulus units are $(dyn/cm^2) \times 10^{-6}$; $\phi = v_f = 0.217$. Narkis and Nicolais (1971).

tensile, and impact behavior must be carefully considered (see Section 12.1.2.4 for discussion of fracture per se). Indeed, the presence of a filler has been shown to have marked and complex effects on the mode of failure and on the yield, tensile, and impact strengths. Although rigorous treatment of these phenomena is not yet available for even unfilled polymers (Leidner and Woodhams, 1974; Nielsen, 1966; Lange, 1974), it is reasonable to briefly examine the evidence.

Using a simple model, Nielsen (1966) has sought to extend additivity predictions to include stress–strain behavior as a function of filler concentration for the cases of (a) perfect adhesion between polymer and filler and (b) no adhesion between them. Since some of the predictions have been verified at least qualitatively in some cases (Trachte and DiBenedetto, 1971; Nicolais and Nicodemo, 1973), the approach is worth reviewing. Although the model neglects many details characteristic of real systems (for example, residual compressive stresses due to cooling from elevated processing temperatures), it does focus attention on the contradictory roles of adhesion in increasing and decreasing ultimate strain. (For an alternate approach to tensile strength, see Section 12.1.2.4.)

For the model shown in Figure 12.5, assuming perfect adhesion, Nielsen calculated the elongation of the plastic per se (in the composite) relative to the overall elongation of the filled specimen to be

$$\varepsilon_p/\varepsilon_c \cong 1/(1 - v_f^{1/3}) \qquad (12.19)$$

where ε_p is the actual elongation of the plastic, ε_c is the overall elongation of the filled system [i.e., $\varepsilon = (L - L_0)/L_0$, where L and L_0 are the final and initial specimen lengths, respectively], and v_f is the volume fraction of filler. [As noted above, a similar relationship was derived by Smith (1959) and

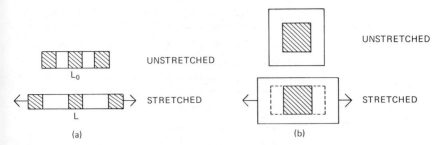

Figure 12.5. Models for filled polymers. (Nielsen, 1966.) (a) Perfect adhesion; (b) no adhesion.

was tested successfully by Narkis and Nicolais (1971).] Assuming that the *polymer* in the composite breaks at the same elongation as the unfilled polymer, the elongation to break for the filled system becomes

$$\varepsilon_B(\text{filled})/\varepsilon_B(\text{unfilled}) = 1 - v_f^{1/3} \tag{12.20}$$

The curve of this function (plotted in Figure 12.6, lower curve) suggests that the deleterious effect of filler on elongation is relatively greater at very low levels of filler as long as adhesion is maintained and aggregation is absent.

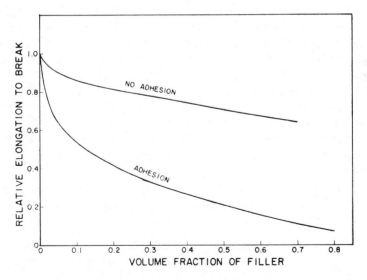

Figure 12.6. Theoretical curves for the elongation to break for the case of perfect adhesion and no adhesion between the filler and polymer phases. (Nielsen, 1966.)

Exceptions to the expected decrease in elongation with increasing filler content (Nielsen, 1966) may, however, occur. Thus Lavengood et al. (1971) have observed increases in toughness and ultimate elongation in glass-sphere-filled poly(phenylene oxide). The explanation proposed is based on the occurrence of crazing in the matrix—a phenomenon not usually found with rigid fillers in matrices. Occurrence of such a deformational mechanism will, of course, make equation (12.19) inapplicable (see also Section 12.1.2.4).

In considering tensile strength σ_B, it is assumed that

$$\sigma_B = E\varepsilon_B \tag{12.21}$$

where ε_B is given by equation (12.20), and E, Young's modulus, is given by equations such as Kerner's (1956b) [equation (12.3)] or Eilers' (1941) [equation (12.7)]. The impact strength σ_I, if crudely assumed to be equal to the area under the stress–strain curve, is given for a glassy system by*

$$\sigma_I = \sigma_B\varepsilon_B/2 = E\varepsilon_B^2/2 \tag{12.22}$$

It may be seen from Figures 12.7 and 12.8 that use of both the Kerner (1956b) and Eilers (1941) relationships for E predicts decreases in tensile and impact strengths at low concentrations of filler. At higher concentrations of filler, each equation predicts a tendency for the tensile strength to increase somewhat as a function of filler content, thus offsetting to various degrees the initial decreases. In an equivalent range of concentration, a similar offsetting of the initial decrease in impact strength is predicted by Eilers', though not Kerner's, equation. Such complex behavior is essentially a consequence of the balancing of the predicted increase in E due to the filler with the predicted decrease in ε.

If, in contrast to the case just treated, the filler does not adhere to the matrix, then the filler particles cannot carry any of the load (Figure 12.5). In addition to the effective porosity thus introduced, stress concentrations around the particles will reduce the strength still further; this effect cannot be calculated simply. Thus Nielsen proposed that the tensile strength should be given approximately by an equation of the following general form†:

$$\sigma_{Bc}/\sigma_{Bp} \cong (1 - K'v_f^{2/3})S' \tag{12.23}$$

---

* Clearly, equation (12.22) can be valid only for systems exhibiting linear stress–strain behavior. In addition, impact strength in real polymers is not equal to the area under a low-strain-rate stress–strain curve.

† In the derivation of equation (12.23) it is assumed that the *area* fraction of filler is equal to $v_f^{2/3}$; in fact, however, area fraction is equivalent to v_f—an identity long used by metallurgists to determine volume fractions of phases from the area fractions observed on polished surfaces (Piggott and Leidner, 1974). [Indeed, the first-power relationship is claimed to give better agreement with some of the data in the literature (Piggott and Leidner, 1974).] However, over much of the range of v_f encountered experimentally, the two dependences are difficult to distinguish.

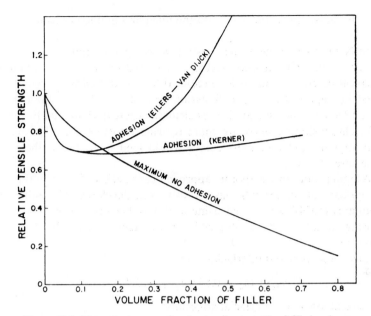

Figure 12.7. Theoretical curves for the tensile strength of filled polymers (Nielsen, 1966). Curve for case of no adhesion was calculated using stress concentration factor $S' = 1$.

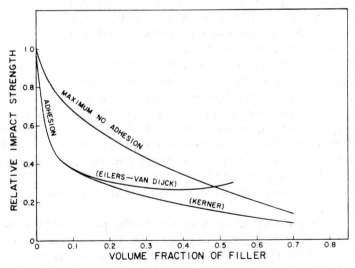

Figure 12.8. Theoretical curves for the impact strength (energy under the stress–strain curve) of filled polymers (Nielsen, 1966). Curve for case of no adhesion was calculated using stress concentration factor $S' = 1$.

where the subscripts c and p refer to the filled and unfilled polymer, respectively.

In Nielsen's treatment, the constant K' is equal to unity, and the value of S', the stress concentration factor, is assumed to be equal to about 0.5 for typical cases. The curves for predicted tensile and impact strengths are plotted in Figures 12.7 and 12.8, assuming that $S' = 1$. [For the curve in Figure 12.8 the value used for E was calculated using the Sato and Furukawa (1963) relationship.] Clearly, for more realistic values of S' (~ 0.5) the predicted strength and toughness curves would tend to lie below the curves for the case of $S' = 1$.

A similar but more elaborate approach to the effect of filler (in this case, spheres) content on strength has recently been developed by Leidner and Woodhams (1974)* and tested using a glass-sphere–polyester system; the following equations were obtained for the tensile strength σ_{Bc} of the composite:

Case of high v_f and interfacial adhesion:

$$\sigma_{Bc} = (\sigma_a + 0.83\tau_p)v_f + \sigma_a S'(1 - v_f) \tag{12.24a}$$

Case of low v_f or interfacial adhesion:

$$\sigma_{Bc} = 0.83p\alpha v_f + K\sigma_p(1 - v_f) \tag{12.24b}$$

Here σ_a and σ_p are the ultimate tensile strengths of the interfacial bond and the matrix, respectively; τ_p is the shear strength of the polymer matrix; S' is the stress concentration factor; K is a parameter dependent on particle diameter; p is the y component of the pressure exerted by the matrix on a sphere; and α is the coefficient of friction.

Experimental results tended to confirm the assumption that equation (12.24b) should be valid in the limit of low values of v_f (say, <0.2), regardless of the degree of interfacial adhesion. According to the analysis, the strength in a well-bonded composite should exhibit a minimum at a particular value of v_f, following equations (12.24b) and (12.24a) at low and high values of v_f, respectively. Though data were somewhat scattered, such a tendency was evident, so that, at least for the larger spheres, curves qualitatively resembled those of Figure 12.7 (Leidner and Woodhams, 1974); small spheres were, however, observed to obey equation (12.24a) to a good approximation.

In any case, experience tends to confirm the qualitative prediction of lower strengths for filled polymers under many circumstances [see Piggott and Leidner (1974) and also the discussion of fracture in Section 12.1.2.4]. For example, Wambach et al. (1968) found that the tensile yield strength of

---

* Based in part on equations developed earlier by Hajo and Toyoshima (1973).

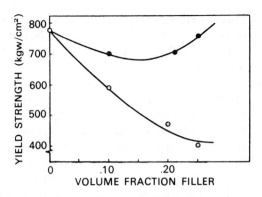

Figure 12.9. Effect of volume fraction of glass beads on yield strength of poly(phenylene oxide). (○) Untreated; (⊗) A-1100 treated. The term A-1100 refers to the silane coupling agent used to enhance adhesion (Wambach *et al.*, 1968). Note that the upper curve corresponds to *ultimate* strength, for fracture occurred prior to yielding.

a glass-bead-filled poly(phenylene oxide) decreased significantly as a function of volume fraction when the glass was untreated (Figure 12.9). On the other hand, as predicted by Nielsen (1966), the ultimate strength passed through a minimum for the silane-treated systems. With poor matrix–filler adhesion, yielding was always observed, as in the pure matrix; on the other hand, with good adhesion, specimens failed before yielding.

In the latter case, results were consistent with an immobilization of the matrix surrounding the filler particles (see also Section 12.3). Using glass-bead–epoxy systems, Sahu and Broutman (1972) reported decreases in flexural and tensile strengths for systems exhibiting good adhesion (i.e., systems in which an appropriate silane coupling agent was applied to the glass spheres). Although the use of finite element analysis in the latter study did not yield very good agreement between predicted and observed values, Nicolais and Nicodemo (1973), using data of Sahu and Broutman (1972), reported excellent agreement for tensile strength (and fairly good agreement for flexural strength) for the case of poor adhesion. For the case of good adhesion, assuming the strength of the composite equals the strength of the polymer itself, Nicolais and Nicodemo (1973) predicted that the strength would be independent of filler content, as observed (an observation in conflict with the prediction of Figure 12.7). With poor adhesion, a variant of equation (12.23) was used in which K' was taken as 1.21; the variant was derived using an approach developed earlier by T. L. Smith (1959) to treat yield strain in a filled polymer.

Applicability of equation (12.23) (with $K' = 1.21$) to several filled polymer systems has been reported (Nicolais and Narkis, 1971).

In any case, it should be noted that the predictions of simple models such as discussed above assume linear stress–strain behavior. In practice, the presence of a filler may not decrease ductility, as mentioned above (Sahu and Broutman, 1972; Wambach *et al.*, 1968), but may increase it instead (Nicolais and Narkis, 1971), depending on the matrix and the degree of adhesion. Such behavior may well be complicated by dewetting or cavitation, which depend strongly on interfacial characteristics (Lange, 1974; Section 12.1.2.1).

Of course it is of interest to consider not only strength per se, but also the effects of temperature, strain rate, and filler size and concentration on yield phenomena and stress–strain curves. Although relatively few experimental studies are available, several trends may be discerned in the literature. In his early studies of glass-bead-filled elastomers, T. L. Smith (1959) observed that the yield stress and yield strain decreased as the filler content was increased—an observation undoubtedly related to dewetting (Section 12.1.4). The yield strain was found to be very dependent on temperature. Similar decreases in yield stress and yield strain with filler content for filled poly(phenylene oxide) have been reported by Moehlenpah et al. (1970, 1971) and Wambach et al. (1968). There may also be an effect of particle size in some cases. In a study of the stress–strain behavior of filled polyurethanes, Nicholas and Freudenthal (1966) showed that the incorporation of NaCl* changed the stress–strain behavior from rubbery to ductile, and that the yield stress varied inversely with the particle size. The yield stress was also studied as a function of strain rate by Ishai and Cohen (1967), who noted that the compressive yield stress for sand-filled epoxy-resin systems was a linear function of the logarithm of the strain rate, with the slope being independent of filler concentration. A similar linear dependence of tensile yield stress on the logarithm of the strain rate has been reported by Moehlenpah et al. (1970, 1971) in a study of tensile yield stress in glass-filled epoxy resins in the region of ductile failure. It was also shown that yield stress isotherms could be rather well superimposed by shifting along the log strain-rate axis (Figure 12.10); the shift factors required did not depend on the filler, but rather on the matrix.

Making use of equation (12.24), Nicolais and Narkis (1971) proposed the following master curve equation to take account of both temperature and filler effects on yield stress as a function of strain rate:

$$\frac{\sigma_{yc}}{1 - 1.21v_f^{2/3}} = A + B \ln(\dot{\varepsilon}a_T) \qquad (12.25)$$

where σ_{yc} is the yield stress of the composite, $\dot{\varepsilon}$ is the strain rate, a_T is the WLF shift factor (Williams et al., 1955), v_f is the volume fraction of filler, and A and B are numerical constants. Reasonable agreement between

---

* Systems comprising an elastomer filled with NaCl are of interest as models for solid propellants, with the elastomer serving as a binder for the inorganic component, which is usually a perchlorate.

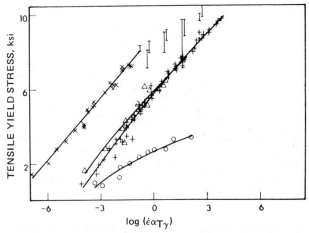

Figure 12.10. Tensile yield stress vs. strain rate for epoxy composites ($T_{ref} = 50°C$) (Moehlenpah *et al.*, 1970, 1971). The term α_{T_y} is the WLF shift factor. ($\times$) Continuous transverse; ($\triangle$) particulate-filled; ($+$) unfilled; ($\bigcirc$) foam; ($\rceil$) brittle failure, range of ultimate strengths.

prediction and theory was reported for glass-bead–(styrene–acrylonitrile) systems in the glassy state (Figure 12.11). In effect, a double shift with respect to both temperature and filler content is performed, as in the case of stress relaxation (Section 12.1.1.2); Narkis and Nicolais, 1971). Since then, Nicolais and DiBenedetto (1973) have also reported that the yield stress and creep of glass-sphere-filled poly(phenylene oxide) can be superimposed to form master curves, and that the shift factors are separable functions of temperature, stress, and composition.

Superposition techniques may also be used to correlate stress–strain behavior in the rubbery state. In their study of styrene–acrylonitrile co-polymers filled with glass beads, Narkis and Nicolais (1971) obtained stress–strain curves at temperatures above T_g. Stress–strain curves were plotted for different fractions of filler, and in terms of both the polymer and composite strain. At a given strain, the stress increased with increasing filler concentration, as expected. It was possible to shift curves of stress vs. polymer strain along the stress axis to produce a master curve (Figure 12.12). In addition to the empirical measurements, an attempt was made to calculate stress–strain curves from the strain-independent relaxation moduli (see Section 1.16 and Chapter 10) by integrating the following equation:

$$f(t) = \dot{\varepsilon} \int_{-\infty}^{\ln t} t E(t)\, d \ln t \qquad (12.26)$$

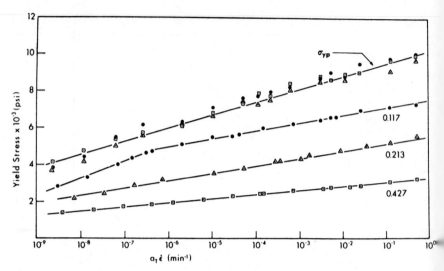

Figure 12.11. Yield stress vs. $a_T\dot{\varepsilon}$ for styrene–acrylonitrile copolymers containing different concentrations of glass beads ($T_{ref} = 24°C$; the numbers on the curves give v_f, the volume fraction of filler). The upper curve corresponds to equation (12.25) with constants A and B equal to 1.0×10^4 and 3×10^2, respectively; it also corresponds closely to estimated values of σ_{yp}, the yield stress of unfilled polymer. (Nicolais and Narkis, 1971.)

where f is the force and $\dot{\varepsilon}$ is the strain rate; the strain-independent moduli were calculated by means of equation (12.18). Shapes of the curves agreed well with prediction, though quantitative agreement with the prediction was only fair.

12.1.2.4. Fracture Toughness

While incorporation of a particulate filler in a plastic invariably results in an increase in modulus, the effect on fracture properties is complex, as was seen in the case of tensile strength. Although there are exceptions, the energy required to induce and propagate a catastrophic crack at normal temperatures is often reduced by the presence of a particulate filler. This is not surprising, for fillers act as stress concentrators and thus supply a composite with potential sites for crack growth, especially if debonding occurs between the filler and matrix (Wambach et al., 1968; Lange, 1974). At the same time, in principle rigid fillers may serve to divert cracks (and thus increase total surface area of fracture) or dissipate energy otherwise associated with crack growth. In practice, such balancing of effects is often not observed, perhaps due to several factors, such as inherent brittleness of the filler itself, and restrictions imposed on mobility (and hence plastic deformation or increased surface area due to filler pullout) of a strongly adherent matrix.

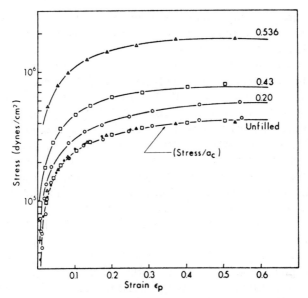

Figure 12.12. Stress vs. polymer strain for styrene–acrylonitrile copolymers containing different concentrations of glass beads; $\dot{\varepsilon}_c = 0.005 \ \mathrm{min}^{-1}$, temperature, 127°C; the numbers on the curves give v_f, the volume fraction of filler. Data for filled systems can be shifted by an empirical factor a_c to give good agreement with curve for unfilled polymer. (Nicolais and Narkis, 1971.)

Of course, the ability to dissipate energy is only one of the characteristics concerned in "toughness," or the ability to resist fracture (see Section 3.2). Thus the modulus itself contributes to the total energy and the stress required for fracture; a particulate filler may, by raising the modulus of a very low-modulus polymer, increase the toughness.

Typical manifestations of embrittlement are obvious in a number of measurements, such as impact strength*; more detailed studies tend to confirm frequent adverse effects on ductility. For example, in the investigation of filled epoxy resins by Moehlenpah *et al.* (1970, 1971) cited previously, it was found that, for both compressive and tensile loading, filling raised the temperature for the brittle–ductile transition at a given strain rate. In other words, the filled material was brittle at higher temperatures and lower strain rates than the unfilled polymer. The effect of filler on the ductile to ductile–rubbery transition was, on the other hand, much less.

---

* See property tables in recent compendia; e.g., *Modern Plastics Encyclopedia* (1974–1975).

There are, of course, many measures of toughness, such as the area under a stress–strain curve, the rate of slow crack growth prior to failure, impact strength, the characteristic surface fracture energy, and the fracture toughness (Andrews, 1968; Broutman and Krock, 1974; Broutman and Sahu, 1971a; DiBenedetto, 1973; DiBenedetto and Wambach, 1972; Eirich, 1975; Fallick et al., 1967; Gent, 1972; Hertzberg et al., 1970a; Irwin, 1956, 1960; Lange, 1974; Manson and Hertzberg, 1973a; Nielsen, 1966; Peretz and DiBenedetto, 1972; Rosen, 1964; Trachte and DiBenedetto, 1971; Wambach et al., 1968). Some correlations between these may be expected, but also some differences, for in general the behavior will depend on the rate and nature of the loading. [Interestingly, however, Wambach et al. (1968) have shown that fracture surface energy* in at least some composites is essentially independent of strain rate over a three-decade range.]

The "fracture toughness," a term defined by Irwin (1956, 1960) to characterize brittleness, provides a measure of the conditions required for catastrophic crack propagation in a material (see Section 1.6). One fracture toughness parameter is the surface fracture energy γ, defined as one-half G_c, the critical strain energy release rate above which catastrophic failure occurs. In turn G_c is related to another convenient toughness parameter, the critical stress intensity factor K_c, a measure of the stress field at the crack tip. For fracture of an isotropic material in a plane strain mode† (Baer, 1964, p. 946):

$$\gamma = \frac{G_c}{2} = K_c^2 \frac{1 - v^2}{2E} \qquad (12.27)$$

where K_c is a measure of the stress intensity near the crack tip at the onset of catastrophic failure, and E and v are Young's modulus‡ and Poisson's ratio, respectively. K_c can be defined in terms of the geometry of a specimen and the stress to break (based on the original cross section); for a plate having two opposite edge notches,

$$K_c = \sigma_B \left[W \left(\tan \frac{\pi(a + r_y)}{W} + 0.1 \sin \frac{2\pi(a + r_y)}{W} \right) \right]^{1/2} \qquad (12.28)$$

where σ_B is the gross stress to break, W is the sample width, a is the half crack

---

* Wambach et al. (1968) use the term "fracture toughness" for the parameter sometimes referred to as the "fracture energy"; many authors refer to Irwin's critical stress field parameter K_c (see below) as "fracture toughness."

† This means that, in a plate having a transversely growing crack, conditions are such that all strain is in the plane of the plate, and that the surfaces of the plate are not pulled inward.

‡ Note that the value of E required is the value just prior to fracture: if dewetting occurs during the test, with the development of small cavities or "pseudovoids," the actual value of E may differ significantly from the value at low strains (Lange, 1974).

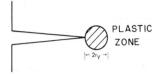

Figure 12.13. Schematic illustration of plastic zone (extensive deformation) at tip of a crack. Actual shape may vary from shape shown.

length just prior to catastrophic failure, and r_y is a measure of the so-called plastic zone size (Figure 12.13). Other expressions are available for other shapes. The parameter r_y takes account of plastic deformation near the crack tip, and is given by

$$r_y = \frac{1}{2\pi}\left(\frac{K_c}{\sigma_y}\right)^2 \tag{12.29}$$

where σ_y is the yield stress.

The effects of several factors on fracture toughness (or on other indexes of toughness) are clearly of interest: temperature; filler type, geometry, size, and content; and interfacial bonding. In recent studies of poly(phenylene oxide)(PPO) and epoxy resins filled with glass beads or short glass or graphite fibers, Trachte, DiBenedetto, and Wambach [Wambach et al., 1968; Trachte and DiBenedetto, 1971; DiBenedetto and Wambach, 1972; DiBenedetto, 1973] showed that Irwin's concepts of fracture mechanics (Irwin, 1956, 1960; DiBenedetto, 1973) could in fact be applied at moderate to fairly elevated temperatures. It was thus possible to gain considerable insight into the roles of several of the factors mentioned.

As shown in Figure 12.14 (Trachte and DiBenedetto, 1971), the values of γ for the PPO composites were nearly independent of temperature up to about 150°C, at which temperature gross yielding was observed, thus invalidating the numbers calculated for γ.* In contrast, the unfilled polymer showed a peak at 120°C. It was suggested that the gross yielding observed in the composite was due to the stress-concentrating effect of the filler particles, with consequent enlargement of the plastic zone. Differences in γ between different filled systems were small at room temperature, of the order of 2 at −50°C. However, a major effect of adhesion was noted. When spheres were treated with a silane coupling agent to promote adhesion, values of γ were decreased in accordance with Nielsen's (1966) predictions. When fibers were so treated, the opposite effect was found; values of γ

---

* The values reported experimentally reflect the energy dissipated in plastic deformation in addition to the intrinsic energy required to form new surfaces during brittle fracture of a glassy material (see Section 1.6). While, strictly speaking, the K_c approach assumes that the specimens exhibit linear elasticity in bulk, it provides useful comparative data. For other approaches, see, for example, Andrews (1968) and Manson and Hertzberg (1973a).

were increased. A plausible explanation for this significant difference between the particulate and fibrous filler was proposed. If improved adhesion is coupled with the possibility of a significant transfer of stress from the matrix to the filler, as with a fiber (see Section 12.1.1.1), γ will increase. On the other hand, if little additional transfer of stress can occur (as with spheres), improved adhesion will inhibit plastic flow at the filler surface, and *decrease* γ.

In any event, the addition of beads reduced γ at 20°C in proportion to the proportion of filler (Figure 12.15). Greater adhesion invariably resulted in a lowering of fracture toughness for the beads. Micrographs of fracture surfaces show (Figures 12.16 and 12.17) that the untreated beads tended to pull out clearly from the resin, while the silane-treated beads tended to lie away from the fracture surface and, if exposed, to retain some adherent resin. Thus it seems plausible that in the latter case constraints due to the filler may inhibit plastic deformation and minimize the overall fracture surface area. The behavior of short fibers was quite different, with a slight maximum in γ at a fiber volume fraction of 0.1 and a relative independence of fiber concentration up to a volume fraction of 0.3. These observations presumably reflect an improved load-bearing ability of fibers in comparison with spheres.

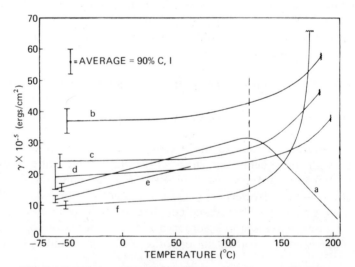

Figure 12.14. Fracture toughness of PPO composites. (a) PPO resin; (b) 20% A-1100 glass-fiber composite; (c) 20% glass-fiber composite; (d) 20% graphite-fiber composite; (e) 20% glass-bead composite; (f) 20% A-1100 glass-bead composite. The term A-1100 refers to the silane used as a coupling agent. (Trachte and DiBenedetto, 1971.)

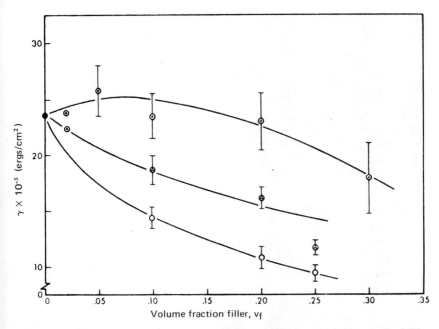

Figure 12.15. Effect of filler content on composite fracture energy at 20°C (Trachte and Di-Benedetto, 1971). (○) Glass fiber/PPO; (△) glass bead/PPO; (○) A-1100 glass bead/PPO (the term A-1100 refers to the silane used as a coupling agent).

Measurements of plastic zone size r_y were also interesting; wide differences between systems were observed. Generally, r_y increased with temperature, but fibers tended to increase r_y whether or not adhesion was improved by use of the coupling agent. Spheres tended to have little effect on r_y of the polymer when adhesion was high, but tended to increase r_y when adhesion was poor—observations consistent with the decrease in γ mentioned above. Microscopic estimates of the plastic zone size were also possible, for crazing and debonding induced density changes that caused whitening; reasonable agreement with calculated values was found. It may be supposed that more energy can be dissipated at a fairly weak interface than at a strong one.

Studies of this type are important in any consideration of toughening mechanisms. The fracture toughness may be expressed explicitly in terms of r_y, σ_y, and E; for a typical polymer having a value for Poisson's ratio of 0.35, γ becomes [from equations (12.27) and (12.29)]

$$\gamma = 2.8 r_y \sigma_y^2 / E \tag{12.30}$$

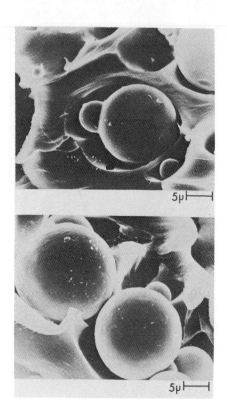

Figure 12.16. Fracture surface of PPO containing untreated glass beads (20 vol %) at 25°C; fast crack region (Trachte and DiBenedetto, 1971). Micrograph by scanning electron microscopy.

If one can increase r_y or σ_y, keeping E constant, one can in principle increase γ. While good adhesion may be desirable for obtaining a high modulus of strength (Section 12.1.2.3) in a particulate composite, it may well be undesirable if a high toughness is wanted, because of the lower value of r_y. One may also have a high toughness, but a low strength. It is always important to choose a suitable compromise between toughness and other mechanical properties. For example, as shown in Table 12.2, silane-treated short fiber composites tended to be superior in most respects, but deficient in ultimate strain. Thus selection of filled systems requires a judicious balancing of properties which may depend on filler characteristics and amounts in different ways.

The trend toward the lowering of fracture toughness by particulate fillers is consistent with considerable experience [see property tables in, e.g., *Modern Plastics Encyclopedia* (1974–1975)] and with results of several other studies of the effects of filler on the area under a stress–strain curve

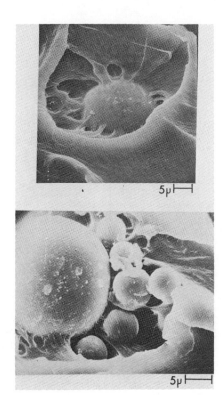

Figure 12.17. Fracture surface of PPO containing silane-treated glass beads (20 vol %) at 25°C; fast crack region (Trachte and DiBenedetto, 1971). Micrograph by scanning electron microscopy. Note interfacial constraint in comparison to Figure 12.16.

Table 12.2
Composite Performance Chart[a]

Design for high→	Ultimate strain ε_u	Ultimate strength σ_u	Elastic modulus E	Fracture energy γ
Unfilled PPO	1	4	6	2
20 vol % glass bead composite	2	6	5	2
20 vol % A-1100 glass bead composite	3	5	5	3
20 vol % glass fiber composite	5	2	2	2
20 vol % A-1100 glass fiber composite	4	1	2	1
20 vol % graphite fiber composite	6	3	1	2

[a] 1 is the highest rating. Trachte and DiBenedetto (1971).

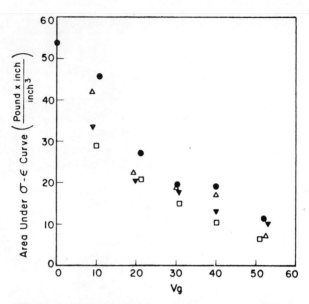

Figure 12.18. Area under flexural stress–strain curves vs. volume percent glass for epoxy composites (Sahu and Broutman, 1972). (●) CPO2; (△) untreated; (▼) DC-20; (□) Z-6076. Code numbers refer to surface treatments used for the glass; CPO2, a mold release agent; DC-20 and Z-6076, coupling agents for improving adhesion.

and on impact strength. In a study of epoxy and polyester resins filled with glass beads, Sahu and Broutman (1972) found that both the areas under the stress–strain curves (Figure 12.18) and unnotched impact strengths were reduced for the epoxy-resin systems in proportion to the volume fraction of filler. Surface treatments to improve adhesion varied in the degree to which they lowered the overall energy to fracture; however, the general effect was detrimental, as noted in the previous study discussed (Trachte and DiBenedetto, 1971). The polyester resin also showed decreases in the same two fracture parameters, though a slight upturn was noted in impact energy at the highest volume fraction of filler, as predicted by the simple theoretical relationship (12.22) derived by Nielsen (1966). As mentioned above, tensile and flexural strengths tend to be reduced by particulate fillers; see Section 12.1.2.3.

Although little research has been reported on the effects of fillers on crack propagation rates, DiBenedetto (1973) and Peretz and DiBenedetto (1972) have investigated the behavior of rapidly growing cracks in epoxy resins filled with glass beads and in a rubber-modified epoxy resin. In both the unfilled and the bead-filled resins, the terminal crack velocity was found to depend only on the matrix properties and to follow the relationship

$$\dot{a}_T = 0.28E/P \qquad (12.31)$$

where $\dot{a}_T$ is the terminal crack velocity, E is Young's modulus, and P is the density of the matrix at the macroscopic strain rate of the experiment. This relationship held regardless of the efficiency of adhesion, which was varied by use of both silane-treated and untreated beads. On the other hand, dispersed rubber particles did slow down the rate of crack growth (see Section 3.2.2.2). The lack of effect of a particulate filler on terminal crack velocity is paralleled by minimal effects of fillers on fatigue in rubbers (Gent, 1972; Manson and Hertzberg, 1972a).

In spite of the general trends toward the reduction of tensile strength, elongation, and fracture toughness by particulate fillers, at least in terms of surface and impact energies, an increasing number of exceptions are becoming apparent. For example, Broutman and Sahu (1971a) did not find a good correlation between crack initiation energy and filler content in glass-sphere–epoxy systems; the values of energy exhibited a maximum in the range of 10–20 vol % filler. DiBenedetto and Wambach (1972) also found significant increases in fracture energy (under some conditions) in glass-bead–epoxy systems (Figures 12.19 and 12.20), and Lavengood et al. (1971) noted a similar effect in the case of glass–PPO composites. Lange and Radford (1971) showed that hydrated alumina increased the fracture energy of an epoxy resin. Interesting exceptions have been described by Fallick et al. (1967), who showed that by coating a particulate filler with a polymer having a modulus between that of the filler and matrix, failure properties of a number of resins could be significantly improved. Similarly, the impact strength of particulate-filled polyethylene can be improved by the use of special coupling agents (Monte and Bruins, 1974). Finally, as discussed in Chapter 10, reinforcing fillers in rubbers provide an important group of exceptions. Clearly, it should not be supposed that rigid fillers *always* pose a fundamental limitation on toughness.

The study by DiBenedetto and Wambach (1972), who observed complex behavior in the surface fracture energy γ as a function of temperature, filler content, and filler surface treatment in epoxy–glass-sphere systems, is illuminating. At low temperatures, at which the epoxy resin was brittle, an increase in bead concentration increased γ monotonically; the tighter the adhesion, the less the effect, though the effects themselves were modest. At higher temperatures, which made the resin more ductile, the beads tended to decrease the fracture energy (see Figures 12.19 and 12.20) when treated with adhesion-promoting silanes. In the latter case, however, a peak in γ was noted when the beads were treated with an adhesion-preventing

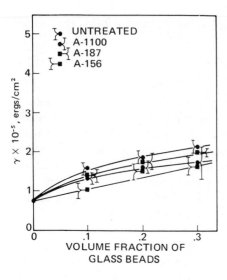

Figure 12.19. Effect of volume fraction of glass beads and bead surface treatment on the gamma-fracture toughness of epoxy composites at 25°C (DiBenedetto and Wambach, 1972). Code numbers refer to various silane surface treatments, with degree of adhesion conferred ranging from strong and positive for A-1100 to negative for A-156.

silicone; plasticization by unreacted curing agent also increased γ. Study of the fracture surfaces (Figure 12.21) showed that γ could be qualitatively correlated with the roughness of the fracture surface—an indication of the work done (per *nominal* surface area) in propagating a crack; a similar correlation was reported by Broutman and Sahu (1971*a*), who also found evidence for subsurface cracking, still another mode of energy dissipation. These results do not, however, necessarily contradict the previous results showing a negative effect of glass beads on γ for glass-bead–PPO systems

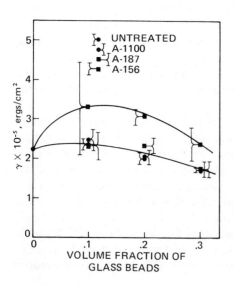

Figure 12.20. Effect of volume fraction of glass beads and bead surface treatment on the gamma-fracture toughness of epoxy composites at 130°C (DiBenedetto and Wambach, 1972). Code numbers refer to various silane surface treatments, with degree of adhesion conferred ranging from strong and positive for A-1100 to negative for A-156.

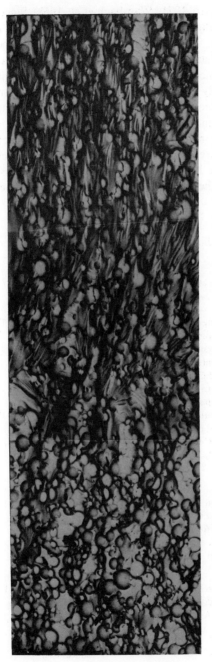

Figure 12.21. Scanning electron micrograph showing fracture surface of an epoxy resin containing 0.3 volume fraction of glass beads treated with a silane to enhance adhesion (DiBenedetto and Wambach, 1972). Specimen fractured at room temperature. Note the surface roughness, which is much greater than in the pure resin (not shown).

(Trachte and DiBenedetto, 1971; Wambach *et al.*, 1968). Since PPO is much more ductile at 25°C than the epoxy resins mentioned, the effects of filler and adhesion promoter on PPO should tend to resemble the effects on an epoxy resin in a *ductile* state (e.g., at 130°C). Indeed this is the case. The point is that a filler tends to increase surface roughness and hence γ in an otherwise *brittle* matrix, especially if the filler–matrix adhesion is poor, but tends to inhibit plastic deformation (by constraints or by simple volume replacement) in an otherwise *ductile* matrix. Such effects are not accounted for in Nielsen's simple treatment (Section 12.1.2.3) and conceivably may occur as competitive mechanisms (see Figure 12.20). A useful summary of such competitive factors is given in Table 12.3 for the glass-bead–epoxy systems (DiBenedetto and Wambach, 1972); the discussion should be relevant to other cases as well.

Table 12.3
Summary of Fracture Toughness Data for Dry Epon 828/Catalyst Z Epoxy Composites[a]

Material	Parameter	Effect on γ of increase in parameter	Reason	Evidence
Unfilled epoxy	Curing agent concentration (catalyst Z concentration varied from 10.38 to 25.65 pph)	Increase	Incompletely reacted curing agent, plasticized epoxy	Decrease in strength and modulus, increase in fracture surface roughness
Unfilled epoxy	Temperature (varied from $-20°C$ to 130°C)	Increase	Increase in ductility	Decrease in strength and modulus; increase in fracture surface roughness
Glass bead/epoxy composites at 25°C (composition range: 0–30% by volume glass)	Adhesion (varied by using A-1100, A-187, and A-156)	None	Constraints on polymer flow due to adhesions are not important when polymer is very brittle	Fractographic plus toughness data
Glass bead/epoxy composites at 100 and 130°C (composition range 0–30% by volume glass)	Adhesion (varied by using A-1100, A-187, and A-156	Decrease	Adhesion at interface constrains polymer flow and significantly reduces energy-absorbing ability of polymer	Fractographic plus toughness data

Table 12.3—continued

Material	Parameter	Effect on γ of increase in parameter	Reason	Evidence
A-156 silane-treated glass bead–epoxy composites at 25, 100, and 130°C	Glass bead concentration (varied from 0–30% by volume)	Increase	Increase in total polymer fracture surface due to increase roughness plus unconstrained polymer flow around beads at fracture surface	Fractographic
Untreated, A-1100 silane-treated, and A-187 silane-treated glass bead–epoxy composites at 25°C	Glass bead concentration (varied from 0–30% by volume)	Increase	Increase in polymer fracture surface roughness counterbalances the supplanting of polymer by beads; polymer ductility is not a dominant factor	Fractographic
Untreated, A-1100 silane-treated, and A-187 silane-treated glass bead–epoxy composites at 100 and 130°C	Glass bead concentration (varied from 0–30% by volume)	Decrease	Increase in total polymer fracture surface is counteracted by constraint of polymer flow near beads	Fractographic

[a] 20 pph catalyst unless stated otherwise, curing cycle of 24 hr at 20°C, 24 hr at 60°C, 24 hr at 100°C, 12 hr at 130°C. DiBenedetto and Wambach (1972).

A rigid filler may be able to improve the fracture toughness of a composite if it can undergo deformation itself (Hertzberg *et al.*, 1970a; Lange, 1970; Lange and Radford, 1971) or interact with growing crack fronts. Evidence for the latter mechanism was found in the study by Lange and Radford (1971) of the fracture energy of an epoxy resin containing hydrated alumina. A maximum in fracture energy was found in a plot of fracture energy against particle size, and explained in terms of a crack interaction model. Crack front interaction with particulates was also observed by Broutman and Sahu (1971a) in their study discussed above.

In summary, the assumptions implied in the predictions of strength and toughness in composites by equations (12.21) and (12.23) do not always hold. Positive deviations from predicted values may occur due to a variety of phenomena involving the matrix, the filler, and the interface, e.g., crazing or subsurface cracking in the matrix, deformation of the filler, increased surface roughness, and interaction with crack fronts. In any case, the effects of adhesion on modulus and elongation at break interact in a complex manner, so that even if fracture energy may be reduced by good adhesion, strength tends to be increased. While some generalizations may be made about effects of particulate fillers, much remains to be learned about the precise relationships between the filler, the matrix, and the interface in particular cases, and the effects of stress fields characteristic of various modes of loading.

12.1.3. Transport Behavior and Other Physical Properties

For many applications of filled polymers, knowledge of properties such as permeability, thermal and electrical conductivities, coefficients of thermal expansion, and density is important. In comparison with the effects of fillers on mechanical behavior, much less attention has been given to such properties of polymeric composites. Fortunately, the laws of transport phenomena for electrical and thermal conductivity, magnetic permeability, and dielectric constants often are similar in form, so that with appropriate changes in nomenclature and allowance for intrinsic differences in detail, a general solution can often be used as a basis for characterizing several types of transport behavior. Useful treatments also exist for density and thermal expansion.

12.1.3.1. Permeability

Although completely rigorous and proven solutions to the problem of the permeation of gases and liquid through polymers have not been developed, typical experimental evidence and some recent approaches to this very important engineering problem are worth presenting. As can be readily seen, the permeation process must be considered with respect to the factors operative in other phenomena in two-component systems, e.g.: the type, shape, and orientation of filler particles; the effects a filler may have on the relaxation characteristics of the matrix resin; and the properties of the interface (Ash et al., 1963; Bardeleben, 1963; Barrer and Chio, 1965; Barrer et al., 1963; Crank and Park, 1969; Funke et al., 1969; Higuchi and Higuchi, 1960; Klute, 1959; Kwei, 1965; Manson and Chiu, 1973a; Maxwell, 1904; Michaels, 1965; Michaels and Bixler, 1971; Nielsen, 1967b; Paul and Kemp, 1972; Perera and Heertjes, 1971; Peterson, 1968; Stallings et al., 1972; Van Amerongen, 1964). In addition to effects characteristic of mechanical response, solubility of the permeant in the matrix, interface, or filler superimposes still another dimension of complexity (Crank and Park, 1969; Nielsen, 1967b).

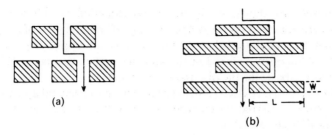

Figure 12.22. Model for the path of a diffusing molecule through a polymer filled with circular or square plates. (a) Cubes; (b) plates. Reprinted from Nielsen 1967b, p. 930, by courtesy of Marcel Dekker, Inc.

In this discussion it will be convenient to discuss "permeability" rather than "diffusivity," since permeability refers to the amount of permeant that passes through a given thickness of a membrane in a given time, and includes the effects of both solubility and inherent diffusivity (Crank and Park, 1969).* Steady-state permeation is also assumed in most cases. Clearly the permeation must be reduced by the presence of an impermeable filler for two related reasons: an increased path length of the permeating molecules (which is equivalent to having a thicker membrane), and a reduced cross-sectional area of matrix available for permeation (Figure 12.22). The first effect is commonly expressed in terms of a "tortuosity factor" τ, which is equivalent to the ratio of the path length traveled through the maze of filler particles to the thickness of the membrane. The second effect is proportional to the volume fraction of filler.

To account for the electrical conductivity of a composite system consisting of noninteracting spheres embedded in a homogeneous matrix, Maxwell (1904) developed an expression of the following form:

$$\tau \simeq 1 + \tfrac{1}{2} v_f \tag{12.32}$$

where v_f is the volume fraction of filler. Though better approximations and more general expressions have, of course, been developed by others (Crank and Park, 1969, Chapter 6), equation (12.32) serves as a useful limiting case. The important point in this discussion is the similarity between laws for the various transport properties. The analogy between electrical conductivity and permeation has long been recognized and applied to permeation in polymers (Crank and Park, 1969; Maxwell, 1904; Nielsen, 1967b).

---

* Thus, in the simplest limiting case of a membrane-permeant system, one in which Henry's law and Fick's laws are observed, the coefficients of permeability P, solubility S, and diffusion D are related as follows: $P = D \times S$.

For example, as shown by Klute (1959), equation (12.32) holds rather well for permeation in polyethylenes having different degrees of crystallinity (assuming the crystallites behave as fillers), in the limiting case of low volume fractions of filler (say, <0.3). For the sake of simplicity, equation (12.32) will be used as the basis for the discussion following.

More recently, Nielsen (1967b) suggested a more general approximation for tortuosity conferred by particles having different ratios of length L to thickness W (Figure 12.22):

$$\tau \simeq 1 + \left(\frac{L}{W}\frac{1}{2}\right)v_f \qquad (12.33)$$

Combination of equation (12.33) with an approximation for the ratio of permeability in the filled (P_c) to the unfilled system (P_p) (Barrer et al., 1963; Michaels and Bixler, 1961),

$$\frac{P_c}{P_p} \cong \frac{v_p}{\tau} \qquad (12.34)$$

gives for the permeation of a gas as a function of particle shape and the volume fractions of polymer v_p and filler v_f

$$\frac{P_c}{P_p} \cong \frac{v_p}{1 + (L/2W)v_f} \qquad (12.35)$$

This relationship has important implications. As shown in Figure 12.23, permeation must be extremely sensitive to particle shape, as expressed in the values of L/W. Thus cubes (or spheres, which are in this sense equivalent) are an order of magnitude less effective than thin plates lying parallel to the plane of the membrane. This trend is in accord with experience (Michaels and Bixler, 1961; Nielsen, 1967b). A second, practical implication is that processing techniques such as extrusion can lead to a significant degree of filler orientation.

Certainly it is general experience that fillers tend to reduce permeation in polymers. Examples include gases or vapors in filled rubbers (Ash et al., 1963; Barrer and Chio, 1968; Barrer et al., 1963; Manson and Chiu, 1973a; Michaels and Bixler, 1971; Van Amerongen, 1964), pigmented paint films (Bardeleben, 1963; Funke et al., 1969; Michaels, 1965; Perera and Heertjes, 1971), and filled epoxy resins (Bardeleben, 1963; Perera and Heertjes, 1971); additional examples are given below. However, there are also many exceptions; for example, in cases in which the filler is not well bonded or uniformly wetted by the resin (Bardeleben, 1963; Funke et al., 1969; Michaels, 1965; Perera and Heertjes, 1971), or in which the filler or interface absorbs the

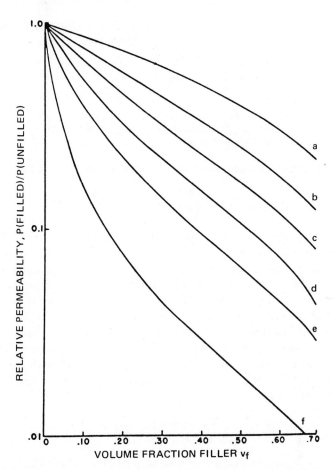

Figure 12.23. Minimum permeability of gases through a polymer filled with plates of different L/W ratio oriented parallel to the surface of the film. (a) $L/W = 1$ (cubes); (b) $L/W = 4$; (c) $L/W = 8$ ($2 \times 2 \times \frac{1}{4}$ plates); (d) $L/W = 15$; (e) $L/W = 27$ ($3 \times 3 \times \frac{1}{9}$ plates); (f) $L/W = 100$. Reprinted from Nielsen 1967b, p. 932, by courtesy of Marcel Dekker, Inc.

penetrant (Funke *et al.*, 1969; Higuchi and Higuchi, 1960; Michaels, 1965; Paul and Kemp, 1972; Perera and Heertjes, 1971; Stallings *et al.*, 1972).

To account for such cases, more general expressions must be used. Nielsen's simple model may be modified to allow for solubility of vapors or liquids in the system, which may vary from the interface to the bulk polymer in real systems (Crank and Park, 1969). With the model shown in Figure 12.24, which indicates separate contributions to permeation from the interface and

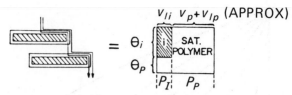

Figure 12.24. Model for the permeability of a liquid through a filled polymer. Reprinted from Nielsen 1967b, p. 933, by courtesy of Marcel Dekker, Inc.

the bulk polymer, Nielsen expressed the total permeation as follows[*]:

$$P_{cl} \cong P_I\left(\frac{v_{li}}{\tau^*}\right) + P'_{pl}\left(\frac{v_p + v_{lp}}{\tau}\right) \tag{12.36}$$

where P_{cl}, P_I, and P'_{pl} are the permeabilities of liquid through the filled polymer, the interfacial regime, and the saturated bulk polymer, respectively; P'_{pl}, of course, must equal the permeability P_{pl} through the unfilled matrix in the absence of changes induced by the filler—an assumption not always valid (Manson and Chiu, 1973a). The tortuosity factor τ^* refers to the interface, while v_{li} and v_{lp} are the volume fractions of liquid collected at the interface and dissolved in the matrix (all volume fractions being taken in the swollen state); v_{li} should be proportional to the surface area of the filler.

If we assume that P_I must imply permeation through both the interface per se, P_i, and through the matrix between interfaces, P_{pl}, it can be shown (Nielsen, 1967b) that

$$P_{cl} \cong \frac{P_iP_{pl}}{P_{pl}\theta_i + P_i\theta_p}\left(\frac{v_{li}}{\tau^*}\right) + P_{pl}\left(\frac{v_p + v_{lp}}{\tau}\right) \tag{12.37}$$

where θ_i and θ_p represent the fractional length of the diffusion paths through the interface and matrix, respectively. The values of θ_i and θ_p are, in general, related to v_f as follows:

$$\theta_i = v_f^n \tag{12.38a}$$

$$\theta_p = 1 - v_f^n \tag{12.38b}$$

Thus the quantity n, where the exponent n denotes the fractional length of the average diffusion path that traverses polymer, depends on particle shape, orientation, and the state of aggregation. Values of n will range from a limit of zero, for very thin plates ($L/W = \infty$) *perpendicular* to the membrane, to unity, for similar plates *parallel* to the membrane. The value of n for spheres falls in between, and is approximately $1/3$.

_____
[*] For the sake of consistency within this book, the subscripts used in the original article (Nielsen, 1967) have been changed.

With the introduction of n, Nielsen (1967b) obtained the following general equation for liquid permeation in a filled system:

$$\frac{P_{cl}}{P_{pl}} \cong \frac{P_i}{P_{pl}v_f^n + P_i(1 - v_f^n)}\left(\frac{v_{li}}{\tau^*}\right) + \left(\frac{v_p + v_{lp}}{\tau}\right) \tag{12.39}$$

The minimum found for equation (12.39) corresponds to the case [$v_{li} = 0$; $v_{lp} = 0$] and to the values predicted by equation (12.34).

Except for n, P_i, and τ^*, the parameters are capable of experimental measurement. However, n may be estimated, at least approximately, for different particle shapes, while τ^* is probably approximately equal to unity in many cases, or at most ~ 2 (Nielsen, 1967b). Although P_i is not conveniently amenable to experimentation, trends in P_{cl} could be predicted in terms of expected trends in P_i; alternatively, of course, curve fitting could be used to estimate P_i from measured values of P_{cl} [assuming equation (12.39) is valid].

Several particular cases were treated in the development of equation (12.39): (1) interface forming channels perpendicular to the membrane ($n = 0$), (2) filler particles composed of porous aggregates (i.e., aggregates containing voids), and (3) uniformly dispersed impermeable particles. Inspection of equation (12.39) shows that interface channels (case 1) may either increase or decrease the permeability of a filled system, depending on the relative magnitudes of the tortuosity and ease of diffusion as implicit in the value of $P_i v_{li}$. With porous aggregates (case 2), it can be shown that aggregates must always increase permeability to liquids (and gases as well). With dispersed, unchanneled particles (case 3), diverse behavior may be expected depending on such factors as the dependence of τ on v_f, the solubility at the interface, and the permeability of the interfacial layer. Figure 12.25 illustrates some of the variations expected for a system having $n = 1/3$, for differences in the relative magnitudes of the permeation parameters. Curves range from one showing a monotonic decrease in P_{cl} as a function of v_f (high L/W ratio, low v_{li}) to one showing a minimum and a tendency to rise above the level characteristic of an unfilled polymer (high v_{li}, medium L/W ratio). Figure 12.26 shows that for a given combination of other permeation parameters, n should be as high as possible if low permeation rates are desired for a filled polymer system, as is the case of platelets parallel to the membrane.

While one should not expect too much in a quantitative sense from such models, they do explicitly predict the effects of factors one intuitively feels to be important.

Other, more sophisticated treatments have, of course, been given (Crank and Park, 1969; Higuchi and Higuchi, 1960). For example, Barrer et al. (1963) and Ash et al. (1963) have developed expressions to account for incomplete wetting of a filler, and Higuchi and Higuchi (1960) have adopted a general

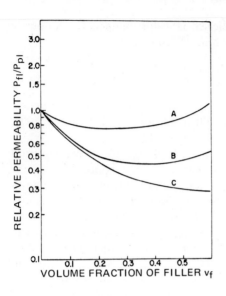

Figure 12.25. Various types of liquid permeability when $n = 1/3$. (A) $P_l/P_{pl} = 10$, $v_{li} = v_f$, $\tau^* = \tau = 1 + 3v_f$, $L/W = 6$. (B) $P_l/P_{pl} = 50$, $v_{li} = 0.5v_f$, $\tau^* = \tau = 1 + 5v_f$, $L/W = 10$. (C) $P_l/P_{pl} = 10$, $v_{li} = 0.3v_f$, $\tau^* = \tau = 1 + 5v_f$, $L/W = 10$. Reprinted from Nielsen 1967b, p. 940, by courtesy of Marcel Dekker, Inc.

expression for dielectric behavior for the case of permeability of two-component systems. Approximate permeability equations for several cases are as follows.

(a) General case:

$$P_c = \frac{3P_pP_fv_f + 2(P_p)^2v_p + P_fP_pv_p}{3P_p + v_p(P_f - P_p)} \tag{12.40}$$

(b) Case for $P_f = 0$:

$$P_c = \frac{2P_p}{(3/v_p) - 1} \tag{12.41}$$

(c) Case for $P_f \gg P_p$:

$$P_c = P_p[1 + (3v_f/v_p)] \tag{12.42}$$

where P_c, P_p, and P_f are the permeabilities of filled polymer, polymers, and filler, respectively, and v_p and v_f are the volume fractions of polymer and filler. Use of more complicated versions of equations (12.40)–(12.42) will, of course, lead to slightly different numerical values (Higuchi and Higuchi, 1960). It is interesting that equation (12.41), which corresponds to the case of a polymer containing a spherical impermeable filler, is precisely equivalent to equation (12.35). Equations (12.40) and (12.42) allow for permeation through the filler phase—a phenomenon to be expected in the case of systems filled with other polymers or porous inorganic materials.

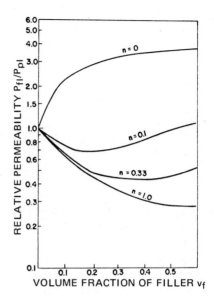

Figure 12.26. Various types of liquid permeability when all variables except n are held constant. $P_l/P_{pl} = 50$, $v_{lp} = 0$, $v_{li} = 0.5v_f$, $\tau^* = \tau = 1 + 5v_f$, $L/W = 10$. Reprinted from Nielsen 1967b, p. 940, by courtesy of Marcel Dekker, Inc.

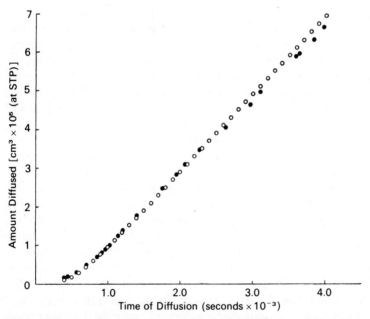

Figure 12.27. Comparison of (○) theoretically predicted diffusion curve based on Higuchi–Higuchi equation (12.40) with (●) experimental results for oxygen permeation through poly(vinyl acetate) containing 0.28 volume fraction of polystyrene latex particles. (Peterson, 1968.)

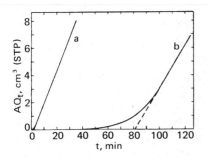

Figure 12.28. Results of transient permeation experiments for (a) pure silicone rubber membranes ($v_d = 0$) and (b) silicone rubber loaded with activated molecular sieves ($v_d = 0.202$). CO_2, $P_2 = 79$ cm Hg. (Paul and Kemp, 1972.)

The Higuchi–Higuchi equation (12.40) was shown by Peterson (1968) to hold very well for the permeation of oxygen through polymer–polymer systems, including various combinations such as poly(vinyl acetate) and poly(vinyl chloride). Typical results are shown in Figure 12.27. Another implication of the Higuchi–Higuchi treatment is important in the case of fillers that can absorb a penetrant: the time lag before a steady state is reached is increased due to the immobilization of those penetrant molecules that penetrate first. This fact leads to the apparent paradox that to protect against any permeation of a given penetrant for a limited time span, one should in principle introduce a filler that absorbs penetrant; the time lag during which permeation is virtually excluded will be increased, though once permeation begins, the steady-state rate may not be much affected (Higuchi and Higuchi, 1960; Paul and Kemp, 1972). An example of this phenomenon may be found in a study by Paul and Kemp (1972), who found that molecular sieves increased the time lag for the permeation of gases such as CO_2 in silicone rubber (see Figure 12.28). Stallings et al. (1972) have also suggested that the nonadditivity of diffusion coefficients as a function of composition in PPO–polystyrene blends may be explained in terms of an immobilization of some penetrant molecules.

As with the predictions of mechanical behavior, predictions of permeability by different models require additional experimental verification to determine the limitations of the expressions proposed. Another interesting point—and one which complicates the quantitative application of permeability expressions—is the possible role of filler–polymer interactions (see also Section 12.3). The mechanism of permeation in polymers is believed to depend upon the ability of segments to move in such a way as to create a hole which can accommodate a penetrant molecule. Thus any restriction or enhancement of mobility would be expected to alter the permeability (Crank and Park, 1969), just as such effects are known to alter the relaxation behavior and hence the glass temperature of a polymer.

Even though fairly large particles ($\sim 1\ \mu$m or more) might be expected to have little immobilizing effect on matrix polymer far removed from the filler, significant specific effects of fillers have been observed by several investigators. For example, Kwei (1965) and Kwei and Kumins (1964), working with a copolymer of vinyl acetate and vinyl chloride, an epoxy resin, and poly(vinyl acetate), found that TiO_2 (particle size $\sim 0.2\ \mu$m) not only raised the T_g of the system (see Section 12.3), but also lowered the sorption of organic vapors. As for permeation itself, the diffusion coefficient and the apparent energy of activation for permeability E_a are often lowered by fillers. For example, Kumins and Roteman (1963) found that the diffusion coefficient for water vapor in a vinyl chloride–vinyl acetate copolymer varied inversely with the volume fraction of TiO_2. Kwei and Arnheim (1965) noted that above T_g, the energy E_a of activation for permeation was decreased significantly ($\sim 40\%$) by as little as 0.1% TiO_2. Curiously, the break in the permeability versus the inverse of temperature curve usually observed for polymers at about T_g is sometimes reduced or absent in filled systems (Kumins and Roteman, 1963; Kwei and Arnheim, 1965). The lower values of E_a were attributed to a decrease in internal pressure or in the volume of activation—each factor being consistent with the ordering of polymer segments. Similarly, as shown in Figure 12.29, Manson and Chiu (1973a,b)

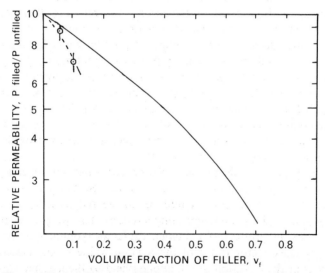

Figure 12.29. Relative permeability to water of a glass-sphere-filled epoxy resin. Solid line represents equation (12.35) and dashed line experimental data taken at 85°C. Similar decreases below the predicted value were noted at other temperatures as well. (Manson and Chiu, 1973a.)

found that glass spheres dispersed in an epoxy resin raised T_g and lowered the value of P (and D) for a given v_f significantly below the value predicted by considerations of volume fraction and tortuosity [equation (12.35)]. Such results may imply an immobilization of matrix resin segments, perhaps due to stresses arising from differential contraction on cooling after curing. For crosslinked matrices that shrink during polymerization (such as epoxy resins), the latter stages of polymerization may induce a long-range state of compression of the polymer about the filler particles. (In fact, a practical reason for including filler in many cases is to reduce shrinkage during fabrication.) It is interesting to speculate whether or not behavior of this type may be consistent with predictions of equation (12.39). In fact, equation (12.39) can predict liquid permeabilities *higher* than for equation (12.35) but not lower values, for the minimum value will be found only if equation (12.39) is reduced to equation (12.35).* Thus, a filler–particle interaction is still probable—a conclusion supported by independent evidence (Section 12.3).

12.1.3.2. Thermal and Electrical Conductivity

The study of thermal and electrical conductivity in composites has an especially long history. By the time Kerner (1965a) reviewed the early literature and proposed his theory, the subject of thermal and electrical conductivity in particulate composites had been studied for over 60 years. Indeed, one of the earliest investigations was by Lord Rayleigh (1892), who analyzed the conductivity of a matrix containing a cubic array of spheres. A little later, Maxwell published an expression for the electrical conductivity of randomly distributed spheres in a continuum; this expression was also used to derive a permeability relationship (Section 12.1.3.1). Later, theories of potential applicability to polymeric composites were developed by such workers as Bruggeman (1935), Fricke (1924), Peterson and Hermans (1969), Behrens (1968), Cheng and Vachon (1969), Tsao (1961), and Hamilton and Crosser (1962). The subject has been recently reviewed by Sundstrom and Chen (1970) and Ashton et al. (1969).

In his approach to the problem, Kerner (who also developed a common additivity law for the modulus of composites; Section 12.1.1.1) assumed that the overall conductivity could be considered as a weighted linear superposition of the component conductivities. The weights in turn were considered as the product of a geometrical factor (volume fraction of filler) and an intensity factor (ratio of the average value of the electrical field component in the direction of the applied field to the average value of the

---

* The complete Higuchi–Higuchi equation gives a still somewhat lower bound to the permea bility, but, at least for the limited evidence available so far (Manson and Chiu, 1973a), still seems to overpredict permeability, at least in the case of a glass-sphere-filled epoxy resin.

field in the bulk). For spheres suspended in a matrix, Kerner obtained the following expression for κ_c, the conductance of the overall composite:

$$\kappa_c = \sum_{i=p,f} \kappa_i v_i \frac{3\kappa_p}{\kappa_i + 2\kappa_p} \bigg/ \sum_{i=p,f} v_i \frac{3\kappa_p}{\kappa_i + 2\kappa_p} \tag{12.43}$$

where κ_c, κ_p, and κ_f are the conductivities of the composite, polymer, and filler, respectively, and v_p and v_f are the volume fractions of polymer and filler, respectively.[*]

Maxwell's (1904) equation for electrical conductivity may be used to predict thermal conductivity as follows:

$$k_c = k_p \left[\frac{2v_f k_p + (3 - 2v_f)k_f}{(3 - v_f)k_p + v_f k_f} \right] \tag{12.44}$$

where k_c, k_f, and k_p are the thermal conductivities of the composite, discrete, and matrix phases, respectively, and v_f is the volume fraction of the discrete phase. This equation, valid for spheres spaced sufficiently far apart that interparticle interactions are negligible, may be adapted for use with other particle shapes (Fricke, 1924; Hamilton and Crosser, 1962). Additional modification is needed if the filler is very much more conductive than the matrix; a ratio of 100 to 1 causes a significant deviation, while the more common ratio of 2 to 1 has a negligible effect.

Several other equations are worth noting, because their validity has been tested for glass sphere–polymer composites. Behrens (1968) derived a general expression for thermal conductivity in two-component systems having orthorhombic symmetry; for spheres in a cubic lattice, the expression becomes

$$k_c = k_p \frac{(\beta + 2) + 2(\beta - 1)v_f}{(\beta + 2) - (\beta - 1)v_f} \tag{12.45}$$

where $\beta = k_f/k_p$ and the other symbols have the same significance as in equation (12.44).

Another expression, which gives results very similar to those from equation (12.45), is based on an approximate solution obtained by Peterson and Hermans (1969) (Kerner, 1956a) for the dielectric constant of two-component systems. In thermal conductivity nomenclature, this becomes

$$k_c = k_p \left[1 + 3\gamma v_f + 3\gamma^2 \left(1 + \frac{\gamma}{4} + \frac{\gamma^2}{256} + \cdots \right) v_f^2 \right] \tag{12.46}$$

where $\gamma = (k_f - k_p)/(2k_p + k_f)$.

For consistency with other sections of this book, the subscripts c, p, and f are used to refer to the composite, polymer, and filler, respectively. In the conductivity literature, however, the subscripts e, c, and d are usually used for the composite, polymeric matrix, and filler, respectively.

A more complex relationship was derived by Cheng and Vachon (1969) based on a theory by Tsao (1961); a parabolic distribution is assumed:

$$\frac{1}{k_c} = \frac{1 - B}{k_p} + \frac{1}{B'C^{1/2}C'} \ln \frac{B' + BC^{1/2}C'/2}{B' - BC^{1/2}C'/2} \tag{12.47}$$

where $B = (3v_f/2)^{1/2}$, $C = 4(2/3v_f)^{1/2}$, and

$$B' = [k_p + B(k_f - k_p)]^{1/2}$$
$$C' = (k_f - k_p)^{1/2}$$

Although the expressions for the thermal conductivity may appear to be exceedingly different, in fact all of them agree fairly well with each other. In a check of equations (12.44)–(12.47), Sundstrom and Chen (1970) found that the Cheng–Vachon equation (12.47) represented the data a little better than the others, at least up to a value for v_f of 0.4 (Figure 12.30). For comparison, the corresponding curve for Kerner's equation (12.43) is plotted for polyethylene; it is seen to fall slightly below the others. Two other points are worth noting for the several polymer systems studied: a relative independence of thermal conductivity on particle size and shape, provided $k_f/k_p < 100$.

Although quantitative tests appear to be limited in number, it seems fair to conclude that the above approximate equations serve to predict transport phenomena in a filled polymer as a function of constituent properties.

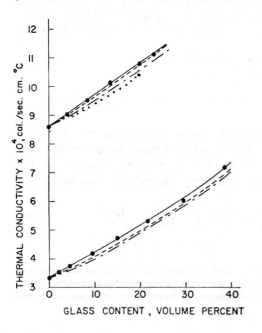

Figure 12.30. Comparison of experimental and predicted thermal conductivities for glass-sphere-filled polymers. The upper curves are for polyethylene, the lower curves for polystyrene. Except for Kerner equation plot (···), curves and data are from Sundstrom and Chen (1970). (--) Maxwell; (—) Cheng-Vachon; (- -) Behrens and Peterson Hermans. (From Sundstrom, D. W. and Chen, S. Y., 1970, J. Compos Mater. 4, 113; courtesy Technomie Publishing Co.)

12.1.3.3. Coefficients of Thermal Expansion and Density

It is a matter of experience that the coefficient of expansion α of a polymeric system may be reduced by introducing an inorganic filler (Holliday and Robinson, 1973; Manabe et al., 1971; Nielsen, 1967a). Prediction of this behavior is important for several reasons.* First, a lowering in α is useful in minimizing dimensional changes in a plastic exposed to changes in temperature during fabrication or use. Second, unequal expansion or contraction of constituents of a composite system may lead to unacceptable stresses, so that matching of α's is frequently desirable. Even if stresses generated do not lead to failure—and they may even be desirable, e.g., in a rubber-modified plastic—they may significantly affect the mechanical behavior of the constituents.

Using concepts of thermoelasticity and the same general averaging approach used to predict electrical conductivities for composites consisting of grains in a continuous medium, Kerner (1965a,b) derived a theoretical expression for the overall volume coefficient of expansion α_c for a composite. For the case of a filled plastic this becomes† (Kerner, 1956b; Nielsen, 1967a)

$$\alpha_c = \alpha_f v_f + \alpha_p v_p - (\alpha_p - \alpha_f) v_p v_f \frac{(1/K_p) - (1/K_f)}{(v_p/K_f) + (v_f/K_p) + (3/4G_p)} \qquad (12.48)$$

where α_f and α_p are the (bulk) coefficients of expansion for the filler and polymer, respectively, v_f and v_p are the respective volume fractions, K_f and K_p are the bulk moduli, and G_p is the shear modulus of the polymer. From the form of this equation, it is clear that a simple additivity relationship ($\alpha_c = \alpha_f v_f + \alpha_p v_p$) can hold only if $K_f = K_p$; otherwise significant deviations may occur which will reach a maximum when $v_p = v_f = 0.5$. The equation as given originally by Kerner can be readily expanded to include additional phases.

An equivalent expression was derived by Manabe et al. (1971) in a slightly different form:

$$\alpha_c = \alpha_f v_f + \alpha_p v_p - \frac{4G_p(K_f - K_p)(\alpha_p - \alpha_f)v_f v_p}{4G_p(K_f - K_p)v_f + K_p(3K_f + 4G_p)} \qquad (12.49)$$

Other equations have been proposed, though testing appears to have been limited. Turner (1946) suggested the following equation for isotropic fillers, where α is independent of size and shape:

$$\alpha_c = \frac{(\alpha_f W_f E_f/d_f) + (\alpha_p W_p E_p/d_p)}{(W_f E_f/d_f) + (W_p E_p/d_f)} \qquad (12.50)$$

---

* Holliday and Robinson (1973) provide an especially comprehensive and useful review.
† In all cases, for the sake of consistency we have converted phase subscripts 1, 2, etc. in the original equations to f, p, etc. Thus f will always refer to the dispersed phase, regardless of its modulus (Kerner, 1956b). For isotropic solids, $\alpha = 3\tilde{\alpha}$, $\tilde{\alpha}$ being the linear coefficient.

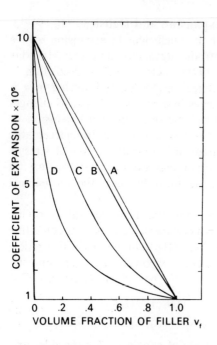

Figure 12.31. Coefficient of thermal expansion of particulate composites as predicted by (A) ideal rule of mixtures, (B) Kerner's equation, (C) Thomas' equation, (D) Turner's equation. The assumed material constants of the components are listed in the text. (From Nielsen, L. E., 1967, *J. Compos. Mater.* **1**, 100; courtesy Technomic Publishing Co.)

where W and E represent the weight fraction and Young's modulus of the given component, respectively, the quantity d represents the density, and the subscripts f and p refer to the filler and polymer, respectively. Thomas (1960) has proposed an alternate equation:

$$\ln \alpha_c = v_p \ln \alpha_p + v_f \ln \alpha_f \qquad (12.51)$$

which is said (Nielsen, 1967a) to represent behavior better than equation (12.50) for some systems.

Deviations of equations (12.48), (12.50), and (12.51) from a linear rule of mixtures are shown in Figure 12.31 for a typical case considered by Nielsen (1967a), who used the following values of properties for the calculations:

$$\alpha_p = 10^{-4} \qquad \alpha_f = 10^{-5}$$
$$K_p = 5 \times 10^{10} \qquad K_f = 1.67 \times 10^{11}$$
$$G_p = 1.07 \times 10^{10} \qquad G_f = 1.25 \times 10^{11}$$
$$E_p = 3 \times 10^{10} \qquad E_f = 3 \times 10^{11}$$
$$v_p = 0.4 \qquad v_f = 0.2$$
$$d_p = 1.0 \qquad d_f = 4.0$$

where the symbols have the same significance as in equations (12.48)–(12.51). It is stated, though without citation, that the scatter in experimental data makes selection of the best equation difficult, but that equations (12.48) and (12.51) are more likely to be generally applicable than equation (12.50).

A recent study by Manabe *et al.* (1971) provides a test of equation (12.49). In this study, the thermal expansion coefficient was determined for various blends of emulsion polymers, e.g., polybutadiene dispersed in polystyrene, and a styrene–butadiene copolymer dispersed in poly(methyl methacrylate). In these examples the "filler" or dispersed phase happens to have a lower modulus than the matrix; this does not affect the argument. As shown in Figures 12.32 and 12.33, experimental data for coefficients of expansion in the glassy state for the two systems mentioned agree rather well with the predictions of equation (12.49) [which is equivalent to Kerner's equation (12.48)]. The data clearly do not agree with the linear relationship for simple additivity on a volume basis. It may be noted in passing that the linear additivity law is very similar to variations in equations (12.48) and (12.49) which account for the possibility of phase inversion, i.e., the low-concentration phases becoming continuous (see Section 3.2 for phase inversion morphology):

$$\alpha_c = \alpha_f v_f + \alpha_p v_p - \frac{4G_f(K_p - K_f)(\alpha_f - \alpha_p)v_f v_p}{4G_f(K_p - K_f)v_p + K_f(3K_p + 4G_f)} \tag{12.52}$$

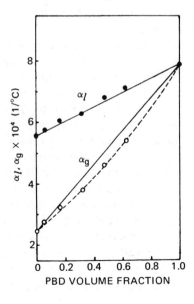

Figure 12.32. Thermal expansion coefficients α_l and α_g vs. volume fraction of polybutadiene (PBD) for the system PS/PBD. α_l and α_g are the values at about 100 and 20°C, respectively. Full line and broken line indicate the values calculated by the additivity law and equation (12.49), respectively, and the circles are observed values. (From Manabe *et al.*, 1971, *Int. J. Polym. Mater.* **1**, 47; courtesy Technomic Publishing Co.)

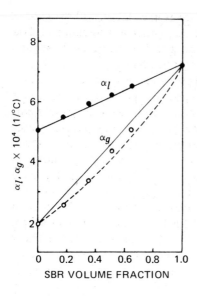

Figure 12.33. Thermal expansion coefficients α_l and α_g vs. volume fraction of SBR for the system of SBR/PMMA. α_l and α_g are the values at about 100 and 20°C, respectively. Full lines and broken line indicate the values calculated by the additivity law and equation (12.49), respectively. Filled and open circles are observed values. (From Manabe *et al.*, 1971, *Int. J. Polym. Mater.* **1**, 47; courtesy Technomic Publishing Co.)

Of course this equation can only be valid when the filler or dispersed phase can in fact undergo a phase inversion; this cannot be the case for a rigid inorganic filler. Above T_g, equations (12.49) and (12.52) give almost the same values for these systems, and a linear additivity relationship holds. This presumably follows from the near equivalence of expansion coefficients so that the right-hand interaction term of equations (12.49) and (12.52) vanishes.

Although these systems appear to be far from a typical inorganic filler–polymer system, the results support the earlier conclusion that equation (12.48) [or (12.49)] should be valid. However, polymer systems may be expected to exhibit additional complications, such as the possibility of complex morphologies, relaxation phenomena in both phases, molecular interactions, and phase inversion (Manabe *et al.*, 1971). Indeed, Manabe *et al.* found that the thermal expansion of a dispersion of polybutadiene in a styrene–acrylonitrile polymer was described by a linear additivity law, and attributed the effect to the ability of the matrix to relax the thermal stresses. In order to account for more complex distributions of phases, the same authors developed more general models and a general expression, of which equations (12.49) and (12.52) are special cases. The models, which treat a two-component system as a core of one phase surrounded by shells, allow for the realistic case of some degree of mixing of the phases by permitting alternate shells of each phase. A fuller discussion of phase mixing and morphology is given in Chapters 2 and 3.

This type of approach would seem to be especially useful in view of the increasing use of polyblends and related materials in which sometimes a rigid phase is used to stiffen a less rigid one. In addition, the models permit prediction of rather complex glass temperature behavior in such systems.

A different approach to the prediction of thermal expansion coefficients was taken by Schapery (1968), who calculated upper and lower bounds for both isotropic and anisotropic composites. The method is applicable to systems containing an arbitrary number of constituents and an arbitrary phase geometry. In some cases, the bounds coincide, and exact solutions may be found; in other cases, approximations only may be derived. In a simple two-component system, Schapery obtained the following expression for the volumetric expansion coefficient α_c of the composite:

$$\alpha_c = \frac{3K_p}{K_c}\left(\frac{K_f - K_c}{K_f - K_p}\right)(\tilde{\alpha}_p - \tilde{\alpha}_f) + 3\tilde{\alpha}_f \qquad (12.53)$$

where K represents the bulk modulus, and the subscripts p and f refer to the matrix and filler phases, respectively. Values of K may be estimated using either an exact predictive relationship (if possible) or upper and lower bound expressions. The expressions should be especially useful in predicting values for anisotropic systems (see Section 12.2); extension to viscoelastic systems has also been indicated (Schapery, 1968). Also, values of K may be calculated from α_c, which is generally easier to measure experimentally than K.

Schapery also related his expressions to other typical relationships. Thus, use of Kerner's equation (12.4), which gives a lower bound *modulus*, can be shown to yield an upper bound to α_c. A comparison of several relationships is given in Figure 12.34 ($\alpha_v/3 \equiv \tilde{\alpha}$ in the text).

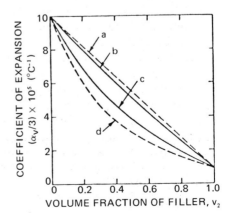

Figure 12.34. Coefficient of thermal expansion for an isotropic, particulate-filled system. (a) Rule of mixtures; (b) upper bound; (c) lower bound; (d) Turner's equation. (From Schapery, R. A., 1968, *J. Compos. Mater.* **2**, 380; courtesy Technomic Publishing Co.)

Still another approach to predicting the thermal expansion coefficient of composites has been taken by Wang and Kwei (1969), who derived expressions to account for deviations from a simple additivity rule due to the development of thermal stresses. Such stresses will arise in practice if a composite is cooled from its fabrication temperature and if the components have different coefficients of expansion. They obtained the following equations to give the linear coefficient of expansion of the composite $\tilde{\alpha}_c$:

$$\tilde{\alpha}_c = \tilde{\alpha}_p[1 - v_f(1 - k_\alpha)C] \tag{12.54}$$

where

$$k_\alpha = \alpha_f/\alpha_p \tag{12.55}$$

$$C = \frac{3\lambda(1 - v_p)}{\lambda[2v_f(1 - 2v_p) + (1 + v_p)] + 2(1 - 2v_f)(1 - v_f)} \tag{12.56}$$

$$\lambda = E_f/E_p \tag{12.57}$$

In these equations, E, $\tilde{\alpha}$, and v are the modulus, linear coefficient of expansion, and Poisson's ratio, respectively (the subscripts f and p referring to filler

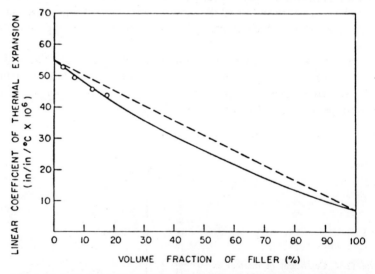

Figure 12.35. Comparison of experimental values of linear coefficient of expansion of a TiO_2-filled epoxy resin with values predicted by equation (12.54). The dashed curve is for the rule of mixture, the solid curve for the predictions of equation (12.54). Temperature range $-50°C$ to $100°C$; $\tilde{\alpha}_f = 7 \times 10^{-6}\,°C$; $E_f = 7.25 \times 10^6$ psi; $v_f = 0.23$; $\tilde{\alpha}_p = 55 \times 10^{-6}\,°C$; $E_p = 7.25 \times 10^5$ psi; $v_p = 0.3$. (Wang and Kwei, 1969.)

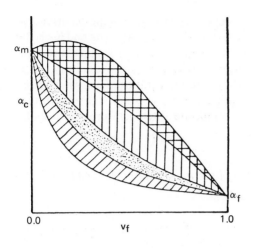

⊞ P.T.F.E. + WIDE RANGE OF INORGANIC POWDERS

⊞ THERMOPLASTICS + WIDE RANGE OF INORGANIC POWDERS

▦ THERMOSETS + GLASS FABRIC

▨ THERMOPLASTICS + GLASS FIBER

Figure 12.36. Volumetric coefficient of expansion α_c of filled polymers as a function of volume fraction filler v_f; schematic diagram. (Holliday and Robinson, 1973.)

and polymer, respectively), and v_f is the volume fraction of filler. As shown in Figure 12.35, good agreement was found between equation (12.54) and experiment for a TiO_2-filled epoxy resin.

One of the most comprehensive comparisons of data for many filled polymer systems is given in the review by Holliday and Robinson (1973), which also provides additional expressions not discussed here. Several points are of special interest. First, equations such as Kerner's (1956a,b) and Wang and Kwei's (1969) agree reasonably well for spherical particles, while Turner's (1946) equation is better for systems in which fillers are fibrous or platelike. Second, fibers and fabrics induce the greatest deviation from additivity (see Figure 12.36). Finally, the behavior of polytetrafluoro-ethylene is strikingly anomalous.

Although reasonable agreement was found in many cases between theory and experiment, a complete picture would also take account of such factors as particle shape and interfacial characteristics, which are usually not explicitly considered.

It may be noted in passing that densities of composites may differ considerably from values predicted by simple additivity. Even if there are no anomalies caused by incomplete wetting or occlusion of air in filler aggregates, there may be deviations due to the occurrence of thermal stresses which affect the coefficient of expansion (Wang and Kwei, 1969). By extending their treatment of differential expansion leading to equation (12.54), Wang and Kwei derived expressions for the effective density ρ_c of the filled polymers at temperatures below the temperature T_0 at which the filled polymer is free from thermal stresses. Since C is unity at temperature T_0, the effective density $\rho_c(T)$ of the composite at temperature T is given by

$$\rho_c(T) = \bar{\rho}_c(T)\phi \tag{12.58}$$

where

$$\bar{\rho}_c(T) = \rho_f(T)v_f(T) + [1 - v_f(T)]\rho_p(T) \tag{12.59}$$

or

$$\bar{\rho}_c(T) = \frac{\rho_f v_f + \rho_p(1 - v_f)}{v_f[1 + \alpha_f(T - 25)] + (1 - v_f)[1 + \alpha_p(T - 25)]} \tag{12.60}$$

and

$$\phi = \frac{1 + \alpha_p(T - 25)[1 - v_f(1 - k_\alpha)]}{\{1 - \alpha_p(T_0 - T)[1 - v_f(1 - k_\alpha)C]\}\{1 + \alpha_p(T_0 - 25)[1 - v_f(1 - k_\alpha)]\}} \tag{12.61}$$

where ρ_f, ρ_p, and v_f are all measured at room temperature (25°C), k_α is given by equation (12.55), α_f and α_p are the volumetric thermal expansion coefficients of the filler and the polymer, respectively, and C is given by equation (12.56).

If α_p is very small compared to unity, the following simplification is obtained:

$$\phi \doteq \frac{1 + \alpha_p(T - 25)[1 - v_f(1 - k_\alpha)]}{1 + \alpha_p(T - 25)\{1 - v_f(1 - k_\alpha)[1 - (C - 1)(T_0 - T)/(T - 25)]\}} \tag{12.62}$$

Testing the equations with TiO_2-filled poly(vinyl acetate), assuming T_0 was equal to the softening temperature (95°C), it was found that the predicted value of density was closer to the experimental value than was the value predicted by additivity, but was slightly high, perhaps due to filler aggregation. Definitive tests of these equations must await further experimentation.

12.2. FIBER-REINFORCED COMPOSITES

It has been seen that particulate fillers tend to increase the stiffness of a matrix resin, but may or may not increase the toughness or tensile strength, depending on the ductility of the matrix and the degree of filler–matrix adhesion. In many cases, the toughness is in fact decreased by spherical or substantially isometric particles, especially if good filler–matrix adhesion exists. On the other hand, even short fibers often confer quite synergistic mechanical properties, such as a combination of high modulus, strength, and toughness.

In this section, major aspects of the mechanical behavior of fiber-reinforced composites (based on both short and continuous fibers) are briefly considered. Since, as with particulate composites, the nature of the interfacial adhesion is important in determining modulus, strength, and toughness, the role of the fiber–matrix interface is also discussed.

12.2.1. Mechanical Properties

As the shape of a filler particle is elongated from a spherical shape to a fiber, the filler stiffens the composite to a greater extent, and begins to carry increasing fractions of the load. Useful fibers may be of many types—glass, boron, graphite, or other polymers; and may exist in various configurations—from discrete whiskers or fibers to woven cloth. With short, discontinuous fibers of a high-modulus material, the mechanical load is shared between the matrix and the filler, and most mechanical properties of the composite are improved to greater or lesser extents in comparison to the matrix. With continuous fibers, on the other hand, the fibers carry most of the mechanical load, while the matrix serves to transfer stresses to the load-bearing fibers, and to protect them against damage (Broutman and Krock, 1967, 1974, Chamis, 1974a; Corten, 1966, 1971; Ferry, 1970; Hattori, 1970; Nicolais, 1974; Parkyn, 1970; Schwartz and Schwartz, 1968; Tsai et al., 1969). In this case, the presence of the matrix makes possible better utilization of the inherently high levels of properties characteristic of fibers than would otherwise be possible. Indeed, fiber–polymer composites are among the strongest and stiffest engineering materials known (see Table 12.4),* especially on a property/density basis.

Although composites of glass cloth with polymeric resins have been common for several decades, the development of aerospace technology in the 1950's and 1960's led to an intensive effort to understand composites better and to optimize their physical behavior. Since weight must be kept to a minimum in aerospace applications, considerable attention was given

---

* Highly oriented crystalline polymers ("molecular composites") clearly also fall in this class (Lindenmeyer, 1975; Porter, 1975; Society of Plastic Engineers, 1975; Halpin, 1975; Kardos and Raisoni, 1975). This is an area of high current interest, for exceptional strengths and moduli can be obtained, often approaching theoretical limits.

Table 12.4
Properties of Oriented Polymeric Chain Structures and Related Engineering Materials[a]

Material	$E_{11}{}^{b} \times 10^{-6}$, psi	$E_{22}{}^{b} \times 10^{-6}$, psi	$\bar{E}_{iso}{}^{b} \times 10^{-6}$, psi	Tensile strength, ksi	Density, lb/in.^{3}
Steels and iron (cast and alloy)	—	—	28–30	13–300	0.25–0.29
Titanium	—	—	19	60–240	0.16
Poly(vinyl alcohol)	36.2	1.54	14.4	~100	0.05
Polyethylene	34	0.7	13.6	~100	0.05
Boron/epoxy	30	2.7	11.89	83	0.07
Aluminum	—	—	10	22–90	0.10
Polytetrafluoroethylene	22.2	—	—	—	0.05
HTS-graphite/epoxy	21	1.7	10	62	0.05
Cellulose I	18.5	—	7.6	50	0.05
E-glass/epoxy	5.6	1.2	2.85	47	0.06
Polypropylene	6.0	0.42	2.5	—	0.05
Poly(ethylene oxide)	1.42	0.56	0.88	—	0.05

[a] Halpin (1974).
[b] E_{11} and E_{22} refer to measurement parallel to and transverse to, respectively, the direction of orientation; E_{iso} refers to isotropic orientation.

to achieving high strength-to-weight and high modulus-to-weight ratios. Table 12.4 shows that, for example, composites based on epoxy resins reinforced with boron or graphite fibers have specific (per unit weight) strengths and moduli far surpassing values for aluminum, titanium, or high-strength steel. Indeed, some of the *absolute* values also attain or exceed those quoted for some of the metals. In spite of the excellent level of strength attained, other properties, such as fatigue behavior, may, however, pose problems in certain cases.

This section will review selected papers on the behavior of composites reinforced with more or less isotropically dispersed short fibers. More detailed technological and theoretical mechanical discussions, as well as additional information on composites containing continuous unidirectional or cross-plied fibers, is given by Ashton et al. (1969), Broutman and Krock (1967, 1974), Chamis (1974a), Corten (1966, 1971), Hattori (1970), Parkyn (1970), Schwartz and Schwartz (1968), Tsai et al. (1969), and Wu (1974), as well as in the general literature.*

* Especially good sources are the *Journal of Composite Materials*, the *Journal of Materials Science*, and proceedings of the annual meetings of the Reinforced Plastics/Composites Institute of the Society of the Plastics Industry.

12.2.1.1. Modulus and Tensile Strength

It can be shown that a matrix can transfer a major portion of an applied stress to fibrous elements only if the ratio of length l to diameter d exceeds a critical value l_c (Broutman and Krock, Chapters 1 and 2):

$$l_c/d = \sigma_f/2\tau \tag{12.63}$$

where σ_f is the maximum stress in the fiber and τ is the matrix or interface shear strength. The reason is that the fiber ends do not carry load; as the fiber length increases, the relatively ineffective portions of the fibers decrease. The average stress in a fiber $\bar{\sigma}_f$ is given by

$$\bar{\sigma}_f = \sigma_f(1 - l_c/2l) \tag{12.64}$$

For the case of a matrix having a yield strength of 8000 psi and a fiber with $\sigma_f = 1 \times 10^6$ psi, the ratio l_c/d would have to be about 60 in order to load the fiber to its maximum stress at the center. Since as a matter of practice l/d ratios are typically in the range 150–5000, it is clear that such a fiber resembles a continuous fiber in its ability to carry load transferred from

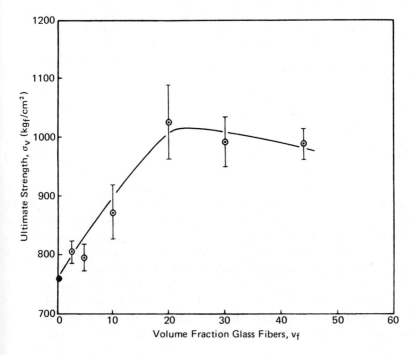

Figure 12.37. Effect of filler content on glass-fiber–PPO composite tensile strength. (Trachte and DiBenedetto, 1971.)

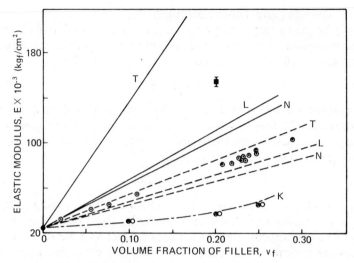

Figure 12.38. Effect of filler content on glass-fiber–PPO composite initial elastic modulus. The term A-1100 refers to the silane coupling agent used with some fillers: (⊙) Glass fiber/PPO; (⊗) A-1100 glass fiber/PPO; (△) glass bead/PPO; (○) A-1100 glass bead/PPO; (■) graphite fiber/PPO. The curves are theoretical: (T) Tsai theory; (L) Lees theory; (N) Nielsen and Chen theory; (K) Kerner theory. (--) Theoretical for both glass fibers/PPO; (–·–) theoretical for both glass beads/PPO; (—) theoretical for graphite fiber/PPO. (Trachte and DiBenedetto, 1971.)

the matrix. Although the role of interfacial adhesion is often complex (Chamis, 1974), good adhesion appears to generally lower the value of l_c/d (Cessna et al., 1969a).

Thus, in contrast to at least some of nonfibrous reinforcements, both tensile moduli and strengths are increased simultaneously by even relatively short fibrous reinforcements (Figures 12.14, 12.37, and 12.38). Of course, the orientation and configuration of the fibers are important. At a given concentration of fiber, isotropic systems will exhibit a lower degree of reinforcement than systems that have preferential alignment in the direction of stressing, while orthogonally oriented systems (fibers at right angles to each other) are less strong, but have good properties in both directions.

With random fibers in a plane a modified rule of mixtures has been proposed by Nielsen (1962) and Nielsen and Chen (1968) to predict the modulus:

$$E_c = KE_f v_f + E_m(1 - v_f) \tag{12.65}$$

where K is an empirical fiber efficiency parameter which depends on the volume fraction of fiber v_f and on the ratio of the fiber to the matrix modulus E_f/E_m. For $100 > E_f/E_m > 10$ and $0.6 > v_f > 0.1$, K lies between 0.15 and 0.56 (Nielsen, 1962). Other, slightly different relationships have been proposed, e.g., by Tsai et al. (1969, p. 233) and Lees (1968). Figure 12.38 gives a comparison of predicted and experimental moduli for carbon and glass fibers dispersed in poly(phenylene oxide) (Tsai et al., 1969). Experimental data fall above the Nielsen–Chen (1968) curve and between the Lees (1968) and Tsai et al. (1969) curves. The one point for a graphite fiber composite also falls between these two relationships, though the absolute level of modulus is higher. It may be seen that randomly oriented short fibers do confer much higher moduli than glass spheres, regardless of the degree of adhesion. The independence from the degree of adhesion may arise from the possibility that shrinkage stresses may provide good mechanical coupling between the fiber and the matrix.

  If fibers are long and oriented in the direction of applied stress, higher levels of reinforcement occur, and the following rule of mixtures is found to hold (Nielsen, 1967a; Ashton et al., 1969, p. 366) in terms of moduli of the constituents:

$$E_{c,\parallel} = E_f v_f + E_m v_m \qquad (12.66)$$

where subscripts c, f, and m refer to the composite, fiber, and matrix, respectively, v_f is the volume fraction of the fiber, and the subscript $\parallel$ means the fiber orientation is parallel to the stress direction.

  It may be recalled from earlier discussions that an equation of this form corresponds to an upper bound to a composite modulus (Section 12.1.1.1). Thus, particulate fillers tend to yield lower-bound values, as predicted by relationships such as Kerner's (1956b), while long, oriented fibers tend to yield upper-bound values of modulus. Short, randomly oriented fibers tend to yield intermediate behavior, but, as pointed out by Brody and Ward (1971), usually lead to moduli closer to the lower than to the upper bound, depending on the modulus of the polymer. On the other hand, Lavengood and Gulbransen (1969) reported moduli closer to upper-bound values for short fibers dispersed in an epoxy resin.

  With long fibers, the advantage gained in the direction of stress is offset by a compensating reduction in the transverse direction as indicated in the following equation (Tsai, 1964):

$$E_{c,\perp} = 2[1 - v_f + (v_f - v_m)v_m] - \frac{M_f(2M_m + G_m) - G_m(M_f - M_m)v_m}{(2M_m + G_m) + 2(M_f - M_m)v_m}$$

$$(12.67)$$

where $E_{c,\perp}$ is the transverse Young's modulus of the composite, M is the area modulus $[M = E/(2 - 2v)]$, v is Poisson's ratio, and the other symbols have the same significance as before. Because of such anisotropy, random fiber systems are often preferred.

We now turn our attention from considerations of stiffness to stress–strain and tensile behavior. Variations in strength and modulus as a function of direction (Broutman and Krock, 1967, Chapter 12) have been treated by several investigators; for example, Tsai (1965) and Brody and Ward (1971). Even though the polymer matrix typically has such a low modulus that it does not contribute much overall to the composite modulus, the matrix can by no means be neglected, because failure often involves catastrophic crack growth in the matrix (see below). Stress–strain curves for unidirectional composites are typically fairly linear up to failure for loading in the direction of the fibers (Broutman and Krock, 1967, p. 370), but quite nonlinear transverse to the fiber direction. The stress to rupture is also very low in the latter case, presumably due to a high concentration of stress in the matrix.

Even in the apparently linear range, the response to stress should be considered as viscoelastic rather than elastic. Most polymers that behave in a linear, viscoelastic manner at small strains ($< 1\%$) behave in a nonlinear fashion at strains of the order of 1 % or more. However, in a fibrous composite, the resin may behave quite differently than it would in bulk. Stress and strain concentrations may exceed the limiting values for linearity in localized regions. Thus the composite may exhibit nonlinearity (Ashton, 1969; Trachte and DiBenedetto, 1968), as is the case with particulate-filled polymers (Section 12.1.2). Although nonlinearity at low strains is characteristic, Halpin and Pagano (1969) have predicted constitutive relations for isotropic linear viscoelastic systems, and verified their prediction using specimens of fiber-reinforced rubbers.

In any case, the effects of high stress concentrations localized at the tips of fibers are extremely important. Several investigators have measured such stresses as a function of the following parameters: critical aspect ratio l_c/d, spacings between fiber ends and between fibers, eccentricity of fibers, and overlap of fiber ends. For example, MacLaughlin (1968) reviewed earlier studies and presented results of a study of this problem by photoelastic techniques. Glass fibers were used in an epoxy matrix; the ratio of fiber to matrix modulus was 40. Maximum shear stresses in the fiber were found to occur for critical aspect ratios of 40–80, with the smaller value for the closer lateral spacing. Maximum shear stresses in the matrix were observed, as expected, at fiber ends; especially high values were observed near fiber ends that were close together. Carrara and McGarry (1968) analyzed matrix and interface stresses as a function of geometry of fiber ends, and showed that a tapered tip should have the lowest stress concentration. Given the high stress concentrations at fiber ends, it seems reasonable to suggest that a matrix should be able to withstand localized yielding, even though the effect of yielding on stress distribution is not known in detail. Unfortunately, typical matrix resins tend to be brittle and relatively incapable of yielding rather than cracking when subjected to a high stress concentration. Brittleness is not desirable per se, but the more brittle resins tend to fulfill the desirable goal of high modulus. It should also be noted that the question of stress concentrations at fiber ends is also relevant to the case of continuous fibers because in such a system some fibers may break at low load levels and thus become discontinuous.

Thus, at least under some circumstances, the matrix (and interface) may play a significant role in determining behavior. Some possible implications are discussed below (see Sections 12.2.1.2 and 12.2.2); the discussion in Section 12.3, though largely concerned with particulate fillers, should also be relevant.

12.2.1.2. Fatigue Behavior

Since many fiber-reinforced plastics are subjected to cyclic loads, an understanding of fatigue behavior is necessary, for both short and continuous fibers. Fatigue has been studied from several viewpoints, such as effects of aspect ratio, frequency, modes of failure, and crack propagation. Several important studies are discussed below; further details may be obtained from the literature (Broutman, 1974; Owen, 1974a, 1974b; Argon, 1974).* Fatigue behavior in plastics themselves has been recently reviewed (Manson and Hertzberg, 1973a).

Working with short boron fibers embedded in an epoxy resin, Lavengood and Gulbransen (1969) determined the number of cycles required for a 20% deflection at a rate of 3 Hz, which was low enough to minimize heating by hysteresis. A significant increase in fatigue life was observed when the aspect ratio equalled 200. Further improvements in fatigue life were small for values of $l_c/d > 200$—a fortunate circumstance in view of the difficulty of fabricating composites containing longer fibers. The failure mechanism was found to be a combination of interfacial fracture and brittle rupture of the matrix at 45° to the fiber axis.

At higher frequencies, a substantial degree of hysteretic heating may occur. Thus, at frequencies up to 40 Hz, Dally and Broutman (1967) observed the generation of surface temperatures up to 265°F for fiber-glass-roving-resin systems (with plies arranged at right angles or at 60° with respect to each other). The heat generated per unit volume per second, q, depends on the

---

* See also the U. S. Government report literature; e.g., Rao and Hofer (1972).

frequency f and tensile strength σ, and is given by

$$q = H\frac{\sigma^2 f}{2E} \qquad (12.68)$$

where E is Young's modulus and H is a constant depending on the loss modulus E''. The number of cycles to failure was somewhat reduced by increasing the frequency, with the effect greater the lower the stress level, and greater for the isotropic material. It was suggested that the increased temperature lowered the modulus, which in turn effectively increased the ineffective fiber length and thus lowered the ability to withstand stressing. In another study, by Cessna *et al.* (1969*b*), the critical role of hysteresis heating in causing failure of both unfilled and fiberglass-filled resins was emphasized. In addition, good matrix-to-fiber coupling was found to be beneficial.

McGarry (1966) and Broutman and Sahu (1969, 1972) separately demonstrated the progressive nature of damage during fatigue loading. Cracks develop at moderate loads even during the first cycle provided there are fibers perpendicular to the loading direction. Cracks in fibers aligned in the stress direction may develop if the stress level exceeds 75% of the ultimate strength. As cycling proceeds, crack lengths increase, and cracks develop along a ply or fiber interface or into a ply or fiber itself. Such damage may be followed by determining the dynamic mechanical behavior (at ultra-sonic or sonic frequencies) of a composite as a function of the number of cycles impressed upon a specimen (DiBenedetto *et al.*, 1972). For example, as shown by DiBenedetto *et al.* (1972), Young's modulus and damping could be correlated with the amount of damage in laminates prepared from glass-fiber-reinforced epoxy and polyester systems and subjected to cyclic loading; the greater the damage, such as interfacial debonding, the lower the modulus and the higher the damping. Such findings [along with results found by others, also described by DiBenedetto *et al.* (1972)] suggest that measurement of the dynamic response of a fatigue-loaded composite may be of considerable use in the development of nondestructive tests for composite integrity.

Studies of fatigue life have also been conducted on a variety of short-fiber reinforced plastics by Theberge (1969). A more detailed study was reported by Dally and Carrillo (1969), who characterized the fatigue behavior of glass-filled nylon, polystyrene, and polyethylene. Nylon was found to be the matrix most resistant to fatigue; in all the materials, debonding at the resin interface was observed. In the long-fiber-reinforced nylon, cracks formed and grew only where the fiber concentration was high, while with the shorter fibers cracks formed in very localized areas. Cracks did not propagate readily into nylon or polyethylene, but did so into polystyrene. These results are consistent with findings of Hertzberg *et al.* (1970), who found crack growth rates to be much higher in polystyrene than in polyethylene or nylon; indeed, crystalline polymers generally exhibit lower rates of fatigue crack growth than glassy polymers (Manson and Hertzberg, 1973a).

The fact that fatigue resistance tends to correlate with static fracture toughness (Manson and Hertzberg, 1973a) should make it possible to more easily select matrix candidates for improved fatigue life, at least for cases in which the matrix limits the strength.

12.2.2. Role of the Matrix and Interface

While it is true that in fibrous composites the fibers bear the major fraction of the load, the matrix and the nature of the fiber–matrix adhesion are often exceedingly important (Chamis, 1974b; Plueddeman, 1974a, 1974b; Scota, 1974; Erickson and Plueddeman, 1974). In fact, the matrix is the strength-limiting variable in the following cases (Goettler et al., 1973): aligned continuous fibers, tested off-axis; cross-plied continuous fibers, tested at all angles; discontinuous fibers more or less aligned, and tested at all angles; and incompletely dispersed random-in-a-plane discontinuous fibers, tested at all angles. Thus, in many common cases, matrix strength and static or dynamic toughness, as well as high modulus, are desired—a combination not often achieved. While the incorporation of rubbery materials in a brittle matrix can increase toughness substantially (McGarry, 1966), the modulus tends to be reduced to some extent. Possibly better approaches to achieve both a high modulus and improved toughness might include the addition of small glass spheres (DiBenedetto and Wambach, 1972), especially if interfacial adhesion is minimized; the use of ductile or crack-stopping fillers (Lange and Radford 1971; Outwater, 1970); the use of high-modulus organic fibers (Blumentritt and Cooper, 1974); the structural modification of epoxy and other resins (Soldatos and Burhans, 1970); and the use of other matrices altogether (Goettler et al., 1973); see Section 12.1.2.

In all these cases, the nature of the interface must be carefully controlled. For example, in contrast to the case of particulate composites (DiBenedetto and Wambach, 1972), good adhesion between a ductile matrix and short glass fibers may enhance crazing (and hence energy dissipation) at the fiber tips. On the other hand, to achieve high impact strength in ABS resin reinforced with potassium titanate fibers, less good adhesion appears to be desirable (Speri and Jenkins, 1973); at relatively low levels of adhesion, the modulus remains high as well, though not the tensile strength.

As with particulate fillers, then, the fracture energy of a fiber-reinforced composite is a complex function of the properties of the reinforcement, the matrix, and the interface (Broutman, 1974; Cooper, 1974). Several modes of energy dissipation may be involved (Allred and Schuster, 1973; Aveston and Kelly, 1973; Beaumont and Harris, 1972; Broutman, 1966, 1970; Broutman and Agarwal, 1973; Cook and Gordon, 1964; DiBenedetto, 1973; Kelly, 1966; Marston *et al.*, 1974; McGarry and Mandell, 1972; Murphy and Outwater, 1973; Outwater and Murphy, 1969; Piggott, 1970, 1974; Williams, T., *et al.*, 1973) (see Table 12.5):

1. Debonding of the fibers from the matrix (Outwater and Murphy, 1969; Murphy and Outwater, 1973) as a crack impinges upon a fiber. Such a debonding may occur when the fracture strain of the fiber is greater than the fracture strain of the matrix, and the extent of debonding will depend on the strength of the interfacial bond relative to that of the matrix (Figure 12.39). A related type of debonding, in which the tensile component in the stress field ahead of the crack can induce interfacial failure, has previously been proposed (Cook and Gordon, 1964).

2. Pulling of broken filaments out of the matrix following fracture (Kelly, 1966).

3. Redistribution of strain energy from a fiber to the matrix after fracture of the fiber (Piggott, 1970).

4. Fracture of the fibers and matrix themselves, the former contributing little to the overall fracture energy (Marston *et al.*, 1974). The case of the interface has already been included in (2); the interfacial fracture energy will of course be added to the strain energy taken up by the fiber, the latter constituting the "debonding" energy per se (Marston *et al.*, 1974).

Table 12.5
Factors Contributing to the Work of Fracture[a]

Type of work	Symbol	Origin	Form of energy dissipation
Fiber internal work	γ_{fb}	Fiber brittle fracture	Stored elastic energy
	γ_{fs}	Fiber bending during pullout	Plastic flow during bending
	γ_{fd}	Fiber ductile fracture	Plastic flow and necking
Interface work	γ_{mf}	Difference in tensile strains across interface	Frictional sliding or plastic shear in matrix
	γ_{fp}	Fiber pullout	Frictional sliding or plastic shear in matrix
Interface and matrix work	γ_{ms}	Splitting of matrix parallel to fibers	Matrix surface energy and fiber–matrix bond energy
Matrix internal work	γ_{m}	Matrix fracture	Matrix surface energy and plastic flow

[a] Piggott (1974).

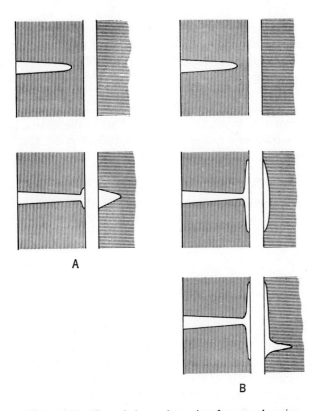

Figure 12.39. Effect of glass–polymer interface on advancing crack. (A) Adhesive strength of bond equals cohesive strength of matrix. (B) Adhesive strength of bond is much less than cohesive strength of matrix. (Broutman, 1966.)

Clearly, all these modes may participate in determining the overall toughness of a composite, though to degrees dependent on the system (Marston *et al.*, 1974; McGarry and Mandell, 1972). Marston *et al.* (1974) and others (Broutman, 1966, 1967; Broutman and Agarwal, 1973) have recently critically analyzed this question, and developed more or less generalized treatments of the overall work required for fracture. Tests with boron–epoxy and carbon–polyester systems showed that such an approach gave improved predictions of fracture toughness (Marston *et al.*, 1974). Such analyses based on fracture mechanics concepts should be useful in balancing the various fracture parameters to obtain a reasonable level of toughness without sacrificing strengths of the order predicted by the rule of mixtures [see equation (12.65). In general, an intermediate value of the interfacial bond strength is desired to optimize toughness; an optimum fiber length (in some cases close to the critical fiber length l_c) also exists, the value depending on the nature of the fibers (Allred and Schuster, 1973; Piggott, 1970).

The interface may also be "tailored" (Kardos, 1974; Kardos *et al.*, 1973) to obtain a gradation in modulus from the fiber to the matrix, and hence a more efficient transfer of stress [compare Fallick *et al.* (1967)]; such a tailoring was effective with annealed graphite and glass-fiber-reinforced polycarbonate.

In any event, greater understanding of the roles of the matrix and interface in determining strength and toughness should help greatly in the selection of matrix resins. The reinforcement may also affect the matrix to a significant degree (see Section 12.3).

12.3. MATRIX AND INTERFACE BEHAVIOR

In Sections 12.1 and 12.2 several important aspects of matrix and interface behavior, such as the effects of an inclusion on modulus and the effect of interfacial adhesion, were described. It is also appropriate to discuss specific molecular effects of rigid inclusions (particulate or fibrous) on a matrix, in order to demonstrate continuity between all types of reinforcements. Effects of environmental exposure on composite behavior are also briefly considered.

12.3.1. Molecular Effects of Rigid Inclusions

In Chapter 10 it was pointed out that certain fillers could dramatically increase the modulus and tensile strength of elastomers to a much greater extent than other fillers. The activity of such so-called "reinforcing" fillers was related to two major factors: small particle size, and the ability to interact strongly in a physicochemical sense with the polymer (as, for example, in the ability to chemisorb and fit between polymer segments). Although the precise mechanisms are not completely understood, the factors mentioned do combine in complex ways to develop a highly efficient filler–polymer network with an improved ability to resist failure (Section 10.7). At the same time, the restrictions on segmental relaxation tend to increase the glass temperature T_g and broaden the spectrum of relaxation times (Sections 10.8, 12.1.1, and 12.1.2). As will be shown, it is reasonable to suppose that the effects of rigid inclusion in a matrix are not restricted to the high-surface-area reinforcing fillers, but may occur with much larger inclusions as well, particularly if the matrix is below its glass transition temperature. Although this discussion deals mainly with evidence drawn from studies of particulate fillers, the conclusions should also be relevant to some aspects of fibrous reinforcements.

The question is of great importance, for typical expressions for composite behavior (Section 12.1) in terms of constituent properties generally assume that the two components are in close contact (in a state of so-called "perfect adhesion") with a singularity at the interface as one passes from one component to the other.

It is now generally recognized that this simple picture is inadequate, and that, probably in general, the interface between an adhesive and adherend or substrate (in this case the matrix and reinforcement, respectively) should be considered not as a singularity but as a layer or "interphase" (National Academy of Sciences, 1974; Sharpe, 1971). Indeed, much evidence for this conclusion has already been presented. As shown in Figure 12.40, then, a filler–matrix system may be considered to possess a region at the interface whose properties differ from those of the bulk matrix. This does not imply that the interphase must itself exhibit sharp phase boundaries; however, the existence of gradients at the boundaries does not affect the general argument. One may envisage several phenomena which may be involved in generation of an interphase (National Academy of Sciences, 1974; Sharpe, 1971): roughness or porosity of the substrate surface, such that the adhesive (or matrix) penetrates beyond the outer bounds of the surface (Bikerman, 1968, p. 1); changes in composition at the interface, as may occur due to selective adsorption of one component, e.g., an amine in an epoxy system (Erickson, 1970; Hirai and Kline, 1972; Kardos, 1974; Patrick *et al.*, 1971); other influences of the substrate on the curing of thermosetting resins

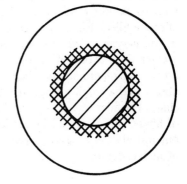

Figure 12.40. Schematic showing filler particle embedded in a matrix resin. Note crosshatched "interphase" region between bulk matrix and outer bound of filler surface. Also note that filler surface may be porous or rough, so that interphase bounds may include filler asperities, and that interphase boundary is not sharp.

(Bascom, 1970; Kwei, 1965); morphological changes induced at an interface with a crystalline polymer during solidification (Fitchman and Newman, 1970; Schonhorn, 1964); the presence of a coupling agent or adhesion promoter, such as a silane, on the substrate (Bascom, 1969; Johannson *et al.*, 1967; Kaas and Kardos, 1971; Plueddeman, 1970; Plueddeman *et al.*, 1962; Sterman and Marsden, 1966); and ordering effects of various kinds, possibly due to annealing effects or thermal stresses (Kardos, 1974; Kardos *et al.*, 1973; Kumins, 1965; Kwei, 1965; Landel, 1958; Manson and Chiu, 1972). Other effects may also be observed or postulated, such as incomplete wetting, air voids (Bascom and Cottington, 1972), and softening due to residual stresses (Lewis and Nielsen, 1970). In any event, a polymeric matrix is not likely to behave like a continuum as one passes from an interface into the bulk. Force fields at short distances (up to 20 Å) are certainly operative (Yim *et al.*, 1973) and long-range effects up to 10^3 Å or more have been postulated (Kumins, 1965; Kwei, 1965; Manson and Chiu, 1972) or directly observed (Trachte and DiBenedetto, 1971).

As a result of such phenomena, observed properties such as relaxation behavior (as seen, for example, in values of T_g and damping behavior), ultimate strength properties, permeability, and solubility may deviate significantly from values predicted on the basis of a singular interface between a filler and a matrix (of course, other causes for deviations, e.g., interaction and agglomeration of filler particles, may exist in a given case, especially with very small particles such as reinforcing carbon blacks—see Section 10.4). In any case, above and beyond effects induced by a given volume of a second phase, effects due to interaction between the filler and the matrix must be considered, both for low- and high-surface-area filler particles.

With high-surface-area fillers, moduli tend to exceed those predicted by a lower-bound additivity equation; indeed, one criterion for a "reinforcing" filler is precisely the exhibition of such a deviation (Broutman and Krock, 1967, Chapter 18); see also Sections 10.11 and 10.12. On the other hand, polymers containing large-particle-size (low-surface-area) fillers tend to exhibit lower-bound behavior (Section 12.1.1.1). Thus, the highly effective fillers, e.g., certain carbon blacks in rubber, differ from other fillers in the *degree* to which they reinforce a polymeric matrix. The degree is greater than expected on the basis of adhesion alone, and, as mentioned above, this phenomenon is believed by many to be related to high adsorptive ability and a small enough size to fit between small segments of a polymer chain. The filler networks formed may also act rather like dispersed collections of fibers. (As discussed in Section 10.12, however, this theory is sufficient, but not always necessary; reinforcement does not always require meeting of such criteria.)

None of these "reinforcing" effects necessarily requires anything other than a purely mechanical effect due to the replacement of volume elements in a polymer by a particulate (or particulate network) phase, as long as "good adhesion" is attained at the interface. In principle, a singular sharply defined interface could suffice (admitting that "good" adhesion might involve adsorption of the closest matrix atoms on the substrate). In such a case, observance of a modulus additivity relationship does not *necessarily* imply an effect of the filler on the viscoelastic state of the matrix, and the relaxation behavior of the composite should be essentially equivalent to that of the bulk matrix (assuming that the filler is very rigid, with low damping capacity). On the other hand, in practice, extensive evidence suggests that the presence of filler particles *does* lead to modification of the state of at least part of the matrix resin, presumably in the vicinity of the filler particles themselves. Effects of this kind have long been recognized in the case of high-surface-area fillers, which can easily be pictured as affecting at least a thin region of matrix resin by adsorption, penetration into cavities, etc. It is not always recognized, however, that such effects may also be observed in low-surface-area fillers (with diameters of 30 μm, for example); indeed, the validity of deductions of this type is still sometimes questioned (Yim *et al.*, 1973).

A careful reading of the literature [coupled, in this case, with experimental work (Manson and Chiu, 1973a)] reveals that *both* small and large filler particles *can* induce significant changes in phenomena such as sorption, permeability, or relaxation behavior of the matrix (notably in the glass transition temperature T_g and in damping characteristics, E'' or tan δ), in addition to the noninteractive increases in modulus predicted by relationships such as Kerner's equation (12.3). The fact that some exceptions and contradictions continue to exist serves to stimulate further study rather than to deny the tendency toward filler–matrix interaction. Typical evidence may now be summarized.

12.3.1.1. Effects of Fillers on Relaxation Behavior and Other Transitions

Because of its obvious significance in determining appropriate temperatures for use, T_g is an extremely important parameter. In considering whether or not presumed changes in T_g are real, one must be careful in interpreting experimental results such as curves of modulus as a function of temperature. As shown in Figure 12.41 (left), the result of a purely mechanical reinforcement is to raise the modulus throughout, though the degree of elevation may vary from the glassy to the rubbery state. The effect is simply one of a vertical shift, as is well known in other systems, such as asphaltic concretes.

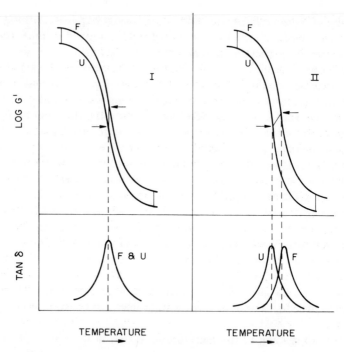

Figure 12.41. Schematic showing storage modulus G' and dissipation factor tan δ as a function of temperature for two cases of filler–matrix systems: I, case of simple volume replacement and stiffening by filler, with no effect on relaxation behavior of filled polymer F; II, case of volume replacement and stiffening *with* an effect of filler on relaxation behavior. (No attempt has been made to illustrate specific effects such as transition broadening, enhanced stiffening in the rubbery region, or additional peaks due to bound resin.)

In such a case, no change in relaxation behavior of the matrix is involved. The T_g, whether indicated by the onset of the decrease in modulus as the temperature is increased or by the temperature corresponding to the mid-point of the transition (e.g., by use of the peak in E''), is unchanged. (It should be noted that it is not legitimate to use the value of T_i, the temperature at which a given modulus is reached, as a measure of T_g in a filled system; this point may easily be neglected.) On the other hand, if the relaxation behavior *is* changed by the filler, some degree of horizontal shift would be observed, or at least a change in slope at the transition. All data discussed below are believed to conform to these restrictions.

A number of studies of high-surface-area fillers (surface area 100 m²/g) reveal detectable increases in T_g or other changes in relaxation behavior.

For example, Payne (1958) noted effects on the shape of relaxation spectra for carbon-black-filled rubber, and a small but significant shift to the transition region to higher frequencies, corresponding to an increase in T_g of up to 5°C. Similarly, Kraus and Gruver (1970) reported modest increases in the T_g of SBR filled with reinforcing carbon black, and Roe et al. (1970) and Waldrop and Kraus (1969) observed shifts in the NMR spin–lattice relaxation time T_1 to higher temperatures in carbon-black-filled polybutadiene and SBR, respectively. Roe et al. (1970) were able to resolve the NMR signals into two components, which were proposed to correspond to the relaxation of two different regions of differing proton mobility; in a similar study of carbon-black-filled rubbers, Kaufman et al. (1971) concluded that an interphase possessing a gradation in properties was present. Significant shifts in T_g have also been observed by Lipatov and Fabuliak (1968) for copolymers of styrene and methyl methacrylate adsorbed on silica and polytetrafluoro-ethylene. Additional confirmation of effects on T_g in several elastomers and polystyrene filled with silica has been reported by Yim et al. (1973). On the other hand, as pointed out by Kraus (1971b) and Baccareda and Butta (1972), effects of carbon blacks on the T_g of elastomers are not necessarily observed in all cases. Section 10.8 contains a further discussion of this point.

However, such effects are not limited to high-surface-area fillers. For example, with a typical pigment, TiO_2 (surface area $\sim 6 \, m^2/g$), several investigators have noted significant increases in T_g of a matrix resin. Thus Kumins and Roteman (1963) found an increase in the lower transition temperature for a copolymer of vinyl chloride and vinyl acetate of 21°C as the volume fraction of filler was increased to 0.2. (The upper transition temperature was decreased, apparently due to complex effects of adsorption.) From measured values of the increase in T_g, they calculated values for a so-called flex-energy parameter $(E_1 - E_2)$. This parameter represents the difference in energy between segmental conformations and hence is a measure of the segmental mobility. Values of $E_1 - E_2$ were increased from 1.38 to 1.54 by addition of the filler. It was suggested that the hindering effects of adsorption could persist outward from the filler surface for distances up to the order of at least $3 \times 10^3 \, \text{Å}$. Results of Galperin (1967) and Dammont and Kwei (1967) with TiO_2-filled epoxy resins, and results of Kwei (1965) and Kwei and Kumins (1964) with a variety of TiO_2-filled resins, are quite consistent; again, significant increases in T_g were observed. As shown by Nielsen et al. (1955), other fillers (perhaps not as finely divided as TiO_2) can increase the T_g of polystyrene by up to 15C.

The earlier observations by Landel (1958) have also been confirmed for several systems containing very large filler particles (up to $\sim 40 \, \mu m$ in diameter). Thus increases in T_g as measured by dilatometry and mechanical studies have been reported by Lipatov et al. (1963, 1974) for polymers such as polystyrene filled with coarse glass powder; increases of up to 24°C were described. As mentioned earlier, Lewis and Nielsen (1970) and Manson and Chiu (1973a) have observed small to moderate (up to 3 and 10°C, respectively) increases in the T_g of different types of epoxy resins filled with glass beads (10–50 μm in diameter), using peaks in damping or E'' to find T_g; see Figures 12.3 and 12.42. Similarly, Droste and DiBenedetto (1969) found small increases in T_g (~5°C) for glass-bead-filled phenoxy resins (Figure 12.43). On the other hand, NMR techniques do not indicate changes in T_1 in such a system, presumably because the segmental motions sensed by the NMR experiment are small compared to those involved in the glass transition per se (Droste et al., 1971). In contrast to these results, some exceptions do exist; for example, as shown by van der Waal et al. (1965) and Yim et al. (1973), sodium chloride appears to be ineffective in raising the T_g of typical elastomers.

While T_g of a matrix resin appears to be often increased by the presence of a filler, the magnitude of the effect does appear to depend on the nature of the surface. Thus Dammont and Kwei (1967) found that at a volume fraction of 0.05, TiO_2 was more effective than Al_2O_3 in raising T_g, albeit by only a few degrees. Surface treatments are also important. Yim et al. (1973) have shown that treatment of silica to *reduce* adhesion eliminates any elevation of T_g in several polymeric matrices. The beads used by Manson and Chiu (1973a), which led to 10°C increases in T_g of an epoxy resin (Figure 12.42), were, on the other hand, treated with a special coupling agent to maximize

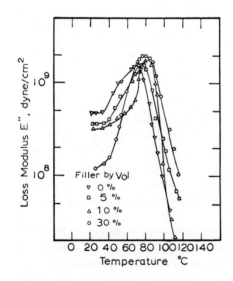

Figure 12.42. Effect of glass beads on loss modulus E'' of an epoxy resin (Manson and Chiu, 1973a). The shift to higher temperatures at higher filler content indicates an increase in T_g—in this case, of about 10°C for a volume fraction of 0.3 (50 wt %).

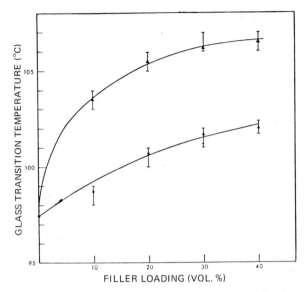

Figure 12.43. Glass transition temperature of poly(phenylene oxide) composite systems as a function of filler concentration. The upper curve is for attapulgite-filled phenoxy, the lower curve for glass-bead-filled phenoxy. (Dróste and DiBenedetto, 1969.)

adhesion. Interestingly, the ΔT_g observed by Yim *et al.* was found to be directly proportional to the polymer–filler interaction energy estimated from heats of adsorption using low-molecular-weight model compounds—an observation of considerable potential importance. These results conflict with those of Lewis and Nielsen (1970), who reported that *good* adhesion resulted in a *negligible* increase in T_g, and that increases were observed only with untreated beads or beads treated to minimize adhesion.

Secondary transitions may also be affected by fillers. Manson and Chiu (1972, 1973a, 1973b) have noted that glass beads in an epoxy resin tend to slightly *lower* the temperature of the low-lying β transition (at about $-60°$C) corresponding to the motion of the glycidyl group. A similar lowering was observed by Kline and Sauer (1962) in the case of Al powder, though not by Hirai and Kline (1972) in the case of carbon and graphite fillers; Jenness (1972) reported a tendency for a β peak to shift to higher temperatures as the volume fraction of filler was increased. In the presence of absorbed water, at least one new peak was observed in the glass-bead system (Figure 12.44) just mentioned (Manson and Chiu, 1973b). New peaks have also been observed in some cases by Yim *et al.* (1973) and attributed to the existence of an absorbed polymer component (Figure 12.45).

Thus it appears that, as long as good bonding can be obtained between a filler and a matrix, relaxation behavior can be significantly affected [see also the discussion by Lipatov *et al.* (1974)]. This is so even for large-particle fillers, for which one would not expect much binding of the matrix by adsorption, as is possible with many small particles, and damping behavior appears to be complex. A filler may increase damping in the matrix, as shown by Lewis and Nielsen (1970) and Nielsen (1969b) for filled epoxy resin and polyethylene. In general, damping is normally greater when the matrix is in the rubbery state, regardless of the degree of adhesion (Morgan, 1974), perhaps due to the occurrence of frictional effects at the interface; this observation is supported by evidence based on a variety of systems (Kardos *et al.*, 1972; Landel, 1958, Ziegel, 1969). In the glassy state, a filler may increase damping associated with the β transition as in the example just mentioned and in the studies by Hirai and Kline (1972) and Jenness (1972), perhaps due to the presence of either excess resin or a separate gel phase at the

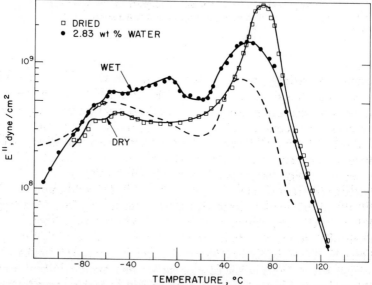

Figure 12.44. Effect of water (2.83 wt %) on damping in glass-bead–epoxy systems ($V_f = 0.3$). Note the appearance of a new peak at 0°C in the presence of water; a shoulder also has developed on the peak near 60°C (Manson and Chiu, 1973b). Dashed line represents unfilled polymer (wet).

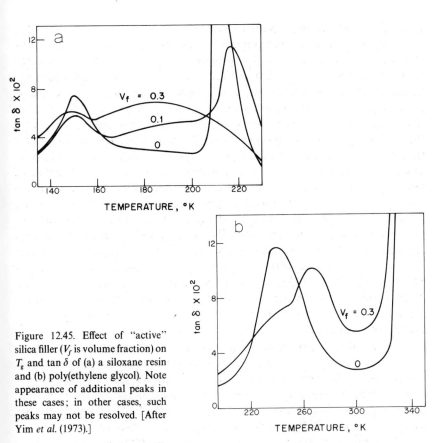

Figure 12.45. Effect of "active" silica filler (V_f is volume fraction) on T_g and $\tan \delta$ of (a) a siloxane resin and (b) poly(ethylene glycol). Note appearance of additional peaks in these cases; in other cases, such peaks may not be resolved. [After Yim *et al.* (1973).]

interface (Lewis and Nielsen, 1970). Generally, however, damping tends to be reduced for glassy matrices, as observed by Galperin and Kwei (1966) for TiO_2-filled poly(vinyl acetate), by Morgan (1974) for silica- and carbon black-filled SBR, and by Hirai and Kline (1972) for the γ transition in a carbon-filled epoxy resin. This tendency is certainly consistent with the existence of some kind of interphase due to constraint by the filler. Morgan (1974) proposed that the relative thickness of the interphase or immobilized layer exhibits a minimum as a function of temperature due to competition between an increasing tendency toward dilatation at lower temperatures and toward increased damping at higher temperatures. Also, the sensitivity to temperature was greater, the less the degree of constraint by the filler. It seems quite possible that the existence of such competitive effects may eventually explain apparent discrepancies in experimental findings.

12.3.1.2. Sorption and Permeability

The sorption of vapors or liquids by a polymer containing a nonsorbing filler will, of course, be reduced in proportion to the volume fraction of the filler (see also Section 12.1.3.1). However, as with the mechanical behavior discussed above, significant interaction between the filler and the matrix (or between a pigment and its binder) may occur (Kumins, 1965), resulting in a value of sorption lower than predicted. Kwei and Kumins (1964) noted that the incorporation of TiO_2 in poly(vinyl acetate) and epoxy resins lowered the ability of the matrix to sorb organic vapors more than predicted on the basis of simple additivity. It was postulated that the TiO_2 immobilized polymer segments close to the surface. An effect of the same type was also observed by Perera and Heertjes (1971) for the sorption of water by red lead–alkyd resin systems, though not for TiO_2–alkyd resin or pigment–epoxy systems; the possibility of a specific interaction between red lead and the alkyd resin was suggested. Specific interaction effects have also been noted with the glass-bead-filled epoxy-resin studies by Manson and Chiu (1973a,b). A small but consistent decrease of the solubility coefficient for water was found in the presence of filler; again, segmental ordering in the presence of the filler was postulated in order to explain the reduction.

Observed values of permeability may also be expected to deviate from predictions based on a simple two-component model of an impermeable, noninteracting filler embedded in a permeable matrix (see Section 12.1.3.1). If an interphase exists, it may be more permeable than the matrix, as in some pigmented paint films (Funke et al., 1969; Michaels, 1965), or less permeable, as in several other filler–polymer systems now to be discussed.

As mentioned above, both permeability and diffusion coefficients are sometimes lowered below predicted values (after correction for filler content); the permeability coefficient includes, of course, a contribution from sorption as well as diffusion per se. In addition, the apparent energy of activation for permeation E_a may be lowered by the presence of a filler. Specific examples have already been given in Section 12.1.3.1 (Kumins, 1965; Kumins and Roteman, 1963; Kwei, 1965; Kwei and Arnheim, 1965; Kwei and Kumins, 1964; Manson and Chiu, 1972, 1973a); data are consistent with the presence of an interphase of some kind, sometimes as thick as 1000 Å or more. Thus the presence of a filler does tend to reduce permeability and diffusivity in a polymer more than expected (at least as long as the interface is tight). Although the possibility of a loosening effect on the polymer close to the surface cannot be ruled out (Morgan, 1974), the fact that equation (12.35) represents an approximate lower bound indicates that some kind of densification or ordering must still occur.

12.3.1.3. Mechanisms of Ordering at the Interface

It has now been shown that recent studies of relaxation, sorptive, and diffusive behavior in many filled polymer systems amply confirm earlier observations of deviations from values predicted by simple additivity (Kumins, 1965). Such effects are not confined to high-surface-area fillers such as certain carbon blacks and fillers (typical "reinforcing" fillers for rubber); they are also observed frequently with low-surface-area fillers, such as pigments and even glass beads with average diameters in the range of tens of micrometers.

Many of these observations, e.g., higher T_g's, lower values of sorption, diffusivity, and mechanical damping (in some cases), are consistent with explanations in terms of adsorption of polymer segments at the interface (Kumins, 1965; Kwei, 1965). The adsorption process would involve a reduction in the number of degrees of freedom, and a consequent decrease in entropy and chain mobility—hence the common use of the term "ordering." It is easy to visualize such a process in the case of small particles, in which even a thin layer of adsorbed and entangled polymer (say, 10–100 Å in thickness) could account for a significant and discrete fraction of the total polymer in the matrix. A thickness of this magnitude is certainly consistent with interactions due to typical surface fields. Some authors suggest much greater zones of influence [up to 10^3 Å or more (Kumins, 1965; Kwei, 1965)], on the assumption that cooperative segmental effects persist relatively far from the surface. A significant role of adsorption is certainly given credence by the recent finding that changes in T_g can (at least in some cases) be correlated with heats of adsorption of model compounds on the filler surfaces (Yim et al., 1973). Studies of this type should be helpful in explaining some discrepancies, e.g., failure to observe changes in T_g in some cases. It has been suggested that enhanced specific polymer–filler interaction may compensate for the lower surface area of large filler particles; this hypothesis has yet to be tested.

It seems likely, however, that the view of the interphase as an adsorbed shell of ordered polymer is excessively simplistic (Erickson and Plueddeman, 1974; Plueddeman, 1974a, 1974b; Schrader, 1974). In some cases, e.g., a silane-treated glass, the matrix may interact strongly with a layer of silane, or even react chemically with it, as can an epoxy resin with an appropriate silane (Plueddeman, 1974). Thus the matrix may exhibit different properties as a function of distance from the filler, due to chemical effects. In many cases, thermal stresses are undoubtedly important (Morgan, 1973). Since most composites are prepared at elevated temperatures and then cooled down, the resin must be under considerable stress after cooling, due to the differences in thermal contraction. Such an effect has been proposed by Dammont and Kwei (1967) as a possible source of segmental immobilization due to the compressive restraints generated. Effects of this kind might well be able to yield layers thicker than the range of surface force fields, and thus explain the ability of low-surface-area fillers to generate interphases such as the 7000-Å-thick oriented layer of epoxy resin observed around a glass bead (Wambach *et al.*, 1968). Birefringence is also characteristic of an epoxy resin between closely spaced glass fibers in a composite (Outwater and Matta, 1962); ordering effects in such cases have been reviewed by Eakins (1966). In an analogous case, evidence has been found for the existence of birefringent layers surrounding filler particles when the matrix was subjected to swelling. Such effects were noted by Sternstein (1972) and Kotani and Sternstein (1971) and by Picot *et al.* (1972), and a theoretical treatment was developed by Sternstein (1972). Indeed, evidence for such dilatational effects has been observed in some cases in terms of modulus and damping as a function of temperature (Lewis and Nielsen, 1970; Morgan, 1973). The coexistence of compressional and dilatational stress fields may well be in part responsible for anomalies such as the existence of positive and negative changes in relative damping (Morgan, 1973; Nielsen, 1969*b*) or the downward shift of the secondary transition and the complex behavior of permeability–temperature curves for glass-bead–epoxy systems (Manson and Chiu, 1973*a,b*).

In conclusion, the picture of an interphase surrounding a filler particle, whether induced by adsorption, by other phenomena such as thermal stresses, or by some combination of these, can help explain specific effects of filler–matrix interactions on behavior. However, the interphase itself may well not be well defined and monolithic, so that a gradation in properties is probably more realistic (Kaufman *et al.*, 1971; Morgan, 1973). In addition, more than one component (e.g., both densified and rarefied) may coexist. Further studies of adsorption per se, the chemistry of cured matrix resins at interfaces, relaxation behavior, and thermal stress effects should be of great value.

12.3.2. Effects of the Environment

As people concerned with such systems as paint, adhesives, and high-performance composites have long known, the effects of environmental exposure, especially to water, may be profound and deleterious. Water may plasticize the matrix and thus change its properties, or, as often happens, it may attack the substrate–matrix interface and cause failure (Halpin, 1969). Indeed, a major reason for the use of silane or other coupling agents on glass is to decrease the sensitivity to moisture (Bascom, 1969, 1970; Broutman and Krock, 1967, Chapters 6 and 13; Corten, 1966; Johannson *et al.*, 1967; Kaas and Kardos, 1971; Plueddeman, 1970, 1974; Plueddeman *et al.*, 1962; Sterman and Marsden, 1966); for further details, see reviews by Bascom (1970, 1974).

Apart from a general tendency for physical properties to deteriorate on exposure to water, specific effects may be quite complicated. For example, DiBenedetto and Wambach (1972) found that short-term water immersion of glass-bead-filled epoxy resins increased ductility and toughness, while long-term immersion resulted in a decrease in toughness. A silane coupling agent minimized the later reduction, thereby increasing the useful life. As shown by Mostovoy and Ripling (1969) and Ripling *et al.* (1971), water can interact with stress applied to, for example, an epoxy resin in an adhesive joint to yield a type of stress corrosion. Values of fracture toughness were shown to increase in some cases, e.g., at the center of the bond, and decrease in others, e.g., at the interface. Outwater and Murphy (1970) have also applied fracture mechanics to analyze the efficacy of various surface treatments on glass in retaining epoxy–glass bond strength in the presence of water. Another interesting study was performed by Galperin *et al.* (1965), who examined the effect of humidity on the tensile behavior of poly(vinyl acetate) and epoxy resins filled with TiO_2. In the case of the epoxy resin, exposure to a high humidity resulted in greater tensile strength and stiffness. However, poly(vinyl acetate) was found to behave differently. As the filler content was increased, strength and modulus were lowered, and elongation increased, especially at lower strain rates. Complex effects of exposure to water have also been noted by Manson and Chiu (1973*b*) for glass-bead-filled epoxy resins; in the presence of water, a secondary relaxation peak was shifted to higher temperature (antiplasticized), while the major peak (corresponding to T_g) was shifted to a lower temperature (plasticized). Such contradictory behavior emphasizes the need for further research on environmental interactions, especially in the presence of mechanical stresses.

APPENDIX A. TYPICAL EQUATIONS USED TO PREDICT MODULI OF PARTICULATE COMPOSITES

In each case, E_c and E_p refer to Young's moduli for the filled and unfilled resin, respectively, G_c and G_p are the corresponding shear moduli, v_f is the volume fraction of filler, and S is the relative sedimentation volume of the filler; S is defined as the ratio of the apparent to true volume of the filler. In all cases the ratio E_c/E_p may be considered as equivalent to G_c/G_p.

1. *Kerner* (*1956b*)

Kerner's equation is

$$\frac{E_c}{E_p} = \frac{G_f v_f/[(7 - 5v)G_p + (8 - 10v)G_f] + v_p/[15(1 - v)]}{G_p v_f/[(7 - 5v)G_p + (8 - 10v)G_f] + v_p/[15(1 - v)]} \quad \text{(A-1)}$$

where G_p and G_f are the shear moduli of the plastic and filler (continuous and discontinuous phase), respectively, v is Poisson's ratio of the plastic, and v_p is the volume fraction of the plastic.

2. *Eilers* (*1941*) (sometimes quoted as the Eilers–Van Dyck equation)

$$E_c/E_p = \{1 + [1.25v_f/(1 - Sv_f)]\}^2 \tag{A-2}$$

[This form is equivalent to equation (12.7).]

The value of S is commonly taken to be in the range 1.2 to 1.3, the theoretical value for close-packed uniform spheres being 1.35 (Nielsen, 1967a).*

3. *Van der Poel* (*1958*)

This equation is not in explicit form.

4. *Guth* (*1944*) *and Smallwood* (*1944*)

$$E_c/E_p = 1 + 2.5v_f + 14.1v_f^2 \tag{A-3}$$

where v_f is the volume fraction of filler. This equation is often used for systems in the rubbery state.

5. *Mooney* (*1951*)

$$\ln\frac{G_c}{G_p} \equiv \ln\frac{\eta_c}{\eta_p} = \frac{2.5v_f}{1 - Sv_f} \tag{A-4}$$

where η_c and η_p are the viscosity coefficients for the filled and unfilled resin. Experimentally, a value of 1.4 is often used for S (Nielsen, 1967a). This equation, often used for systems in the rubbery state, is equivalent to equation (12.8).

6. *Tsai and Halpin* (*Ashton et al., 1968*)

$$\frac{E_c}{E_p} = \frac{G_c}{G_p} = \frac{1 + ABv_F}{1 - Bv_F} \tag{A-5}$$

where A and B are constants for a given composite. In this generalized equation, the constant A is a function of several factors, such as the Poisson's ratio of the polymer and the shape of the filler, while the constant B is defined in terms of the filler and polymer elastic moduli E_f and E_p, respectively:

---

* S is the inverse of ϕ_m, the maximum packing fraction.

$$B = \frac{(E_f/E_p) - 1}{(E_f/E_p) + A} \tag{A-6}$$

Many equations for composite moduli can be put in this form.

7. *Nielsen* (*1970b*)

$$\frac{E_c}{E_p} = \frac{G_c}{G_p} = \frac{KBv_f}{1 - B\psi v_f} \tag{A-7}$$

In this extension of the Tsai–Halpin equation (Brassell and Wischmann, 1974), K is a generalized Einstein (1905, 1906, 1911) coefficient equal to 2.5 for a suspension of rigid spheres in a matrix having a Poisson's ratio equal to 0.5, and given approximately for other values of v in the reference cited. Note that $K = A + 1$ [see equation (12.5)]. The constant B is defined as in equation (A-6) and ψ is given by functions such as

$$\psi = \left[1 + \frac{1 - \phi_m}{\phi_m^2}\right]v_f \quad \text{or} \quad \psi\phi_f = 1 - \exp\frac{-v_f}{1 - (v_f/\phi_m)} \tag{A-8}$$

where ϕ_m is the maximum volumetric packing fraction.

As mentioned in the text, many equations approach similar limits at low values of v_f. Some hold better for glassy than elastomeric systems, and vice versa, and many variations and simplifications are possible in certain cases. For detailed applications, the appropriate references in this chapter should be consulted.

13

A Peek into the Future

This chapter will be devoted to the consideration of nascent research undertakings, possible new materials, and as yet unanswered research questions. First, an effort will be made to codify and classify composite materials and polyblends with respect to topological considerations. We will examine what other ways may possibly exist to make new combinations of two types of polymeric molecules. How many ways exist to mix two kinds of polymer molecules? What relationships can be developed among such diverse materials as particulate and fiber-reinforced plastics, polymer-impregnated concrete, and foams, paint films, etc.?

Another portion of this chapter will treat some polymer mixing problems. Brief consideration will be given to possible polymer/polymer eutectic systems, which have never been made. The status of our knowledge of polymer/polymer mixing at phase boundaries will be reviewed, with an attempt to emphasize unknown or poorly understood factors.

The chapter concludes with a section briefly exploring the polyblend and composite characteristics of paints and adhesives, followed by a brief examination of environmental and economic problems.

13.1. CLASSIFICATION SCHEME FOR COMPOSITE MATERIALS

Broadly defined, composite materials include all two-component structures; however, discussions in this monograph have been limited to composites containing at least one polymeric component. Even so, many important subclasses of polymer-based composites were either treated extremely cursorily or omitted entirely. It is valuable, however, to classify the known types of polymer-based composites, since such a scheme might promote the discovery or development of yet new types. Such a classification scheme should have a topological foundation (Chinn and Steenrod, 1966),

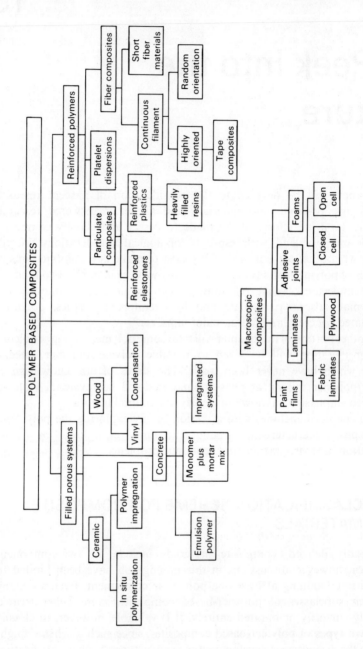

Figure 13.1. Polymer composite classification scheme. (Sperling, 1974d.)

since we are really asking: In how many significantly different ways can a polymer be mixed with a nonpolymer? Figure 13.1 displays one such possible classification mode. The classification in Figure 13.1 should be compared to Figure 13.2 (Section 13.2). In Figure 13.1 under filled porous systems are sketched the principal composites where the nonpolymeric component forms the more continuous phase. Ceramics, concrete, and wood form important substrates, which can be further subdivided in terms of the mode of polymer addition. (Wood, however, is obviously polymeric, as discussed in Section 9.8.)

An important feature of filled porous systems relates to the continuity of the polymeric component. Considering the reinforcing of concrete, addition of an emulsion polymer latex to the mix in the preparative stage clearly may lead to a continuous or discontinuous polymeric component, depending on composition. However, adding monomer to the mix or impregnating a set concrete with monomer may yield two different morphologies, each with the possibility of the polymer forming a more or less continuous phase. (See Chapter 11.)

When the polymeric component forms the continuous phase, spheres, cylinders, or platelets may be added, as illustrated under reinforced polymers. The fiber composites are the most highly researched, as far as different modes of mixing are considered. The filaments may be continuous or discontinuous, or oriented or random in the matrix, with many subclasses of partial orientation possible (not shown). The tape composites are interesting since in some quarters these may be considered a two-dimensional analog of the highly oriented, continuous fibers embedded in a plastic matrix. The reinforced elastomers differ from the reinforced plastics in two ways: the mechanical properties of the polymeric substrate, and the size of the reinforcing particles with respect to polymer chain dimensions. Because of the poor properties often obtained, it is rare to see a research paper on large particles dispersed in an elastomer.

A broad class of composites not discussed in detail in this monograph may be designated as the macroscopic composites. Laminates, adhesive joints, foams, and paint films fall in this category. The reader may also observe that paint films are composites in terms of (1) their being laminates and (2) their composition, which usually includes several types of pigments (see Section 13.6). Closed- and open-cell foams constitute topologically different forms, since the continuity of the gaseous components is in question. Fabric laminates could also be classified under fiber composites, but the fabric itself is considered to have a two-dimensional structure (Sperling, 1974d).

Thus there are many different classes of polymer composites, each having characteristic morphologies.

13.2. CLASSIFICATION OF POLYMER BLENDS

Polymer blends may be classified according to use, overall composition, morphology, or method of synthesis (Ore, 1963; Sperling, 1974c,e). For example, the frontispiece classifies blends with respect to morphology and plastic/rubber content. In the several chapters of this monograph concerned with polymer blends, the materials have been grouped by use or structure. This section will be concerned with a systematic classification of polymer blends according to topology and graph theory (Ore, 1963). An important secondary objective will be to suggest as yet unsynthesized types of polymer blends.

The polymer blend classification scheme shown in Figure 13.2 groups materials in the form of sets and subsets (Sperling, 1974c). For example, the block copolymers are considered to be a special case of the graft copolymers because a block is in effect a graft that always appears on the end mer of the "backbone" polymer. The IEN's are a subset of the latex blends, which in turn are a subset of all of the kinds of blends involving mixing of two previously polymerized homopolymers. The general set of two kinds of mers also conveniently allows for classification of the random and alternating copolymers. Although the ionomers occur as random copolymers, they are grouped with the block copolymers, of which they are also a limiting subset. Star block copolymers are another structure (Bi and Fetters, 1975).

Figure 13.2 makes some (though admittedly incomplete) allowance for the distinguishability of compatible and incompatible pairs, as well as crystalline and amorphous polymer pairs. For example, most of the polymer pairs shown may be both amorphous, both crystalline, or one of each type, yielding four possible combinations. The relative quantities of the two components (not shown in Figure 13.2) are also important. Thus inversion of the plastic/rubber compositions yields a different material. Take, for example, the five major forms observed in block copolymers: elastomeric spheres or cylinders dispersed within a plastic matrix; plastic spheres or cylinders within a rubbery matrix; and alternating plastic/rubber lamellae.* With the same method of synthesis, both components may be plastic, both rubbery, or one of each type. If these various possible permutations are superimposed on the ~ 25 odd blend types enumerated in Figure 13.2, at least 100 clearly different blending possibilities exist.

---

* One may ask if any other block-copolymer structures are possible. Physically, one dimension is constrained to remain smaller than twice the contour length of the constituent blocks. The spheres, cylinders, and lamellae are topologically similar, and if we are limited to shapes without holes or other constructions, they appear to be the only shapes possible. However, even though entropic considerations may render some structures improbable, we may consider such other forms as closed cylinders (toroids) hooked together like chain links, like the catenanes (Frisch and Klempner, 1970b).

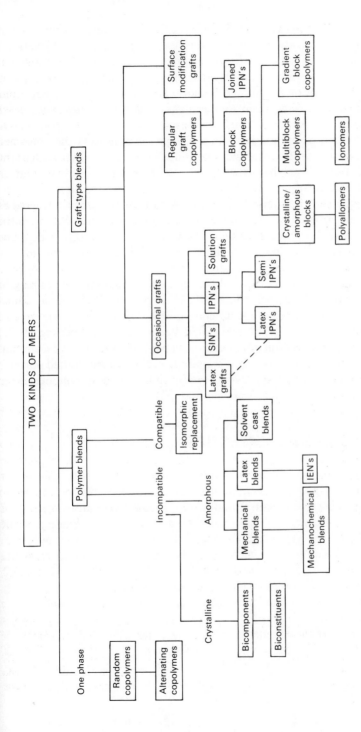

Figure 13.2. Polymer blend classification scheme. Random copolymers, block copolymers, graft copolymers, IPN's, a mechanical blends of various types have a definite relationship to each other. (Sperling, 1974c.)

While all the materials in Figure 13.2 are defined elsewhere in this monograph, a few clarifications are in order. Regular graft copolymers are distinguished from the occasionally grafted materials on the basis that specific grafting sites have been established for the regular graft copolymers in such a way as to discourage homopolymer formation, or that some definite effort has been made to eliminate the formation or existence of homopolymers. The occasionally grafted copolymers, on the other hand, usually have initiation and termination steps separate from the presumed grafting sites, and grafting depends on random, statistical events, such as the polymerization of polymer II through an indigenous double bond in polymer I and the occurrence of chain transfer or other reactions. Thus the block copolymers and some types of the joined IPN's may be considered as regular graft polymers, while the solution and latex materials belong in the occasionally grafted category. The term surface-modification graft should be restricted likewise to materials that specifically undergo initiation or termination on the surface in question.

Yet more exciting is the possibility that major classes of blends remain undiscovered. Recently, Sperling (1974c,e) presented a mathematical approach, employing group theory concepts, to classify the known types of polymer blends, blocks, grafts, and IPN's. It was suggested (Sperling, 1974c) that new morphologies and topologies could be synthesized by application of inverse reactions. For example, the degrafting of a previously grafted polymer pair, or decrosslinking of a network, employed in a series of synthetic steps, could lead to novel polymer combinations.

13.3. POLYMERIC EUTECTICS

Among the many types of blends discussed, an obvious deficiency is the absence of polymer/polymer eutectics from the polymer literature. Here we imagine two polymers that form a true solution in the melt,* and coprecipitate in crystalline form to make a type of crystalline/crystalline polyblend.

Known situations that meet some, but not all, of the above criteria include:

1. Copolymer melting-point depression curves (Edgar and Hill, 1952), which result from one mer's disruption of the crystalline structure of the second mer (see Figure 13.3).

2. Isomorphous polymer crystals, where the individual polymers are capable of replacing each other in the crystals. This class of materials,

---

* Even this simple prerequisite exists only in rare cases.

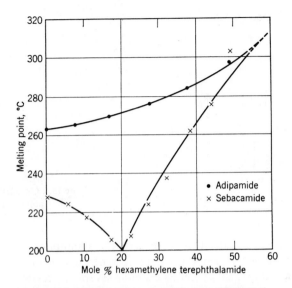

Figure 13.3. Melting points of copolymers of hexamethylene adipamide and terephthalamide, and of hexamethylene sebacamide and terephthalamide (Billmeyer, 1962, pp. 213–214; Edgar and Hill, 1952). Adipamide copolymer with terephthalamide forms isomorphous crystals, whereas sebacamide–terephthalamide copolymers result in more normal melting-point depression curves, which superficially resemble curves characteristic of eutectic behavior.

already discussed in Section 9.7.2, is distinct from the class in which individual mers are isomorphously replaced by others in random copolymers, as illustrated by the adipamide–terephthalamide pair in Figure 13.3.

  3. The biconstituent and bicomponent polymer pairs, discussed in Section 9.2.

  The interesting features of a true polymer/polymer eutectic might include a highly controllable fine structure, especially in fiber formation. In this respect, the materials might be made to behave like the unidirectionally solidified metallic eutectics of Kraft (Sheffler *et al.*, 1969).

13.4. THE POLYMER MIXING PROBLEM

  Starting with recent studies by Meier (1969, 1970) on block copolymers (see Section 4.6.3), we have begun to learn about the phase relationships and degree of mixing at the phase boundaries in polymer blends. In fact,

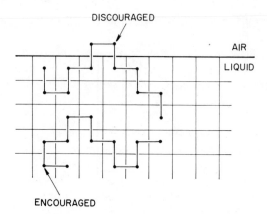

DISCOURAGED

AIR

LIQUID

ENCOURAGED

Figure 13.4. Quasilattice model for
dilute polymer solutions.

many workers still make the naive classical assumption that the phase
boundary is infinitely sharp, even for styrenes that exhibit considerable
mixing of the components.* Let us examine several aspects of this problem.

13.4.1. Dilute Polymer Solution–Air Interface

The free energy of mixing may, of course, be written

$$\Delta F_M = \Delta H_M - T \Delta S_M \qquad (13.1)$$

where ΔH_M is the heat and ΔS_M the entropy of mixing. Let us examine the
effects of polymer chain loops extending above the liquid surface into the
air (see Figure 13.4). Obviously, the entropy of the system increases, since
more conformations are possible. However, the enthalpy of mixing of air
(mostly empty space) with polymer segments is very positive (related to the
heat of vaporization, per segment), strongly discouraging such mixing.
According to Flory-type mixing statistics [see the volume exclusion problem
(Flory, 1953, pp. 423, 521, 530)] in such "forbidden" circumstances we should
erase the molecule and pick a new set of conformations at random. The net
result is that surface (or wall) regions may have an abnormally low polymer
chain concentration, a conclusion consistent with surface tension experiments.
However, quantitative calculations of expected surface polymer concentrations have been slow in forthcoming. This solution–air interface model
problem sets the stage for the polymer–polymer mixing discussion developed
below.

---

*The polymer–filler interface (Chapter 12) involves a similar problem. The interface is not
smooth or flat; the presence of asperities in fillers assures a more complex joining of phases.

13.4.2. Emulsion Polymer Conformations

Some aspects of the core–shell model of Grancio and Williams (1970) have already been examined in relation to graft-type ABS plastics (Section 3.1.2.2). Here, too, the unfavorable heat of mixing between polymer loops and water seems to be controlling. A more detailed view of these experiments will now be presented.

By studying the kinetics of emulsion polymerization of homopolymers, Grancio and Williams (1970) demonstrated that the actual polymerization locus in the latex particles was within an essentially pure monomer phase up to about 60% total conversion. To locate the monomer and the actual sites of polymerization, Grancio and Williams (1970) added trace quantities of butadiene to a partly polymerized styrene latex system. A cross section of a thin section of a latex particle (stained with OsO_4) is shown in Figure 13.5. Most of the butadiene mers (stained dark) are located in a shell surrounding a polystyrene (unstained) core. The shell–core appearance of this essentially homopolymer composition shows that polymer incompatibility per se in latex grafts does not fully explain the latex graft shell–core morphology (see Figure 13.5). Figure 13.5 shows that polymer I forms in or migrates to the core, leaving monomer II (either the same as or different from monomer I) in the shell. At higher conversions, the remaining monomer polymerizers.

While a quantitative molecular explanation of overcoating in latex polymerization is lacking, a qualitative picture can be synthesized with the aid of the polymer solution statistical mechanics developed by Flory (1953) and the phase separation statistics of Meier (1969, 1970). Let us examine

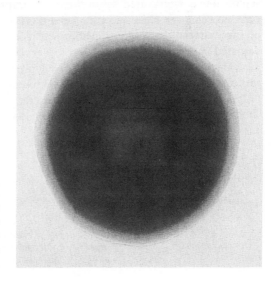

Figure 13.5. Center section of a composite butadiene–styrene particle, 250,000X, showing dark shell and light core. OsO_4 vapor staining technique. (Grancio and Williams, 1970; Williams, 1971, pp. 123–126.)

the conditions for thermodynamic equilibrium in a monomer-swollen latex particle:

1. The conformational entropy of the individual polymer molecules must be maximized. (See Figure 13.6.) For partly polymerized latex particles whose diameter is the same order of magnitude as the end-to-end distance of individual chains (~ 500–800 Å), the chains will prefer to have loops and ends dangling in the aqueous phase (Sperling *et al.*, 1972), as illustrated in Figure 13.6A.

2. The condition depicted in Figure 13.6A becomes unrealistic when the heat of mixing between polymer molecules and water is considered. This heat of mixing should be positive enough to severely restrict or forbid protrusion of loops and ends into the water "nonsolvent."

The Flory-type mixing statistics (Flory, 1953) suggests that whenever a polymer molecule is placed in an unrealistic or impossible position, the entire molecule must be lifted out and again placed into the medium, each new placement being on a statistical basis. For monomer-swollen latex particles, the polymeric portion will tend to be concentrated in the center of each latex particle so as to avoid polymer–water mixing. The result is that the monomer is predicted to take up the remaining sites mostly near the surface (Figure 13.6B) and, on polymerization, will form a more or less permanent shell. According to the model depicted in Figure 13.6B, the shell–core boundary need not be sharp, but may be graded. Such features can be controlled by crosslinking, as in latex IPN's (Section 8.8.2), or by varying compatibility.

If the polymers are chemically distinct, of course, the degree of phase separation is aggravated by ordinary incompatibility. Incompatibility not only prevents the shell and core from mixing, but causes phase separation of the material polymerized within the core, yielding the microstructure depicted in Figure 3.6.

There remains also an unresolved problem in rubber elasticity with respect to latexes. It is easy for the end-to-end conformations of individual polymer molecules to exceed the diameter of the containing particle, as shown in Figure 13.7. A necessary consequence is that these molecules may

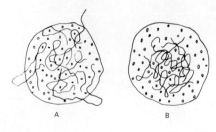

A   B

Figure 13.6. Models for partly polymerized polymer latex particles. (Sperling *et al.*, 1972.)

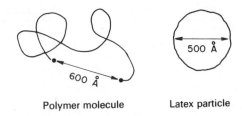

Figure 13.7. The θ-condition end-to-end distance of a medium-molecular-weight polymer molecule (left) may exceed the diameter of the containing latex particle (right).

Polymer molecule   Latex particle

be under considerable compression and hence may have higher free energy than fully relaxed molecules. Such an increase in free energy might encourage the sintering of latex particles during drying, as well as molecular mixing in graft copolymer latexes.

13.4.3. Polyblend Phase Boundaries

In this case the molecules on both sides of the phase boundary are both polymeric. The heat of mixing will still be positive, but much less so. In fact, in many such cases extensive (but incomplete) solution does take place.

In any case, the phase boundaries must be very diffuse. An important factor is the huge size of the polymer chains compared to small molecules. As shown in Figure 13.8, loops and ends may cross the nominal phase

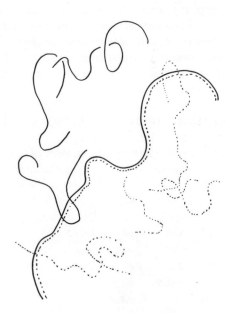

Figure 13.8. Nominal phase boundary in a polyblend, showing crossover by occasional molecules. (—) Polymer I; ($\cdots$) polymer II; ($-\cdot-$) nominal phase boundary.

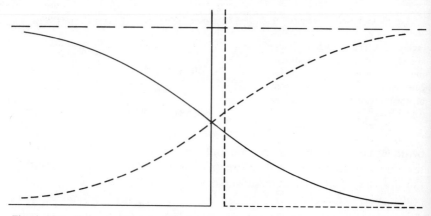

Figure 13.9. Concentration of molecules as a function of distance from nominal phase boundary, based on molecular representation in Figure 13.8. Note slight real molecular solubility far from nominal phase boundary. (==) Nominal phase boundary; (−−) bulk density.

boundary under the influence of thermal forces. The gain in entropy is eventually offset by loss in enthalpy to provide a broad range where very extensive mixing may take place (Figure 13.9). Uncalculated are the depth and extent of such mixing as functions of solubility parameters. For instance, for very small phases (~ 100–500 Å) the phase boundary dimensions may be of the same order of magnitude as the phase domains themselves.

It should be pointed out that, interestingly enough, in the diffusion theory of adhesion (Voyutskii and Vakula, 1969), polymer molecular diffusion is regarded as the mechanism behind the creation of an adhesive bond, and the strength of the bond is attributed to intermolecular forces.

13.4.4. Interfacial Tensions of Polymers

Before leaving the subject of interfacial behavior in polymers, it is instructive to consider the interfacial tension, and resulting interfacial density profiles. Making effective use of the Flory interaction parameter χ, Helfand and Tagami (1972), Gaines (1972), Wu (1974), and others estimated the interfacial surface tension between incompatible polymer pairs (see Table 13.1). Also shown in Table 13.1 are theoretically estimated values of χ. (See Section 4.7 and especially Sections 4.7.3 and 9.6 for related discussion.) Helfand and Tagami found that the characteristic thickness of the interface is proportional to $\chi - \frac{1}{2}$ for small χ. For a polystyrene/poly(methyl methacrylate) system, the value of χ leads to an estimated interfacial thickness of 50 Å. This value is much less than that estimated by Voyutskii and Vakula

Table 13.1
Tabulation of Typical Values of Interfacial Tensions between Polymers[a]

Polymer pairs[b]	Interfacial tension, dyn/cm			$-(\partial\gamma/\partial T)$
	100°C	140°C	180°C	
Polyethylene vs. others				
PE/PP	—	1.1	—	—
L-PE/PS	6.7	5.9	5.1	0.020
B-PE/PCP	4.0	3.7	3.4	0.0075
L-PE/PVAc	12.4	11.3	10.2	0.027
B-PE/PEVAc, 25 wt % VAc	1.6	1.4	1.2	0.005
L-PE/PMMA	10.4	9.7	9.0	0.018
L-PE/PnBMA	5.9	5.3	4.7	0.015
B-PE/PiBMA	4.7	4.3	3.9	0.010
B-PE/PtBMA	5.2	4.8	4.4	0.009
B-PE/PEO	10.3	9.7	9.1	0.016
B-PE/PTMO	4.5	4.2	3.9	0.007
B-PE/PDMS	5.2	5.1	5.0	0.002
Polypropylene vs. others				
PP/PS	—	5.1	—	—
PP/PDMS	3.0	2.9	2.8	0.002
Polyisobutylene vs. others				
PIB/PVAc	8.3	7.5	6.7	0.020
PIB/PDMS	4.4	4.2	4.0	0.006
Polystyrene vs. others				
PS/PCP	0.6	0.5	0.4	0.0014
PS/PVAc	3.9	3.7	3.5	0.0044
PS/PMMA	2.2	1.6	1.1	0.013
PS/PEVAc, 38.7 wt % VAc	—	5.6	—	—
PS/PDMS	6.1	6.1	6.1	0.000
Polychloroprene vs. others				
PCP/PnBMA	1.8	1.6	1.4	0.0047
PCP/PDMS	6.7	6.5	6.3	0.0050
Poly(vinyl acetate) vs. others				
PVAc/PnBMA	4.6	4.2	3.8	0.010
PVAc/PDMS	7.7	7.4	7.1	0.0081
PVAc/PTMO	4.9	4.6	4.3	0.0081
PVAc/PEVAc, 25 wt % VAc	6.0	5.8	5.6	0.0043
Poly(methyl methacrylate) vs. others				
PMMA/PnBMA	2.4	2.0	1.5	0.012
PMMA/PtBMA	2.5	2.3	2.1	0.005
Polydimethylsiloxane vs. others				
PDMS/PnBMA	3.9	3.8	3.6	0.0037
PDMS/PtBMA	3.4	3.3	3.2	0.0025
PDMS/PEO	10.2	9.9	9.6	0.0078
PDMS/PTMO	6.3	6.3	6.2	0.0012

continued

Table 13.1—continued

Polymer pairs[b]	Interfacial tension, dyn/cm			$-(\partial\gamma/\partial T)$
	100°C	140°C	180°C	
Polyethers vs. others				
PEO/PEVAc, 25 wt % VAc	6.0	5.8	5.6	0.0045
PEO/PTMO	4.1	3.9	3.7	0.0051
PTMO/PEVAc, 25 wt % VAc	1.3	1.2	1.1	0.0023

[a] Wu (1974).
[b] PCP = polychloroprene, PDMS = polydimethylsiloxane, PE = polyethylene, B-PE = branched polyethylene, L-PE = linear polyethylene, PEO = poly(ethylene oxide), PEVAc = poly(ethylene-co-vinyl acetate), PIB = polyisobutylene, PMMA = poly(methyl methacrylate), PnBMA = poly(n-butyl methacrylate), PiBMA = poly(isobutyl methacrylate), PtBMA = poly(t-butyl methacrylate), PP = polypropylene, PS = polystyrene, PTMO = poly(tetramethylene oxide) or polytetrahydrofuran, PVAc = poly(vinyl acetate).

(1969). Most surprisingly, Helfand and Tagami (1972) found evidence for a significant rarefication of matter at the interface of incompatible blends. The density profile for the polystyrene/poly(methyl methacrylate) interface is shown in Figure 13.10. They found that the shape of the density profiles depends on both energetic and entropic factors. For the case of macromolecules near the interface, a minimum in free energy must be struck

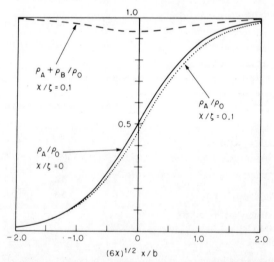

Figure 13.10. Density profiles through the polymer–polymer interface. Results for both an incompressible and a moderately compressible system are shown. (Helfand and Tagami, 1972.)

between those conformations of high energy that penetrate the opposite phase, and the rapidly rising number of conformations available when excursions into these higher energy regions are possible.

The subject of interfacial surface tensions was reviewed recently by Gaines (1972) and Wu (1974).

13.5. NOISE AND VIBRATION DAMPING

At the glass–rubber transition temperature, the rate of coordinated motion of polymer chain segments is of the same order of magnitude as the rate of the experiment used to detect the transition. Mechanical energy supplied at the temperature and frequency corresponding to the transition will be absorbed by the system and reduced to heat. In other words, in the region of the glass temperature, in between conditions of mobility (rubbery state) and immobility (glassy state), chain segments follow only with difficulty an oscillating force field (mechanical or dielectric) and a maximum amount of energy is dissipated as heat.

Thus, near their glass transition temperatures, polymers display maximum values of loss moduli and tan δ, which are measures of energy dissipation. More simply, their "lossiness" is at a maximum. Through application of such polymers in resonant vibrating systems, unwanted noise and vibration may be attenuated (Ball and Salyer, 1966, 1968; Beranek, 1971; Pisarenko and Shemegan, 1972; Plunkett, 1959; Warson, 1972). However, for homopolymers the useful temperature range (at constant frequency) is usually quite narrow, corresponding to a range of about 20–30°C about their glass transition temperature.

By application of the time–temperature superposition principle, a decade of frequency can be shown to correspond to a 6 or 7°C shift in T_g. Noting that the normal acoustical range goes from 20 to 20,000 Hz, or three decades, it can be seen that the equivalent temperature range is 18–20°C. We then conclude that a properly chosen homopolymer can just damp all acoustical frequencies at a single use temperature.

13.5.1. Vibration-Damping Blends

Many different efforts have been made in recent years to broaden the useful temperature range of damping. Since blends of two polymers possessing different glass temperatures immediately come to mind, many such combinations have been tried. By and large, however, incompatible polymer blends

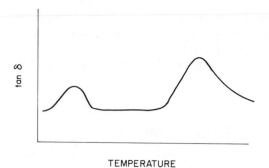

TEMPERATURE

Figure 13.11. Schematic damping behavior of an incompatible blend.

(Mizumachi, 1969, 1970), grafts (Albert *et al.*, 1966; Eustice, 1972; Oberst and Schommer, 1965; Oberst *et al.*, 1961, 1970, 1971), and blocks containing an elastomeric and a plastic component fail; one gets damping over two narrow temperature ranges corresponding to the two transitions, with little in between (see Figure 13.11). This is because of the limited molecular mixing at the phase boundaries (Section 13.4.3). Another interesting development relates to several layers of polymers applied to the vibrating surface consecutively. The layers differ systematically in T_g, the upper layers having the higher T_g's for the best effect. (See Section 13.5.2.) A random copolymer of varying overall composition fulfills the required condition and also allows good interlayer adhesion. The result is a modest amount of damping over the required temperature range (Albert *et al.*, 1966).

An improved approach involves the preparation of polymer blends, blocks, grafts, or IPN's with controlled compatibility and subsequent degrees of mixing. In the semicompatible range, where mixing is extensive but incomplete, a true synergism arises, with better than expected damping qualities over a range spanning the two polymer transitions. We may speculate that the improved damping arises from the frictional dissipation of energy caused by the motion of flexible chains over stiff ones when they are partly mixed.

The characteristic damping behavior of a semicompatible mechanical blend (Mizumachi, 1969, 1970) is illustrated in Figure 13.12. This behavior may be compared with the behavior of the semicompatible latex IPN already discussed in Section 8.8.2.

13.5.2. Extensional and Constrained Layer Damping

Two types of damping configuration are in common use. Application of a single layer of polymer on the vibrating surface is called extensional

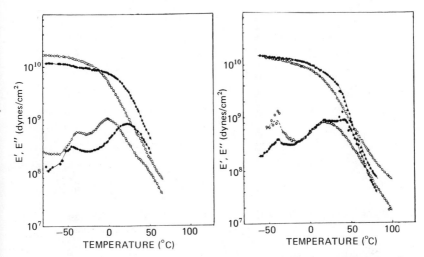

Figure 13.12. Semicompatible polymer blends as vibration dampers (110 Hz). In left diagram: (●) NBR(30)/PVAc·VC(50·50) (1:1); (○) NBR(30)/PVAc/PVC (2:1:1). In right diagram: (●) NBR(30)/PVAc/PVC (1:1:1); (○) NBR(30)/PVC (1:1). (Mizumachi, 1969, 1970.)

damping. The quantity E'' controls damping level attained in extensional damping systems.

A further improvement in damping may be attained by covering the damping layer with a stiff constraining layer (Beranek, 1971; Botsford, 1970; Mizumachi, 1969, 1970; Plunkett, 1959; Wollek, 1965). When the substrate material vibrates, the constraining layer will, in general, respond with vibrations that are out of phase with the substrate. The resulting shearing action on the intermediate damping layer causes enhanced degradation of the mechanical energy to heat. The quantity tan δ controls the damping level in constrained layer configurations.

A common configuration involves sandwiching the polymer between two sheets of metal to make a true composite material. While such composites exhibit optimum damping characteristics, they necessarily have limited formability. Alternatively, damping tapes (Wollek, 1965) have found important applications. In these systems, the adhesive serves also as the damping layer, and aluminum foil as the backing; multiple layers may be applied with good effect.

13.5.3. Gradient IPN's

These structures differ from homogeneous sequential IPN's in that the macroscopic composition varies in a systematic way. For example, the ratio

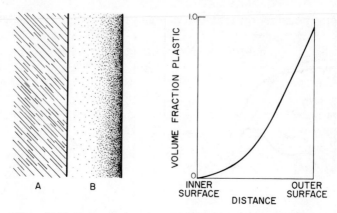

Figure 13.13. Proposed use of gradient IPN's as constrained layer damping materials. Left, A is the substrate to be damped, B is a gradient IPN with the plastic component increasing in concentration toward the outside, as shown on the right.

of network I to network II can vary across the thickness of a sheet. Such gradient IPN's can be synthesized by polymerizing a second monomer that has been nonuniformly swollen into a first polymer network. If an elastomeric layer is swollen from one side with a plastic-forming layer (or vice versa), a constrained layer damping system with a gradient IPN structure will be formed. Of course, the material must be applied rubber-side-down (see Figure 13.13). The compositional gradient causes increased temperature width of the damping spectrum (Sperling and Thomas, 1974).

13.5.4. Constrained Layer Coatings

A stiff polymer coating, such as a reinforced plastic composite, may also serve as a constraining layer. For example, a semicompatible latex IPN paintlike material can be overcoated with a stiff plastic-former to attain a

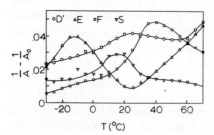

Figure 13.14. Damping as a function of temperature. The quantity A is the amplitude of the damped reed, and A_0 is the undamped reed. E is an epoxy material, F is a polyvinyl acetate material, S is an acrylic house paint, and D' is the prototype "Silent Paint." (Sperling and Thomas, 1974; Sperling et al., 1974.)

broad temperature spectrum of damping over arbitrary configurations (Sperling and Thomas, 1974; Sperling *et al.*, 1974). Figure 13.14 compares such a coating system to two single-layer, extensional damping homopolymer materials and a latex house paint, the last produced solely for decorative and protective properties. While the damping peaks of the homopolymers are somewhat higher in the vicinity of their respective glass transition temperatures, the considerably greater temperature coverage of the constrained coating is apparent. Due to the large amount of filler in the house paint, damping activity is considerably restricted. Damping may be increased also with the judicious use of fillers (see Section 12.3.1.1).

13.6. PAINTS AND ADHESIVES

The great bulk of this monograph has emphasized toughened plastics and reinforced elastomers. While paints and adhesives as such are beyond the scope of this work, it is important to mention some of the features (and research problems) that paints and adhesives have in common with plastics and elastomers.

Nearly all paints and adhesives contain various fillers (Bikerman, 1968; Martens, 1968, Chapters 3 and 4; Patrick, 1973), ranging from titanium dioxide through ferric salts to clay. It is common in these industries to refer to the filler as the pigment and the polymer as the binder, which already brings to the fore the problems associated with bonding the filler to the polymer. The fillers not only serve to extend the polymer, but also to raise its modulus and toughen the final product.

Both paints and adhesives are commonly formulated as polymer blends or grafts. In fact, some compositions resemble semi-IPN's or AB crosslinked copolymers (Section 8.7). For example, epoxy adhesive resins are often cured with polyamides (Bikerman, 1968). The product is tougher than materials cured with low-molecular-weight amines, possibly because of a separate amide phase in this AB crosslinked copolymer.* A more complex molecular architecture is exhibited by the alkyd resins common in oil-based paints (Martens, 1968, Chapters 3 and 4). The major component is a polyester, which often forms a network structure on drying. The polyester component is reacted with various drying oils, such as linseed oil or tung oil (Martens, 1968, Chapters 3 and 4). These oils form an ester link to the polyester structures and also polymerize through their multiple double bonds. Latex paints always contain thickeners, such as cellulosics, poly(acrylic acid), casein,

---

* Nylon homopolymers are also sometimes added to the epoxies for increased toughness.

sodium alginate, or starch, which are dissolved in the aqueous phase (Martens, 1964). Upon drying, a polymer blend is formed between the thickener and the latex.

Thus we observe that most paints and adhesives are complex-composite–polymer blend systems. Upon reviewing the literature (Bikerman, 1968; Martens, 1964, 1968; National Academy of Sciences, 1974; Patrick, 1973), however, it appears that less is known about both the composite and polyblend characteristics of paints and adhesives than is known in the corresponding fields of plastics and elastomers. It appears that some very interesting research lies ahead.

13.7. AN OLD ART IN A CHANGING WORLD

While composites of various kinds have been used since ancient times, the purpose was generally to overcome some deficiency in one of the components, for example, to strengthen mud bricks by adding straw (*Exodus* 5: 7–19). Today, on the other hand, it is possible to develop entirely new properties or much higher levels of properties by judiciously combining two materials (Deanin *et al.*, 1974; Keely, 1967). Indeed, a recurrent theme in this monograph has been the principle of combined action, or synergism. It has become quite clear that the control of heterogeneity and the relationships between phases, especially on the microscopic or submicroscopic scale, is often the key to implementation of this principle. Greater understanding, both empirical and theoretical, has made it possible to transform an ancient intuitive art to an increasingly sophisticated modern science.

In turn, the development of principles of polymer and composite behavior, coupled with the availability of instrumentation to characterize both morphology and physical properties, has made possible a much better scientific understanding of polymer composites. Further research on phase and interfacial characteristics continues to be productive, especially now that newer instruments permit quantitative elemental analysis of exceedingly small volume elements (Hercules, 1972). The knowledge gained should greatly improve the now rather undeveloped state of polymer selection and design.

Where, then, do polymer composites stand in the world today, and where are they likely to stand in the future? Certainly in recent years such materials have comprised an increasingly important share of the rapidly growing volume of polymers produced, the overall volume of the latter now being of the order of that for all metals, ferrous and nonferrous (Eirich and Williams, 1973; National Commission on Materials Policy, 1973). In addition to the volume produced, the number of different types of polymeric materials is believed to now exceed those of steel; the range of polymeric blends and composites must surely contribute largely to this trend (Hercules, 1972).

The major reasons for the growing interest in these materials for engineering purposes may be summarized as follows: a combination of low cost and low weight with superior mechanical and other properties.

Beginning in the early 1970's, however, questions have been raised about the future of synthetic polymers as materials, particularly in view of increasing recognition of problems with the future supply of energy and materials (Hercules, 1972). Since most synthetic polymers are derived from petroleum, the conservation of which is of great concern, the question is certainly important in both technological and political senses. For example, the automotive industry, a major consumer of polymeric materials, has reassessed its status in the light of increased costs and potential availability and concluded that plastics and their composites, such as ABS and sheet or bulk molding compounds, will probably continue to play an increasing role in automotive technology (Anon, 1974). To put this problem in perspective, several points may be noted. First, the production of all petrochemicals, including feedstocks for monomers, constitutes a small fraction (probably $< 10\%$) of all petroleum production (Guillet, 1973). Second, when properly designed and used, polymer blends and composites can serve the cause of materials conservation quite well. Inherently, corrosion-resistant composites can provide an optimum level of performance per unit cost. Since as energy costs rise, so do the costs of all materials, the relative cost-effectiveness of polymer blends and composites seems likely to be retained. Indeed, the use of petroleum feedstocks to produce materials with improved lifetimes would seem to be inherently preferable to the use of petroleum as a source of energy per se. From the standpoint of ecology, the major problem with composites is probably the difficulty of disposal or recycling—matters requiring considerable attention.

Thus, in spite of the difficulties mentioned, it may be expected that the virtues of polymer blends and composites will continue to outweigh current problems. Uses in diverse applications, such as impact-resistant plastics, toughened elastomers, stronger adhesives, energy-absorbing systems, corrosion-resistant structural elements, new fibers, and protective coatings, seem likely to develop still further, especially as materials selection and design become increasingly sophisticated.

References

Adler, E. (1961), *Papier* **15**, 604.

Aggarwal, S. L. (1969), presented before the American Physical Society Meeting, Philadelphia, Pennsylvania, February 1969.

Aggarwal, S. L., Ed. (1970), *Block Polymers*, New York, Plenum.

Aignesberger, A., Rey, T., and Schraemli, W. (1969), *Zement-Kalk-Gips* **58**(7), 297.

Aklonis, J. J., MacKnight, W. J., and Shen, M. (1972), *Introduction to Polymer Viscoelasticity*, Wiley–Interscience.

Albert, W., Bohn, L., and Ebigt, J. (1966), U. S. patent 3,271,188.

Alfrey, T., and Gurnee, E. F. (1967), *Organic Polymers*, New York, Prentice-Hall.

Alfrey, Jr., T., Gurnee, E. F., and Schrenk, W. J. (1969), *Polym. Eng. Sci.* **9**, 400.

Allegra, G. (1970), private communication, June 17, 1970.

Allegra, G., and Bassi, I. W. (1969), *Adv. Polym. Sci.* **6**, 549.

Allen, G., Gee, G., and Nicholson, J. P. (1956), *Polymer* **1**, 8.

Allen, G., Bowden, M. J., Blundell, D. J., Hutchinson, F. G., Jeffs, G. M., and Vyvoda, J. (1973a), *Polymer* **14**, 597.

Allen, G., Bowden, M. J., Blundell, D. J., Jeffs, G. M., Vyvoda, J., and White, T. (1973b), *Polymer* **14**, 604.

Allen, G., Bowden, M. J., Lewis, G., Blundell, D. J., and Jeffs, G. M. (1974a), *Polymer* **15**, 13.

Allen, G., Bowden, M. J., Lewis, G., Blundell, D. J., Jeffs, G. M., and Vyvoda, J. (1974b), *Polymer* **15**, 19.

Allen, G., Bowden, M. J., Todd, S. M., Blundell, D. J., Jeffs, G. M., and Davies, W. E. A. (1974c), *Polymer* **15**, 28.

Allied Chemical Corp. (n.d.), Technical literature on "Source."

Allred, R. E., and Schuster, D. M. (1973), *J. Mater. Sci.* **8**, 245.

Amagi, Y., Ohya, M., Shuki, Z., and Yusa, H. (1972), U. S. patent 3,671,610.

American Chemical Society (1972), Thermomechanical Analysis of Polymer Solids, Symposium held at annual meeting, Polymer Division, American Chemical Society, New York, September 1972; see *Polym. Preprints* **13**, 1124 (1972) and *J. Macromol. Sci. Phys.* **B9**, No. 2 (1974).

American Chemical Society (1973), Symposium on Polymer-Impregnated Solids, Polymer Division, American Chemical Society, Chicago, Illinois, August 1973; see *Polym. Preprints* **14**, No. 2 (1973).

American Concrete Institute (1972), Meeting at Hollywood, Florida, November 1972.

American Concrete Institute (1973a), Polymers in Concrete, American Concrete Institute Publication SP-40.

American Concrete Institute (1973b), Meeting at Atlantic City, New Jersey, March 1973.

American Concrete Institute (1975), Polymers in Concrete: A State-of-the-Art Review, report of ACI Committee 548, in press.

Amos, J. L. (1974), *Polym. Eng. Sci.* **14**, 1.

Amrhein, E. (1967), *Kolloid Z. Z. Polym.* **216**, 38.

Anderson, F. R. (1964), *J. Appl. Phys.* **35**, 65.

Andrews, E. H. (1968), *Fracture in Polymers*, American Elsevier.

Andrews, E. H. (1969), in *Testing of Polymers*, Vol. IV, Brown, W., ed., New York, Interscience, p. 267.

Andrews, E. H. (1972), in *Macromolecular Science* (MTP International Review of Science, Physical Chemistry Series One, Vol. 8), Bawn, C. E. H., ed., London, Butterworths and Baltimore, University Park Press, p. 227.

Andrews, R. D., Hofman-Bang, D., and Tobolsky, A. V. (1948), *J. Polym. Sci.* **3** 669.

Anon. (1962), *Nucleonics* **20**(3), 94.

Anon. (1974), *Chem. Eng. News* **52**(20), 10.

Argon, A. S. (1974), in *Fracture and Fatigue* (*Composite Materials*, Vol. 5), Broutman, L. J., ed., Academic, p. 154.

Arnold, K. R., and Meier, D. J. (1968), *J. Polym. Sci.* **26C**, 37.

Arthur, Jr., J. C. (1971), in *Multicomponent Polymer Systems* (Adv. Chem. Series 99), Gould, R. F., ed., American Chemical Society, Chapter 21.

Artificial Heart Program (1968), Annual report, The Artificial Heart Program of the National Health Institute, Department of Health, Education, and Welfare.

Ash, R., Barrer, R. M., and Petropolous, J. (1963), *Br. J. Appl. Phys.* **14**, 854.

Ashton, J. E. (1969), *J. Compos. Mater.* **3**, 116.

Ashton, J. E., Halpin, J. C., and Petit, P. H. (1969), *Primer on Composite Materials: Analysis*, Stamford, Connecticut, Technomic Publishing Co.

Auer, E. E. (1964), in *Encyclopedia of Polymer Science and Technology*, Mark, H. F., Gaylord, N. G., and Bikales, N. M., eds., Interscience, Vol. I, p. 22.

Auskern, A. (1970), Report BNL-14575R1, Brookhaven National Laboratory, March 1970.

Auskern, A. (1973), paper presented at American Concrete Institute Meeting, Atlantic City, New Jersey, March 1973.

Auskern, A., and Horn, W. (1974), *Cement Concrete Res.* **4**, 785.

Auskern, A., and Horn, W. (1971), *J. Am. Ceram. Soc.* **54**, 282.

Aveston, J., and Kelly, A. (1973), *J. Mater. Sci.* **8**, 352.

Avrami, M. (1939), *J. Chem. Phys.* **7**, 1103.

Avrami, M. (1940), *J. Chem. Phys.* **8**, 212.

Avrami, M. (1941), *J. Chem. Phys.* **9**, 177.

Baccareda, M., and Butta, E. (1962), *J. Polym. Sci.* **57**, 617.

Bachmann, J. H., Sellers, J. W., Wagner, M. P., and Wolf, R. W. (1959), *Rubber Chem. Technol.* **32**, 1286.

Baer, E., ed. (1964), *Engineering Design for Plastics*, Reinhold.

Baer, M. (1962*a*), U. S. patent 3,041,306.

Baer, M. (1962*b*), U. S. patent 3,041,308.

Baer, M. (1962*c*), U. S. patent 3,041,309.

Baer, M. (1964), *J. Polym. Sci.* A-2, **2**, 417.

Baer, M. (1972), *J. Appl. Polym. Sci.* **16**, 1109, 1125.

Ball, G. L., and Salyer, I. (1966), *J. Acoust. Soc. Am.* **39**, 663.

Ball, G. L., and Salyer, I. (1968), U. S. patent 3,399,104.

Bamford, C. H., and Eastmond, G. C. (1974), in *Recent Advances in Polymer Blends, Grafts, and Blocks*, Sperling, L. H., ed., New York, Plenum.

Bamford, C. H., Eastwood, G. C., and Whittle, D. (1971), *Polymer* **12**, 247.

Bank, M., Leffingwell, J., and Thies, C. (1969), *J. Polym. Sci.* A-2 **7**, 795.

Bardeleben, J. (1963), *Kunststoffe* **53**, 162.

Barrer, R. M., and Chio, H. T. (1965), *J. Polym. Sci.* **10C**, 111.

Barrer, R. M., Barrie, J. A., and Rogers, M. G. (1963), *J. Polym. Sci.* *A* **1**, 2565.

Bartkus, E. J., and Kroekel, C. H. (1970), in *Polyblends and Composites*, Bruins, P. F., ed., New York, Wiley.

Bartung, N. G., Fear, R. H., and Wagner, M. P. (1964), *Rubber Age* **96**, 405.

Bascom, W. D. (1969), The Function of Adhesive Promoters in Adhesive Bonding, Naval Research Laboratory Report 6916, June 1969.

Bascom, W. D. (1970), *J. Adhesion* **2**, 161.

Bascom, W. D. (1974), in *Interfaces in Polymer Matrix Composites* (*Composite Materials*, Vol. 6), Plueddemann, E. P., ed., Academic, p. 79.

Bascom, W. D., and Cottington, R. L. (1972), *J. Adhesion* **4**, 193.

Bateman, L., ed. (1963), *The Chemistry and Physics of Rubber-Like Substances*, New York, Wiley.

Battaerd, H. A. J., and Tregear, G. W. (1967), *Graft Copolymers*, Interscience.

Bauer, P., Henning, J., and Schreyer, G. (1970), *Angew. Makromol. Chem.* **11**, 145.

Baumann, G. F. (1969a), in *Engineering Plastics and Their Commercial Development*, Gould, R. F., ed., American Chemical Society, pp. 40–41.

Baumann, G. F. (1969b), *Solid Polyurethane Elastomers*, Wright, Haigh, and Hochland, Ltd.

Bawn, C. E. H., ed. (1972), *Macromolecular Science* (MTP International Review of Science, Physical Chemistry Series One, Vol. 8), London, Butterworths, and Baltimore, Maryland, University Park Press.

Beaumont, P. R. W., and Harris, B. (1972), *J. Mater. Sci.* **7**, 1265.

Becker, G. W., and Oberst, H. (1956), *Kolloid-Z.* **148**, 6.

Beecher, J. F., Marker, L., Bradford, R. D., and Aggarwal, S. L. (1969), *J. Polym. Sci.* **26C**, 117.

Behrens, E. (1968), *J. Compos. Mater.* **2**, 2.

Beranek, L., ed. (1971), *Noise and Vibration Control*, New York, McGraw-Hill.

Bergen, Jr., R. L. (1968), *Appl. Polym. Symp.* **7**, 41.

Berry, G. C. (1966), *J. Chem. Phys.* **44**, 4550.

Berry, G. C. (1967), *J. Chem. Phys.* **46**, 1338.

Berry, G. C. (1971), *J. Polym. Sci. A-2* **9**, 687.

Berry, G. C., and Fox, T. G. (1964), *J. Am. Chem. Soc.* **86**, 3540.

Berry, J. P. (1961), *J. Polym. Sci.* **50**, 107.

Berry, J. P. (1964), in *Fracture Processes in Polymeric Solids*, Rosen, B., ed., New York, Wiley.

Bi, L. K., and Fetters, L. J. (1975), *Macromolecules* **8**, 90.

Biglione, G., Baer, E., and Radcliffe, S. V. (1969), in *Proc. Int. Symp. on Fracture, Brighton, England*, London, Chapman and Hall, paper no. 44, p. 503.

Bikerman, J. J. (1968), *The Science of Adhesive Joints*, 2nd ed., New York, Academic.

Billmeyer, Jr., F. W. (1962), *Textbook of Polymer Science*, New York, Wiley.

Billmeyer, Jr., F. W. (1971), *Textbook of Polymer Science*, 2nd ed., New York, Wiley.

Bills, K. W., Sweeney, K. H., and Salcedo, F. S. (1960), *J. Appl. Polym. Sci.* **4**, 259.

Blankenhorn, P. R., Weyers, R. E., Kline, D. E., and Cady, P. D. (1975), *Transp. Eng. J.*, ASCE, No. TE1, **101**, 65.

Blatz, P. H. (1956), *Ind. Eng. Chem.* **48**, 727.

Blumentritt, B. F., and Cooper, S. L. (1974), in *Proc. 32nd Ann. Tech. Conf. SPE, San Francisco, California, May 1974*, p. 317.

Blundell, D. J., Longman, G. W., Wignall, G. D., and Bowden, M. J. (1974), *Polymer* **15**, 33.

Bohn, L. (1966), *Kolloid Z. Z. Polym.* **213**, 55.

Bohn, L. (1968), *Rubber Chem. Technol.* **41**, 495.

Bonard, B. B. (1957), *Kolloid Z. Z. Polym.* **231**, 438.

Bonart, R. (1968), *J. Macromol. Sci. B* **2**, 115.

Boni, K. A., and Sliemes, Jr., F. A. (1968), *Battelle Tech. Rev.* (May–June) 2.

Bonotto, S., and Bonner, E. F. (1968), *Polym. Preprints* **9**, 537.

Bonvicini, A., Monaci, A., and Cappuccio, V. (1963), U. S. patent, 3,073,667.

Boonstra, B. B. (1965), in *Reinforcement of Elastomers*, Kraus, G., ed., New York, Interscience.

Boonstra, B. B. (1970), in *Polyblends and Composites*, Bruins, P. F., ed., New York, Interscience.

Bostick, E. E. (1970), in *Block Polymers*, Aggarwal, S. L., ed., New York, Plenum.

Botsford, J. H. (1970), U. S. patent 3,534,882.

Bovey, F. A., Kolthoff, I. M., Medalia, A. I., and Meehan, E. J. (1955), *Emulsion Polymerization* (High Polymer Series IX), Interscience.

Boyer, R. F. (1968), *Polym. Eng. Sci.* **8**, 161.

Boyle, W. R., Winston, A. W., and Loos, W. E. (1971), Preparation of Wood–Plastic Combinations Using Gamma Irradiation to Induce Polymerization, Final Report (ORO-2945-10), Contract AT(40-1)-2945, for the U. S. Atomic Energy Commission, West Virginia University, May 10, 1971.

Bragaw, C. G. (1970), *Polym. Preprints* **11**, 368.

Bragaw, C. G. (1971), in *Multicomponent Polymer Systems* (Adv. Chem. Series 99), Platzer, N. A. J., ed., American Chemical Society.

Bramfitt, B. L., and Marder, A. R. (1968), in *IMS Proc.*, p. 48.

Brandrup, J., and Immergut, E. H. (1966), *Polymer Handbook*, Interscience.

Brassell, G. W., and Wischmann, K. B. (1974), *J. Mater. Sci.* **9**, 307.

Brennen, J. J., Dannenberg, E. M., and Rigbi, Z. (1969), in *Proc. Int. Rubber Conf.*, *1967*, New York, Gordon and Breach, p. 123.

Breuers, W., Hild, W., and Wolff, H. (1954), *Plaste Kautsch.* **1**, 170.

Bridgman, P. W. (1952), *Studies in Large Plastic Flow and Fracture with Special Emphasis on the Effects of Hydrostatic Pressure*, 1st ed., New York, McGraw-Hill.

Brody, H., and Ward, I. M. (1971), *Polym. Eng. Sci.* **11**, 139.

Broutman, L. J. (1966), *Polym. Eng. Sci.* **6**, 263.

Broutman, L. J. (1970), in *Proc. 25th Tech. Conf., Reinforced Plastics/Composites Div., SPI*, Section 13-B, p. 1.

Broutman, L. J., ed. (1974), *Fracture and Fatigue (Composite Materials*, Vol. 5), Academic.

Broutman, L. J., and Agarwal, B. D. (1974), *Polymer Eng. Sci.* **14**, 581.

Broutman, L. J., and Kobayashi, T. (1971), Report AD 736859, U. S. Department of Defense, Washington, D.C.

Broutman, L. J., and Kobayashi, T. (1972), paper presented at Int. Conf. on Fracture Mech., Lehigh University.

Broutman, L. J., and Krock, R. H. (1967), *Modern Composite Materials*, Addison-Wesley, Reading, Massachusetts.

Broutman, L. J., and Krock, R. H., eds. (1974), *Composite Materials*, 8 vols., New York, Academic.

Broutman, L. J., and Sahu, S. (1969), in *Proc. 24th Ann. Conf. Reinforced Plastics Div. SPI*, Section 11-D, p. 1.

Broutman, L. J., and Sahu, S. (1971), in *Proc. 26th Ann. Conf. Reinforced Plastics Div., SPI*, Section 14-C, February 1971, p. 1.

Broutman, L. J., and Sahu, S. (1972), ASTM Spec. Tech. Publ. 497, p. 170.

Brown, H. P., Panshin, A. J., and Forsaith, C. C. (1949), *Textbook of Wood Technology*, New York, McGraw-Hill.

Bruggeman, D. A. G. (1935), *Ann. Phys. (Leipzig)* **24**, 635.

Bruins, P. F., ed. (1970), *Polyblends and Composites* (J. Appl. Polym. Sci. Applied Polymer Symp. No. 15), Interscience.

Bryant, W. M. D. (1947), *J. Polym. Sci.* **2**, 547.

Buchdahl, R., and Nielsen, L. E. (1950), *J. Appl. Phys.* **21**, 482.

Buckley, R. A., and Phillips, R. J. (1969), *Chem. Eng. Prog.* **65**(10), 41.

Bucknall, C. B. (1967a), *Br. Plastics* **40**(11),118.

Bucknall, C. B. (1967b), *Br. Plastics* **40**(12), 84.

Bucknall, C. B. (1969), *J. Mater. Sci.* **4**, 214.

Bucknall, C. B. (1972), paper presented at the American Chemical Society Biennial Polymer Symp., Ann Arbor, Michigan, June 1972.

Bucknall, C. B., and Clayton, D. (1972), *J. Mater. Sci.* **7**, 202.

Bucknall, C. B., and Drinkwater, I. C. (1973), *J. Mater. Sci.* **8**, 1800.

Bucknall, C. B., and Hall, M. M. (1971), *J. Mater. Sci.* **6**, 95.

Bucknall, C. B., and Smith, R. R. (1965), *Polymer* **6**, 437.

Bucknall, C. B., and Street, D. G. (1967), *Soc. Chem. Ind. Monograph No. 26.* London, p. 272.

Bucknall, C. B., Clayton, D., and Keast, W. (1972a), *J. Mater. Sci.* **7**, 1443.

Bucknall, C. B., Drinkwater, I. C., and Keast, W. (1972b), *Polymer* **13**, 115.

Bucknall, C. B., Clayton, D., and Keast, W. (1973), *J. Mater. Sci.* **8**, 514.

Bueche, A. M. (1956), *J. Polym. Sci.* **19**, 297.

Bueche, F. (1965), in *Reinforcement of Elastomers,* Kraus, G., ed., Interscience.

Burke, J. J., and Weiss, V., eds. (1973), *Block and Graft Copolymerization,* Syracuse, New York.

Burlant, W. J., and Hoffman, A. S. (1960), *Block and Graft Polymers,* New York, Reinhold.

Cahn, J. W., and Charles, R. J. (1965), *Phys. Chem. Glasses* **6**, 181.

Callan, J. E., Hess, W. M., and Scott, C. E. (1971), *Rubber Chem. Technol.* **44**, 814.

Campos-Lopez, E., McIntyre, D., and Fetter, L. J. (1973), *Macromolecules* **6**, 415.

Capet-Antonini, F. C., Grimard, M., and Tamenasse, J. (1973), *Thromb. Res.* **2**(6), 479.

Carbonell, R. G., and Kostin, M. D. (1972), *AIChE J.* **18**, 1.

Carrara, A. S., and McGarry, F. J. (1968), *J. Compos. Mater.* **2**, 222.

Casale, A., and Porter, R. S. (1971), *Rubber Chem. Technol.* **44**, 534.

Casale, A., Porter, R. S., and Johnson, J. F. (1971), *Rubber Chem. Technol.* **44**, 534.

Catsiff, E., and Tobolsky, A. V. (1955), *J. Colloid Sci.* **10**, 375.

Catsiff, E., and Tobolsky, A. V. (1956), *J. Polym. Sci.* **19**, 111.

Ceresa, R. J. (1962), *Block and Graft Copolymers,* London, Butterworths.

Ceresa, R. J., ed. (1973), *Block and Graft Copolymerization,* Wiley-Interscience, Vol. 1.

Cessna, L. C., Levens, J. A., and Thomson, J. B. (1969a), *Polym. Eng. Sci.* **9**, 339.

Cessna, L. C., Thomson, J. B., and Hanna, R. D. (1969b), *SPE J.,* **25** (10), 35.

Chahal, R. S., and St. Pierre, L. E. (1969), *Macromolecules* **2**, 193.

Chamis, C. C. (1974a), in *Fracture and Fatigue (Composite Materials,* Vol. 5), Broutman, L. J., ed., Academic, p. 94.

Chamis, C. C. (1974b), in *Interfaces in Polymer Matrix Composites (Composite Materials,* Vol. 6), Plueddemann, E. P., ed., Academic, p. 32.

Chander, N. (1971), M. S. Report, Department of Chemical Engineering, Lehigh University, June 1971.

Chapiro, A. (1962), *Radiation Chemistry of Polymeric Systems* (High Polymer Series, Vol. XV), New York, Interscience.

Charlesby, A. (1960), *Atomic Radiation and Polymers,* New York, Pergamon.

Charlesby, A., and Wycherly, V. (1957), *Int. J. Appl. Radiat. Isot.* **2**, 26.

Chen, W. F., and Dahl-Jorgensen, E. (1973), in *Polymers in Concrete,* American Concrete Institute Publication SP-40, paper No. 17, p. 347.

Cheng, S. C., and Vachon, R. I. (1969), *Int. J. Heat Mass Transfer* **12**, 249.

Chinn, W. G., and Steenrod, N. E. (1966), *First Concepts of Topology,* Random House.

Chiu, E. H. (1973), Ph.D. Thesis, Lehigh University, June 1973.

Cizak, E. P. (1968), U. S. patent 3,383,435.

Cohan, L. (1947), *India Rubber World* **117**, 343

Conaghan, B. F., and Rosen, S. L. (1972), *Polym. Eng. Sci.* **12**, 134.

Cook, J., and Gordon, J. E. (1964), *Proc. R. Soc. London* **A282**, 508.

Cooper, S. L., and Tobolsky, A. V. (1966a), *J. Appl. Polym. Sci.* **10**, 1837.

Cooper, S. L., and Tobolsky, A. V. (1966b), *Text. Res. J.* **36**, 800.

Coover, Jr., H. W., McConnell, R. L., Joyner, F. B., Slonaker, D. F., and Combs, R. L. (1966), *J. Polym. Sci. A-1* **4**, 2563.

Cornish, P. J., and Powell, B. D. W. (1974), *Rubber Chem. Technol.* **47**, 481.

Corten, H. T. (1966), in *Engineering Properties of Plastics*, Baer, E., ed., New York, Reinhold–Van Nostrand.

Corten, H. T., ed. (1971), *Conf. on Composite Materials: Testing and Design*, ASTM Special Technical Publication 497.

Coughlin, R. W., Hartman, M., Gebauer, J. (1970), Corrosion Protection by Inhibitor Cross-Linking and Surface Coating Grafting, Final Report, Grant No. 14-01-0001-1640, Office of Saline Water, U. S. Department of the Interior, August 31, 1970 (Lehigh University).

Counto, U. J. (1964), *Mag. Concr. Res.* **16**, 129.

Cox, R. G. (1969), *J. Fluid Mech.* **37**, 601.

Crank, G. S., and Park, W. R. (1969), *Diffusion in Polymers*, London, Academic.

Cresentini, L. (1971), U. S. patent 3,595,935.

Cross, A., and Haward, R. N. (1973), *J. Polym. Sci. A-2* **11**, 2423.

Crystal, R. G. (1971), in *Colloidal and Morphological Behavior of Block and Graft Copolymers*, Molau, G. E., ed., New York, Plenum, p. 279.

Crystal, R. G., O'Malley, J. J., and Erhardt, P. F. (1969), *Polym. Preprints* **10**, 804.

Crystal, R. G., Erhardt, P. F., and O'Malley, J. J. (1970), in *Block Copolymers*, Aggarwal, S. L., ed., New York, Plenum, p. 179.

Curtius, A. J., Covitch, M. J., Thomas, D. A., Sperling, L. H. (1972), *Polym. Eng. Sci.* **12**, 101.

D'Agostino, V., and Lee, J. (1972), AFML-TR-72-13, Final Technical Report to Wright-Patterson Air Force Base, Ohio, Section 2.6.

Dahl-Jorgensen, E., and Chen, W. F. (1973), Stress–Strain Tests of Polymer Modified Concrete, Lehigh University, Fritz Engineering Laboratory Report No. 390.1, February 1973.

Dahl-Jorgensen, E., and Chen, W. F. (1974), *Mag. Concr. Res.* **26**, 16.

Dahl-Jorgensen, E., Chen, W. F., Manson, J. A., Vanderhoff, J. W., and Liu, Y. N. (1975), *Transp. Eng. J., ASCE* **101**, No. TE1, 29.

Dally, J. W., and Broutman, L. J. (1967), *J. Compos. Mater.* **1**, 424.

Dally, J. W., and Carrillo, D. H. (1969), *Polym. Eng. Sci.* **9**, 434.

Dammont, F. R., and Kwei, T. K. (1967), *Polym. Preprints* **8**, 920.

Dannenberg, E. M. (1966), *Trans. IRI* **42**, T. 26.

Dannis, M. L. (1963), *J. Appl. Polym. Sci.* **7**, 231.

Davenport, N. E., Hubbard, L. W., and Pettit, M. R. (1959), *Br. Plastics* **32**, 549.

Davies, W. E. A. (1971a), *J. Phys. D* **4**, 1176.

Davies, W. E. A. (1971b), *J. Phys. D* **4**, 1325.

Davis, H. A., Longworth, R., and Vaughan, D. J. (1968), *Polym. Preprints* **9** (1), 515.

Deanin, R. D. (1972), *Polymer Structure, Properties, and Applications*, Boston, Cahners.

Deanin, R. D. (1973), *Polym. Preprints* **14**(2), 728.

Deanin, R. D., Deanin, A. A., and Sjoblom, T. (1974), in *Recent Advances in Polymer Blends, Grafts, and Blocks*, Sperling, L. H., ed., New York, Plenum.

Delft, B. W., and McKnight, W. J. (1969), *Macromolecules* **2**, 309.

DePuy, G. W., Kukacka, L. E., Auskern, A., Cowan, W. C., Colombo, P., Lockman, W T., Romano, A. J., Smoak, W. G., Steinberg, M., and Causey, F. E. (1973), Concrete-Polymer Materials, Fifth Topical Report, BNL 50390 and USBR REC-ERC-73-12, December.

DeVries, K. L., Wilde, T. B., and Williams, M. L. (1973), *Polym. Preprints* **14**, 463.

DiBenedetto, A. T. (1973), *J. Macromol. Sci. Phys.* **B7**, 657.

DiBenedetto, A. T., and Wambach, A. D. (1972), *Int. J. Polym. Mater.* **1**, 159.

DiBenedetto, A. T., Gauchel, J. V., Thomas, R. L., and Barlow, J. W. (1972), *J. Mater.* **7**, 211.

Dickie, R. A. (1973), *J. Appl. Polym. Sci.* **17**, 45.

Dickie, R. A., and Cheung, M.-F. (1973), *J. Appl. Polym. Sci.* **17**, 79.

Dickie, R. A., Cheung, M.-F., and Newman, S. (1973), *J. Appl. Polym. Sci.* **17**, 65.

Dietrich, K., and Tolsdorf, S. (1970), *Faserforsch. Textiltech.* **21**(7), 295.

Dikeou, J. (1973), paper presented at the American Concrete Institute Meeting, Atlantic City, New Jersey, March 1973.

Dikeou, J. T., Kukacka, L. E., Backstrom, J. E., and Steinberg, M. (1969), *Am. Concr. Inst. J.* **66**, 829.

Dikeou, J. T., Steinberg, M., Cowan, W. C., Kukacka, L. E., DePuy, G. W., Auskern, A., Smoak, W. G., Colombo, P., Wallace, G. B., Hendrie, J. M., and Manowitz, B. (1971), Concrete-Polymer Materials, Third Topical Report, Brookhaven National Laboratory, Report BNL 50275, January 1971.

Dikeou, J. T., Cowan, W. C., DePuy, G. W., Smoak, W. G., Wallace, G. B., Steinberg, M., Kukacka, L. E., Auskern, A., Colombo, P., Hendrie, J. M., and Manowitz, B. (1972), Concrete Polymer Materials, Fourth Topical Report, Brookhaven National Laboratory, Report BNL 50328, January 1972.

Dobry, A., and Boyer-Kawenoki, F. (1947), *J. Polym. Sci.* **2**, 90.

Donatelli, A. A., Thomas, D. A., and Sperling, L. H. (1974*a*), in *Recent Advances in Polymer Blends, Grafts, and Blocks*, Sperling, L. H., ed., New York, Plenum, p. 375.

Donatelli, A. A., Thomas, D. A., and Sperling, L. H. (1974*b*), presented at the 177th National AIChE Meeting, Pittsburgh, Pennsylvania, June 1974.

Donatelli, A. A., Thomas, D. A., and Sperling, L. H. (1975) to be published.

Doppert, H. L., and Overdiep, W. S. (1971), in *Multicomponent Polymer Systems* (Adv. Chem. Series 99), Platzer, N. A. J., ed., Chapter 5.

Dow Chemical Co. (n.d.), technical service literature.

Dresner, L. (1972), *Desalination* **10**, 293.

Droste, D. H., and DiBenedetto, A. T. (1969), *J. Appl. Polym. Sci.* **13**, 2419.

Droste, D. H., DiBenedetto, A. T., and Stejskal, E. D. (1971), *J. Polym. Sci. A-2* **9**, 187.

Duck, E. W. (1970), *Br. Polym. J.* **2**, 60.

Dunn, A. S., and Melville, H. W. (1952), *Nature* **169**, 699.

DuPre, D. B., Samulski, E. T., and Tobolsky, A. V. (1971), in *Polymer Science and Materials*, Tobolsky, A. V., and Mark, H. F., eds., Wiley–Interscience.

Duran, J. A., and Meyer, J. A. (1972), *Wood Sci. Technol.* **6**, 59.

Eakins, W. J. (1966), Initiation of Failure Mechanisms in Glass–Resin Composites, NASA Report CR-518, August 1966.

Eash, D. R., and Shafer, H. H. (1975), *Trans. Res. Rec.*, in press.

Eastman Chemical Products (n.d.), General Information on Tenite® Polyallomer, Publication No. PO-12(03-69), Eastman Chemical Products, Inc., Kingsport, Tennessee.

Edgar, O. B., and Hill, R. (1952), *J. Polym. Sci.* **8**, 1.

Eilers, H. (1941), *Kolloid-Z.* 313.

Einstein, A. (1905), *Ann. Phys.* **17**, 549.

Einstein, A. (1906), *Ann. Phys.* **19**, 289.

Einstein, A. (1911), *Ann. Phys.* **34**, 591.

Eirich, F. R. (1965), *Appl. Polym. Symp.* **1**, 271.

Eirich, F. R., and Williams, M. L., eds. (1973), *Polymer Engineering and Its Relevance to National Materials Development*, Report on National Science Foundation Workshop, Washington, D. C., December 1972, UTEC Do 73-013, Inst. of Materials Research, University of Utah, June 1973.

Eisenberg, A. (1970), *Macromolecules* **3**, 147.

Eisenberg, A. (1973), *Polym. Preprints* **14**(2), 871.

Eisenberg, A., and King, M. (1971), *Macromolecules* **4**, 204.

Eisenberg, A., and King, M. (1972), *Rubber Chem. Technol.* **45**, 908.

Eisenberg, A., and Navratil, M. (1972), *J. Polym. Sci. B* **10**, 537.

Eppe, R., Fischer, E. W., and Stuart, H. S. (1959), *J. Polym. Sci.* **34**, 721.

Erhardt, P. F., O'Malley, J. J., and Crystal, R. G. (1969), *Polym. Preprints* **10**, 812.

Erhardt, P. F., O'Malley, J. J., and Crystal, R. G. (1970), in *Block Copolymers*, Aggarwal, S. L., ed., New York, Plenum, p. 195.

Erickson, P. W. (1970), *J. Adhesion* **2**, 131.

Erickson, P. W., and Plueddemann, E. P. (1974), in *Interfaces in Polymer Matrix Composites* (*Composite Materials*, Vol. 6), Plueddemann, E. P., ed., Academic, p. 2.

Estes, G. M., Seymour, R. W., Huh, D. S., and Cooper, S. L. (1969), *Polym. Eng. Sci.* **9**, 383.

Estes, G. M., Huh, D. S., and Cooper, S. L. (1970), in *Block Copolymers*, Aggarwal, S. L., ed., New York, Plenum, p. 225.

Eustice, A. L. (1972), U. S. patent 3,636,158.

Fallick, G. J., Bixler, H. J., Marsella, R. A., Garner, F. R., and Fettes, E. M. (1967), in *Proc. 22nd Ann. Tech. Conf. Reinforced Plastics Div. SPI*, Section 17-E, p. 1.

Fava, R. A. (1969), *Br. Polym. J.* **1**, 59.

Fedors, R. F., and Landel, R. F. (1967), AIAA Paper 67-491, Am. Inst. Aeronautics and Astronautics.

Feibush, A. M. (1963), U. S. Atomic Energy Commission, Report NYO-3334.

Fekete, F., and McNally, J. S. (1969), Fr. patent 1,567,700; *Chem. Abs.* **71**, 125445w.

Fels, M., and Huang, R. (1970), *J. Appl. Polym. Sci.* **14**, 537.

Ferry, J. D. (1961), *Viscoelastic Properties of Polymers*, Wiley, pp. 218–224.

Ferry, J. D. (1970), *Viscoelastic Properties of Polymers*, 2nd ed., Wiley.

Fetters, L. J., and Morton, M. (1969), *Macromolecules* **2**, 453.

Finaz, G., Skoulios, A., and Sadron, C. (1961), *C. R. Acad. Sci.* **253**, 265.

Finaz, G., Rempp, P., and Parrod, J. (1962), *Bull. Soc. Chim. Fr.* 262.

Fischer, E. W. (1957), *Z. Naturforsch.* **12a**, 753.

Fitchman, D. R., and Newman, S. (1970), *J. Polym. Sci. A-2* **8**, 1545.

Flajsman, F., Cahn, D. S., and Phillips, J. C. (1971), *J. Am. Ceram. Soc.* **54**, 129.

Fletcher, W. P., and Gent, A. N. (1953), *Trans. Inst. Rubber Ind.* **29**, 266.

Flory, P. J. (1947), *J. Chem. Phys.* **15**, 684.

Flory, P. J. (1950), *J. Chem. Phys.* **18**, 108.

Flory, P. J. (1953), *Principles of Polymer Chemistry*, Ithaca, New York, Cornell University Press.

Flory, P. J., and Rehner, J. (1943), *J. Chem. Phys.* **11**, 521.

Flory, P. J., Ciferri, A., and Hoeve, C. A. J. (1960), *J. Polym. Sci.* **45**, 235.

Ford, W. E., Callan, J. E., and Hess, W. M. (1963), *Rubber Age* **92**, 738.

Fordham, S. (1960), *Silicones*, London, George Newnes, pp. 180–181.

Fowler, D. W., Houston, J. T., and Paul, D. R. (1973), paper presented at the American Concrete Institute Meeting, Atlantic City, New Jersey, March 1973.

Fowler, D. W., Paul, D. R., and Shoch, E. P., (1975), *Transp. Res. Rec.*, in press.

Franta, E., Skoulios, A., Rempp, P., and Benoit, H. (1965), *Makromol. Chem.* **87**, 271.

Freudenberg, K. (1966), in *Lignin Structures and Reactions* (Adv. Chem. Series 59), Martin, J., ed., American Chemical Society.

Fricke, H. (1924), *Phys. Rev.* **24**, 575.

Frisch, H. L., and Klempner, D. (1970a), in *Advances in Macromolecular Chemistry*, Vol. 2, p. 149.

Frisch, H. L., and Klempner, D. (1970b), *Macromol. Rev.* **1**, 149.

Frisch, K. C., Frisch, H. L., and Klempner, D. (1972), Ger. patent 2,153,987.

Frisch, K. C., Klempner, D., Frisch, H. L., and Ghiradella, H. (1974a), in *Recent Advances in Polymer Blends, Grafts, and Blocks*, Sperling, L. H., ed., New York, Plenum.

Frisch, K. C., *et al.* (1974b), *J. Appl. Polym. Sci.* **18**, 683, 689.

Fritsche, A. K., and Price, F. P. (1969), *Polym. Preprints* **10**, 893.

Fritsche, A. K., and Price, F. P. (1970), in *Block Polymers*, Aggarwal, S. L., ed., New York, Plenum.

Fritsche, A. K., and Price, F. P. (1971), in *Colloidal and Morphological Behavior of Block and Graft Copolymers*, Molau, G. E., ed., New York, Plenum.

Fujino, K., Ogawa, Y., and Kawai, H. (1964), *J. Appl. Polym. Sci.* **8**, 2147.

Fukuma, N. (1971), U. S. patent 3,607,610.

Funke, W., Zorel, U., and Murthy, B. K. G. (1969), *J. Paint Technol.* **41**, 210.

Furukawa, J. (1970), *Bull. Japan Petroleum Inst.* **12**, 40.

Furukawa, J., Iseda, Y., and Kobayashi, E. (1971), *Polym. J.* **2**, 337.

Gaines, G. L. (1972), *Polym. Eng. Sci.* **12**, 1.

Galanti, A. V., and Sperling, L. H. (1970a), *J. Appl. Polym. Sci.* **14**, 2785.

Galanti, A. V., and Sperling, L. H. (1970b), *Polym. Lett.* **8**, 115.

Galanti, A. V., and Sperling, L. H. (1970c), *Polym. Eng. Sci.* **10**, 177.

Gallot, Y., Franta, E., Rempp, P., and Benoit, H. (1963), *J. Polym. Sci.* **4C**, 473.

Galperin, I. (1967), *J. Appl. Polym. Sci.* **11**, 1475.

Galperin, I., and Kwei, T. K. (1966), *J. Appl. Polym. Sci.* **10**, 673.

Galperin, I., Kwei, T. K., and Arnheim, W. M. (1965), *J. Appl. Polym. Sci.* **9**, 3215.

Gamow, G. (1967), *One, Two, Three, . . . Infinity*, 6th ed., Bantam, p. 56.

Gauri, R. L. (1970), *Nature* **228**, 882.

Gebauer, J., and Coughlin, R. W. (1971), *Cement Concrete Res.* **1**, 187

Gebauer, J., and Manson, J. A. (1971), unpublished work, Lehigh University.

Gebauer, J., Hasselman, D. P. H., and Thomas, D. A. (1972a), *J. Am. Ceram. Soc.* **55**, 175.

Gebauer, J., Hasselman, D. P. H., and Long, R. G. (1972b), *Am. Ceram. Soc. Bull.* **51**, 471.

Gee, G. (1966), *Polymer* **7**, 373.

Geil, P. H. (1960), *J. Polym. Sci.* **44**, 449.

Geil, P. H. (1963), *Polymer Single Crystals*, New York, Interscience.

Gent, A. N. (1970), *J. Mater. Sci.* **5**, 925.

Gent, A. N. (1972), Fracture of Elastomers, in *Fracture—An Advanced Treatise*, Vol. VII, Liebowitz, H., ed., New York, Academic.

Gervais, M., and Gallot, Y. (1970), *C. R. Acad. Sci.* **270**, 784.

Gesner, B. D. (1965), *J. Appl. Polym. Sci.* **9**, 3701.

Gesner, B. D. (1967), *J. Appl. Polym. Sci.* **11**, 2499.

Gibbs, J. H., and DiMarzio, E. A. (1958), *J. Chem. Phys.* **28**, 373, 807.

Gibbs, J. H., and DiMarzio, E. A. (1959), *J. Polym. Sci.* **40**, 121.

Girolamo, M., and Urwin, J. R. (1971), *Eur. Polym. J.* **7**, 693.

Gladstone, H. M., Loan, L. D., and Dawes, L. D. (1971), private communication to Szymczak, T. J. (1971).

Godard, P., Delmon, B., and Mercier, J. P. (1974), *J. Appl. Polym. Sci.* **18**, 1477.

Goettler, L. A., Lewis, T. B., and Nielsen, L. E. (1973), *Polym. Preprints* **14**(1), 436.

Goldsmith, H. L., and Mason, S. G. (1967), in *Rheology*, Vol. 4, Eirich, F. R., ed., New York, Academic.

Goodrich, B. F. (1954), Br. Patent 707,425.

Grancio, M. R. (1971), *Polym. Preprints* **12**, 68, 681.

Grancio, M. R., and Williams, D. J. (1970), *J. Polym. Sci. A-1* **8**, 2617.

Greensmith, H. W., Mullins, L., and Thomas, A. G. (1963), in *The Chemistry and Physics of Rubber-Like Substances*, Bateman, L. B., ed., Maclaren and Sons, Chapter 19.

Griffith, A. A. (1921), *Phil. Trans. R. Soc.* A **221**, 163.

Grosius, P., Gallot, Y., and Skoulios, A. (1970), *Eur. Polym. J.* **6**, 355.

Guillet, J., ed. (1973), *Polymers and Ecological Problems*, New York, Plenum, Introduction.

Guinier, A., and Fournet, G. (1955), *Small-Angle Scattering of X-Rays* (transl. by C. B. Walker), New York, Wiley.

Guth, E. (1944), *J. Appl. Phys.* **15**, 758.

Guth, E. (1945), *J. Appl. Phys.* **16**, 20.

Hagenmeyer, Jr., H. J., and Edwards, M. B. (1966), *J. Polym. Sci.* **4C**, 731.

Hagenmeyer, Jr., H. J., and Edwards, M. B. (1970), U. S. patent 3,529,037.

Hajo, H., and Toyoshima, W. (1973), in *31st ANTEC, SPE*, Montreal, Canada, p. 163.

Hall, C. E. (1966), *Introduction to Electron Microscopy*, 2nd ed., New York, McGraw-Hill.

Halley, W., Blackburn, D. H., Wagstaff, F. E., and Charles, R. J. (1970), *J. Am. Ceram. Soc.* **53**, 34.

Halpin, J. C. (1969), Effects of Environmental Factors on Composite Materials, Report AD 692 481, June 1969, U. S. Department of Defense, Washington, D.C.

Halpin, J. C. (1975), *Polym. Eng. Sci.*, **15**, 132.

Halpin, J. C., and Bueche, F. (1964), *J. Appl. Phys.* **35**, 3142.

Halpin, J. C., and Kardos, J. L. (1972), *J. Appl. Phys.* **43**, 2235.

Halpin, J. C., and Pagano, N. (1969), *J. Compos. Mater.* **3**, 720.

Halpin, J. C., and Polley, H. W. (1967), *J. Compos. Mater.* **1**, 64.

Ham, G. E., ed. (1967), *Kinetics and Mechanisms of Polymerization*, Vol. I, *Vinyl Polymerization*, New York, Marcel Dekker.

Hamilton, R. L., and Crosser, O. K. (1962), *Ind. Eng. Chem. Fund.* **1**, 187.

Hammon, H. G. (1958), *SPE J.* **14**(3), 40.

Han, C. D. (1973), *J. Appl. Polym. Sci.* **17**, 1289.

Harmer, D. E. (1962), in *Radiation Chemistry of Polymer Systems* (High Polymer Series, Vol. XV), Chapiro, A., ed., New York, Interscience, pp. 670–673.

Harmer, D. (1967), in *Irradiation of Polymers* (Adv. Chem. Series No. 66), Gould, R. F., ed. American Chemical Society.

Harrell, Jr., L. L. (1969), *Macromolecules* **2**, 607.

Harrell, Jr., L. L. (1970), in *Block Polymers*, Aggarwal, S. L., ed., New York, Plenum, p. 213.

Harwood, J. A. C., Mullins, L., and Payne, A. R. (1967), *J. IRI* **1** (January/February), 17.

Hashin, Z., and Shtrikman, S. (1963), *J. Mech. Phys. Solids* **11**, 127.

Hasselman, D. P. H. (1962), *J. Am. Ceram. Soc.* **45**, 452.

Hasselman, D. P. H., and Gebauer, J. (1973), *Polym. Preprints* **14**(2), 1193.

Hasselman, D. P. H., and Penty, R. A. (1973), *J. Am. Ceram. Soc.* **56**, 105.

Hasselman, D. P. H., Gebauer, J., and Manson, J. A. (1972), *J. Am. Ceram. Soc.* **55**, 588.

Hasselberger, F., Allen, B., Parachuri, E. K., Charles, M., and Coughlin, R. W. (1974), *Biochem. Biophys. Res. Comm.* **57**(4), 1054.

Hattori, K. (1970), in *Encyclopedia of Polymer Science and Technology*, Mark, H. F., and Atlas, S., eds., Vol. 12.

Havlik, A. J., and Smith, T. L. (1964), *J. Polym. Sci. A*, **2**, 539.

Haward, R. N. (1970), *Br. Polym. J.* **2**, 209.

Haward, R. N., and Mann, J. (1964), *Proc. R. Soc.* **282A**, 120.

Hayes, B. T. (1969), *Chem. Eng. Prog.* **65**(10), 50.

Hearle, J. W. S. (1963), *J. Polym. Sci.* **7**, 1175.

Hearmon, R. F. S. (1953), in *Mechanical Properties of Wood and Paper*, Meredith, R., ed. New York, Interscience, Chapter 2.

Heidemann, G., Jellinek, G., and Ringens, W. (1967), *Kolloid Z. Z. Polym.* **221**, 119.

Heijboer, J. (1968), *J. Polym. Sci.* **16C**, 3755.

Heins, C. F. (1973), *Polym. Preprints* **14**(2), 1130.

Helfand, E., and Tagami, Y. (1972), *J. Chem. Phys.* **56**, 3592.

Henderson, J. F., and Swarc, M. (1968), *Macromol. Rev.* **3**, 317.

Hercules, D. M. (1972), *Anal. Chem.* **44**(5), 106.

Hertzberg, R. W., Manson, J. A., and Nordberg, H. (1970), *J. Mater. Sci.* **5**, 521.

Hertzberg, R. W., Manson, J. A., and Wu, W. C. (1973), ASTM Spec. Tech. Publ., **536**, 30.

Hess, W. M. (1965), in *Reinforcement of Elastomers*, Kraus, G., ed., New York, Interscience.

Higuchi, W. I., and Higuchi, T. (1960), *J. Am. Pharm. Assoc. Sci. Ed.* **49**, 598.

Hills, P. R., Barrett, R. L., and Pateman, R. J. (1969), U. K. Atomic Energy Agency, Report AERE-R-6090 (available through U. S. National Technical Information Service).

Hirai, T. (1970), *Japan Plastics* **4**(October), 22.

Hirai, T., and Kline, D. E. (1973), *J. Compos. Mater.* **7**, 160.

Hobbs, D. W. (1969), Report TRA 437, Cement and Concrete Association, London, December.

Holden, G., Bishop, E. T., and Legge, N. R. (1969a), *J. Polym. Sci.* **26C**, 37.

Holden, G., Bishop, E. T., and Legge, N. R. (1969b), in *The Proceedings of the International Rubber Conference, 1967*, Maclaren.

Holliday, L., and Robinson, J. (1973), *J. Mater. Sci.* **8**, 301.

Hopfenberg, H. B., Kimura, F., Rigney, P. T., and Stannett, V. (1969), *J. Polym. Sci.* **28C**, 243.

Horie, K., Mita, I., and Kambe, H. (1967), *J. Appl. Polym. Sci.* **11**, 57.

Horino, T., Ogawa, Y., Soen, T., and Kawai, H. (1965), *J. Appl. Polym. Sci.* **9**, 2261.

Hosemann, R. (1962), *Polymer* **3**, 349.

Hosemann, R., and Bonart, B. (1957), *Kolloid Z. Z. Polym.* **152**, 53.

Huang, R. Y. M., and Kanitz, P. J. F. (1969), *J. Appl. Polym. Sci.* **13**, 669.

Huelck, V., and Covitch, M. (1971), unpublished, Lehigh University.

Huelck, V., Thomas, D. A., and Sperling, L. H. (1972), *Macromolecules* **5**, 340, 348.

Hughes, L. J., and Brown, G. L. (1961), *J. Appl. Polym. Sci.* **5**, 580.

Hughes, L. J., and Brown, G. L. (1963), *J. Appl. Polym. Sci.* **7**, 59.

Huglin, M. B., and Johnson, B. L. (1972), *Eur. Polym. J.* **8**, 911.

Huh, D. S., and Cooper, S. L. (1971), *Polym. Eng. Sci.* **11**, 369.

Hull, D. (1970), *J. Mater. Sci.* **5**, 357.

Hull, D., and Owen, T. W. (1973), *J. Polym. Sci.* **11**, 2039.

Husson, F., Hertz, J., and Luzzati, V. (1961), *C. R. Acad. Sci.* **252**, 3290, 3462.

Hustad, G. O., Richardson, T., and Olson, N. F. (1973), *J. Dairy Sci.* **56**(9), 118.

Idorn, G. M., and Fördös, Z. (1973), Cement Polymer Materials, Concrete Research Laboratory, Aktieselskabet Aalborg Portland-Cement-Fabrik, Karlstrup, Report BFL 312.

Imasawa, Y., and Matsuo, M. (1970), *Polym. Eng. Sci.* **10**, 261.

Inoue, T., Soen, T., Hashimoto, T., and Kawai, H. (1969), *J. Polym. Sci. A-2* **7**, 1283.

Inoue, T., Soen, T., Hashimoto, T., and Kawai, H. (1970a), *Macromolecules* **3**, 87.

Inoue, T., Soen, T., Hashimoto, T., and Kawai, H. (1970b), in *Block Copolymers*, Aggarwal, S. L., ed., New York, Plenum, p. 53.

International Atomic Energy Agency (1963), *Report of Conference on the Application of Large Radiation Sources in Industry, Salzburg, 1963*, Vienna, IAEA, 2 vols.

International Atomic Energy Agency (1968), *Report of Study Group on Impregnated Fibrous Materials, Bangkok, 1967*, Vienna, IAEA.

International Conference on Plastics and Paper (1973), London, May 1973.

Probst, S. A., and Manson, J. A. (1970), *Polym. Preprints* **11** (2), 765.

Probst, S. A., and Manson, J. A. (1972), *J. Polym. Sci. A-1* **10**, 179.

Irwin, G. R. (1956), U. S. Naval Research Laboratory Report 4763.

Irwin, G. R. (1960), U. S. Naval Research Laboratory Report 5486.

Isenburg, J. E., and Vanderhoff, J. W. (1974), *J. Am. Ceram. Soc.* **57**, 242.

Isenburg, J. E., Sutton, E. J., Rapp, D. E., and Vanderhoff, J. W. (1971), *Highway Research Record* No. 370, 75.

Ishai, O., and Cohen, L. H. (1967), *Int. J. Mech. Sci.* **9**, 539.

Jenckel, E., and Herwig, H. U. (1963), *Kolloid-Z.* **148**, 57.

Jenkins, A. D., ed. (1972), *Polymer Science. A Materials Science Handbook*, Amsterdam, North-Holland.

Jenness, J. R. (1972), Ph. D. Thesis, The Pennsylvania State University, December 1972.

Johannson, O. K., Start, F. O., Vogel, G. E., and Fleischmann, R. M. (1967), *J. Compos. Mater.* **1**, 278.

Johnson, F. A., and Radon, J. C. (1972), *Eng. Fract. Mech.* **4**, 555.

Johnson, O. B., and Labana, S. S. (1972), U. S. patent 3,659,003.

Kaas, R. L., and Kardos, J. L. (1971), *Polym. Eng. Sci.* **11**, 11.

Kallaur, M. (1969), Fr. patent 1,567,022; *Chem. Abs.* **71**, 125443u.

Kambe, H., and Kato, T. (1971), in *Proc. 5th Int. Cong. Rheol.*

Kambour, R. P. (1966), *J. Polym. Sci. A-2* **4**, 17, 349, 359.

Kambour, R. P. (1968), *Appl. Polym. Symp.* **7**, 215.

Kambour, R. P. (1970), in *Block Polymers*, Aggarwal, S. L., ed., New York, Plenum, p. 263.

Kambour, R. P. (1973), *Macromol. Rev.* **7**, 1.

Kambour, R. P., and Barker, Jr., R. E. (1966), *J. Polym. Sci. A-2* **4**, 359.

Kambour, R. P., and Kopp, R. W. (1969), *J. Polym. Sci. A-2* **7**, 183.

Kambour, R. P., and Robertson, R. E. (1972), in *Polymer Science. A Materials Science Handbook*, Jenkins, A. D., ed., Amsterdam, North-Holland, Chapter 11.

Kambour, R. P., and Russell, R. R. (1971), *Polymer* **12**, 237.

Kardos, J. L. (1974), Interface Modification in Reinforced Plastics, presented at the Conference on Adhesion, Polymer Conference Series, University of Utah, Salt Lake City, June 1974.

Kardos, J., and Raisoni, J. (1975), *Polym. Eng. Sci.*, **15**, 183.

Kardos, J. L., McDonnell, W. L., and Raisoni, J. (1972), *J. Macromol. Sci. Phys.* **B6**, 397.

Kardos, J. L., Cheng, F. S., and Tolbert, T. L. (1973), *Polym. Eng. Sci.* **13**, 455.

Karpov, V. L., Malinsky, Y. M., Serenkov, V. I., Klimanova, V. R., and Freiden, R. S. (1960), *Nucleonics* **18**(3), 88.

Kato, K. (1966), *Polym. Lett.* **4**, 35.

Kato, K. (1968), *Japan Plastics* **2**(April), 6.

Katz, D., and Tobolsky, A. V. (1964), *J. Polym. Sci. A-2*, **2**, 1587.

Kaufman, M. (1968), *Giant Molecules*, Doubleday.

Kaufman, S., Slichter, W. P., and Davis, D. D. (1971), *J. Polym. Sci. A-2* **9**, 829.

Kawai, H., and Inoue, T. (1970), *Japan Plastics* **4**(July), 12.

Kay, D. H. (1965), *Techniques for Electron Microscopy*, 2nd ed., Philadelphia, Davis.

Kay, G. (1968), *Process Biochem.* **36**(August).

Kedem, O., and Katchalsky, A. (1963), *Trans. Faraday Soc.* **59**, 1918, 1931, 1941.

Keely, A. (1967), in *Materials*, San Francisco, California, W. H. Freeman.

Keith, H. D. (1969), *Kolloid Z. Z. Polym.* **231**, 421.

Keller, A. (1957), *Phil. Mag.* **2**, 1171.

Keller, A. (1971), *Polymer* **12**, 22.

Keller, A., Pedemonte, E., and Willmouth, F. M. (1970), *Kolloid Z. Z. Polym.* **238**, 385.

Kelly, A. (1966), in *Strong Solids*, Oxford University Press.

Kenga, D. L., Fennessey, J. P., and Stannett, V. T. (1962), *For. Prod. J.* **4**, 161.

Kennedy, J. P. (1971), in *XXII IUPAC Meeting, Boston, Massachusetts, July 1971, Preprints* Vol. I, p. 105.

Kennedy, J. P., and Baldwin, F. P. (1969), Fr. patent 1,564,485; *Chem. Abs.* **71**, 102926g.

Kennedy, J. P., and Smith, R. R. (1974), in *Recent Advances in Polymer Blends, Grafts, and Blocks*, Sperling, L. H., ed., New York, Plenum, p. 303.

Kent, J. A., Taylor, G. B., Boyle, W. R., and Winston, A. W. (1968), *Chem. Eng. Prog. Symp. Series* **64**(83), 137.

Kenyon, A. S., and Duffey, H. J. (1967), *Polym. Eng. Sci.* **7**, 189.

Kenyon, A., and Nielsen, L. E. (1969), *J. Macromol. Sci. Chem.* **A3**, 275.

Kerner, E. H. (1956a), *Proc. Phys. Soc. London* **69B**, 802.

Kerner, E. H. (1956b), *Proc. Phys. Soc. London* **69B**, 808.

Keskkula, H., ed. (1968), *Polymer Modification of Rubbers and Plastics* (J. Appl. Polym. Sci. Applied Polymer Symp. No. 7), Interscience.

Keskkula, H., and Traylor, P. A. (1967), *J. Appl. Polym. Sci.* **11**, 2361.

Keusch, P., and Williams, D. J. (1973), *J. Polym. Sci. Polym. Chem. Ed.* **11**, 301.

Keyser, C. A. (1968), *Materials Science and Engineering*, Columbus, Ohio, Merrill, Chapters 6 and 7.

Khanderia, J., and Sperling, L. H. (1974), *J. Appl. Polym. Sci.* **18**, 913.

Kinell, P.-Ö., and Aagard, P. (1968), *Report of Study Group on Impregnated Fibrous Materials, Bangkok, 1967*, IAEA, Vienna, 1968.

Kinsey, R. H. (1969), *Appl. Polym. Symp.* **11**, 77.

Klempner, D., Frisch, H. L., and Frisch, K. C. (1970), *J. Polym. Sci. A-2* **8**, 921.

Kline, D. E., and Sauer, J. A. (1962), *SPE Trans.* **2**, 21.

Klute, C. H. (1959), *J. Appl. Polym. Sci.* **1**, 340.

Kohler, J., Riess, G., and Banderet, A. (1968), *Eur. Polym. J.* **4**, 173.

Koleske, J. V., and Lundberg, R. D. (1969), *J. Polym. Sci. A-2* **7**, 795.

Kollman, F. F. P., and Cote, Jr., W. A. (1968), *Principles of Wood Science and Technology*, Springer-Verlag.

Koozu, H., Oodan, K., and Sacki, S. (1963), *Shikizai Kyokaishi* 362.

Korshak, V. V., Silin, A. A., Frunze, T. M., Kragel'skii, I. V., Dukhovskii, E. A., and Baranov, E. L. (1968), *Mekh. Polim.* **4**, 559 (*Polym. Mech. USSR* **4**, 432).

Kotaka, T., Tanaka, T., and Inagaki, H. (1972), *Polym. J.* **3**, 327, 338.

Kotani, T., and Sternstein, S. S. (1971), in *Polymer Networks: Structure and Mechanical Properties*, Chompff, A., and Newman, S., eds., New York, Plenum, p. 273.

Koutsky, J. A., Hien, N. V., and Cooper S. L. (1970), *Polym. Lett.* **8**, 353.

Kovacs, A. J. (1967), *Chim. Ind. Genie Chim.* **97**, 315.

Kovacs, A. J., and Manson, J. A., with Levy, D. (1966), *Kolloid-Z.* **214**, 1.

Kraus, G. (1965a), *Rubber. Chem. Technol.* **38**, 1070.

Kraus, G., ed. (1965b), *Reinforcement of Elastomers*, Interscience.

Kraus, G. (1965c), in *Reinforcement of Elastomers*, Kraus, G., ed., Interscience, Chapter 4.

Kraus, G. (1971), *Fortschr. Hochpolym.-Forsch.* **8**, 156.

Kraus, G., and Gruver, J. T. (1970), *J. Polym. Sci. A-2* **8**, 571.

Kraus, G., Rollmann, K. W., and Gruver, J. T. (1970), *Macromolecules* **3**, 92.

Krause, S. (1969), *J. Polym. Sci. A-2* **7**, 249.

Krause, S. (1970), *Macromolecules* **3**, 84.

Krause, S. (1971), in *Colloidal and Morphological Behavior of Block and Graft Copolymers*, Molau, G. E., ed., New York, Plenum.

Krause, S. (1972), *J. Macromol. Sci. Rev. Macromol. Chem.* **C7**, 251.

Krauss, W. (1963), Ph. D. Dissertation, California Institute of Technology.

Krigbaum, W. R., and Goodwin, R. G. (1965), *J. Chem. Phys.* **43**, 4523.

Krigbaum, W. R., Roe, R. J., and Smith, K. J. (1964), *Polymer* **5**, 533.

Krock, R. H. (1966), *J. Mater. (ASTM)* **1** (2), 278.

Krock, R. H., and Broutman, L. J. (1967), in *Modern Composite Materials*, Broutman, L. J., and Krock, R. H., eds., Reading, Massachusetts, Addison-Wesley, Chapter 1.

Kukacka, L. E., Romano, A. J., Reid, M., Auskern, A., Colombo, P., Klamut, C. J., Pike, R. G., and Steinberg, M. (1972), Concrete-Polymer Materials for Highway Applications, Progress Report No. 2, Brookhaven National Laboratory, report BNL 50348.

Kukacka, L. E., Fontana, J., Romano, A. J., Steinberg, M., and Pike, R. G. (1973), Concrete-Polymer Materials for Highway Applications, Progress Report No. 3, BNL 50417 and FHWA-RD-74-17, December.

Kuhn, R. (1968), private communication.

Kuhn, R., Cantow, J. J., and Burchhard, W. (1968a), Incompatible Nature of a Ternary System Polymer (1)/Polymer (2)/Solvent, presented at the Macromolecular IUPAC Symposium, Toronto, 1968.

Kuhn, R., Cantow, J. J., and Burchhard, W. (1968b), *Angew. Makromol. Chem.* **2**, 146, 157.

Kuhn, W., Ramel, A., Walters, O., and Kuhn, H. (1960), *Fortschr. Hochpolym.-Forsch.* **1**, 540.

Kumins, C. A. (1965), *Off. Digest* **37**, 1314.

Kumins, C. A., and Roteman, J. (1963), *J. Polym. Sci. A* **1**, 527.

Kwei, T. K. (1965), *J. Polym. Sci. A* **3**, 3229.

Kwei, T. K., and Arnheim, W. M., (1965), *J. Polym. Sci. C* **10**, 103.

Kwei, T. K., and Kumins, C. A. (1964), *J. Appl. Polym. Sci.* **8**, 1483.

Laible, R. C., Sultan, J. N., and McGarry, F. J. (1974), in *Proc. 32nd Ann. Tech. Conf.*, SPE, San Francisco, California, May 1974, p. 113.

Lake, G. J., Lindley, P. B., and Thomas, A. G. (1969), in *Proc. 2nd Int. Conf. on Fracture*, Brighton, England, Pratt, P. L., ed., Chapman & Hall.

Landel, R. F. (1958), *Trans. Soc. Rheol.* **2**, 53.

Landel, R. F., and Fedors, R. F. (1965), in *Proc. 1st Int. Conf. on Fracture*, Sendai, Japan, Yokobori, T., Kawasaki, T., and Swedow, J. L., eds., Vol. 2, p. 1247.

Landel, R. F., Moser, B. G., and Bauman, A. (1965), in *Proc. 4th Int. Cong. Rheology*, Lee, E. H., ed., New York, Interscience, Part 2, p. 663.

Lange, F. F. (1970), *Phil. Mag.* **22**, 983.

Lange, F. F. (1974), in *Fracture and Fatigue* (*Composite Materials*, Vol. 5), Broutman, L. J., ed., Academic, p. 2.

Lange, F. F., and Radford, K. C. (1971), *J. Mater. Sci.* **6**, 1197.

Langwig, J. E., Meyer, J. A., and Davidson, R. W. (1968), *For. Prod. J.* **18**(17), 33.

Langwig, J. E., Meyer, J. A., and Davidson, R. W. (1969), *For. Prod. J.* **19**(11), 57.

Lannon, D. A. (1967), in *Encyclopedia of Polymer Science and Technology*, Mark, H. F., Gaylord, N. G., and Bikales, N. M., eds., New York, Interscience, Vol. 7, p. 574.

Lannon, D. A., and Hoskins, E. J. (1965), in *Physics of Plastics*, Richie, P. D., ed., London, Plastics Institute, p. 381.

Lavenwood, R. E., and Gulbransen, L. E. (1969), *Polym. Eng. Sci.* **9**, 365.

Lavengood, R. R., Nicolais, L., and Narkis, M. (1971), Technical Report AD 891,254, NTIS, November 1971.

Learmonth, G. S., and Pritchard, G. (1969), *Br. Polym. J.* **1**, 88.

Lee, T. C. P., Sperling, L. H., and Tobolsky, A. V. (1966), *J. Appl. Polym. Sci.* **10**, 1831.

Lees, J. K. (1968), *Polym. Eng. Sci.* **8**, 184.

Leidner, J., and Woodhams, R. T. (1974), *J. Appl. Polym. Sci.* **18**, 1639.

Leitz, F. B., and McRae, W. A. (1972), *Desalination* **10**, 293.

Leitz, F. B., and Shorr, J. (1972), Research on Piezodialysis—Third Report, Research and Development Progress Report No. 775, Office of Saline Water, U. S. Department of the Interior, May 1972.

Lenschow, R., Hofsoy, A., and Sopler, B. (1971), Polymer Concrete Materials, Project C 149, Report No. 2, Cement and Concrete Research Institute, Technical University of Norway, January 1971.

Lenz, R. W. (1967), *Organic Chemistry of Synthetic High Polymers*, New York, Interscience

Lewis, T. B., and Nielsen, L. E. (1970), *J. Appl. Polym. Sci.* **14**, 1449.

Lim, C. K., and Tschoegl, N. W. (1969), in *Proc. JANNAF Mechanical Behavior Working Group, 8th Meeting*, p. 153.

Lindenmeyer, P. F. (1965), *Science* **147**, 1256.

Lindenmeyer, P. F. (1975), *Polym. Eng. Sci.* **15**, 236.

Lindley, P. B. (1964), *Engineering Design with Natural Rubber*, NRPRA Tech. Bull. No. 8, Arrowsmith, Ltd.

Lipatov, Y. S., and Fabuliak, F. G. (1968), paper presented at the IUPAC Int. Symp. Macromol. Chem., Toronto.

Lipatov, Y. S., Lipatova, T. E., Vasilenko, Y. P., and Sergeva, L. M. (1963), *Vysokomol. Soedin.* **5**, 290.

Lipatov, Y. S., Babich, V. F., and Rosovzky, V. F. (1974), *J. Appl. Polym. Sci.* **18**, 1213.

Litt, M., and Herz, J. (1969), *Polym. Preprints* **10**, 905.

Litt, M., and Tobolsky, A. V. (1967), *J. Macromol. Sci. Phys.* **B1**, 433.

Liu, Y. N., and Manson, J. A. (1975), unpublished, Lehigh University.

Longworth, R., and Vaughan, D. J. (1968), *Nature* **218**, 85.

Lord Rayleigh (1892), *Phil. Mag.* **34**, 481.

Lotz, B. (1963), Thesis (3rd cycle), Centre des Recherches sur les Macromolécules, Strasbourg.

Lotz, B., and Kovacs, A. J. (1966), *Kolloid-Z.* **209**, 97.

Lotz, B., and Kovacs, A. J. (1969), *Polym. Preprints* **10**, 820.

Lotz, B., Kovacs, A. J., Bassett, G. A., and Keller, A. (1966), *Kolloid-Z.* **209**, 115.

Lusis, J., Woodhams, R. T., and Xanthos, M. (1973), *Polym. Eng. Sci.* **13**, 139.

Lyman, D. J. (1966), *Rev. Macromol. Chem.* **1**, 355.

Lyons, B. J. (1960), *Nature* **185**, 6041.

Mabis, A. J. (1962), *Acta Cryst.* **15**, 1152.

MacKnight, W. J., Stoelting, J., and Karasz, F. E. (1971), in *Multicomponent Polymer Systems* (Adv. Chem. Series No. 99), Platzer, N. A. J., ed., Chapter 3.

MacLaughlin, T. F. (1968), *J. Compos. Mater.* **2**, 44.

Maewaka, E., Nakao, M., and Ninomaya, K. (1967), *Nippon Gomu Kyokaishi* **40**, 549.

Manabe, S., Murakami, R., and Takayanagi, M. (1969), *Mem. Fac. Eng., Kyushu Univ.* **28**, 295.

Manabe, S., Murakami, R., Takayanagi, M., and Vemura, S. (1971), *Int. J. Polym. Mater.* **1**, 47.

Mandekern, L. (1964), *Crystallization of Polymers*, McGraw-Hill, Chapter 8.

Manning, D. G., and Hope, B. B. (1971), *J. Cement Concrete Res.* **1**, 631.

Manning, D. G., and Hope, B. B. (1973), in *Polymers in Concrete*, American Concrete Institute Publ. SP-40, paper No. 9, p. 191.

Manson, J. A., and Chiu, E. H. (1972), *Am. Chem. Soc., Div. Org. Coat. Plast. Prepr.* **32**(2), 162.

Manson, J. A., and Chiu, E. H. (1973a), *J. Polym. Sci. Symp.* **41**, 95.

Manson, J. A., and Chiu, E. H. (1973b), *Polym. Preprints* **14**, 469.

Manson, J. A., and Hertzberg, R. W. (1973a), *CRC Crit. Rev. Macromol. Sci.* **1**, 433.

Manson, J. A., and Hertzberg, R. W. (1973b), *J. Polym. Sci., Polym. Phys. Ed.* **11**, 2483.

Manson, J. A., and Kovacs, A. J. (1965), paper presented at IUPAC Congross, Prague, Preprint No. 556.

Manson, J. A., Iobst, J. A., and Acosta, R. (1974), *J. Macromol. Sci., Phys.* **B9**, 301.

Manson, J. A., Chen, W. F., Vanderhoff, J. W., Cady, P. D., Kline, D. E., and Blankenhorn, P. R. (1975b), paper given at VI Interamerican Congress of Chemical Engineering, Caracas, July, 1975.

Manson, J. A., Chen, W. F., Vanderhoff, J. W., Liu, Y.-N., Dahl-Jorgenson, E., and Mehta, H. (1973), *Polym. Preprints* **14**(2), 1203.

Manson, J. A., Chen, W. F., Vanderhoff, J. W., Cady, P. D., Kline, D. E., and Blankenhorn, P. R. (1975), paper given at the International Congress on Polymer Concretes, London, May, 1975.

Mark, H. (1943), in *Cellulose and Cellulose Derivatives*, Ott, E., ed., Interscience, pp. 999–1004.

Mark, H. F. (1966), *Giant Molecules*, Time, Inc.

Mark, H. (1971), in *Polymer Science and Material* Tobolsky, A. V., and Mark, H., eds., New York, Wiley–Interscience, Chapter 11.

Mark, H. F., Gaylor, N. G., and Bikales, N. M., eds. (1967), *Encyclopedia of Polymer Science and Technology*, Interscience.

Mark, J. E. (1972a), *J. Am. Chem. Soc.* **94**, 6645.

Mark, J. E. (1972b), *J. Chem. Phys.* **57**, 2541.

Mark, J. E. (1973), *J. Polym. Sci., Polym. Phys. Ed.* **11**, 1375.

Marston, T. U., Atkins, A. G., and Felbeck, D. K. (1974), *J. Mater. Sci.* **9**, 447.

Martens, C. R. (1964), *Emulsion and Water Soluble Paints and Coatings*, Reinhold.

Martens, C. R., ed. (1968), *Technology of Paints, Varnishes, and Lacquers*, Reinhold.

Marx, C. L., Caulfield, D. F., and Cooper, S. L. (1973), *Macromolecules* **6**, 344.

Matsuo, M. (1966), *Polymer* **7**, 421.

Matsuo, M. (1968), *Japan Plastics* **2**(July), 6.

Matsuo, M. (1969a), *Polym. Eng. Sci.* **9**, 206.

Matsuo, M. (1969b), private communication.

Matsuo, M., Ueno, T., Horino, H., Chujyo, S., and Asai, H. (1968), *Polymer* **9**, 425.

Matsuo, M., Nozaki, C., and Jyo, Y. (1969a), *Polym. Eng. Sci.* **9**, 197.

Matsuo, M., Sagae, S., and Asai, H. (1969b), *Polymer* **10**, 79.

Matsuo, M., Ueda, A., and Kondo, Y. (1970a), *Polym. Eng. Sci.* **10**, 261.

Matsuo, M., Kwei, T. K., Klempner, D., and Frisch, H. L. (1970b), *Polym. Eng. Sci.* **10**, 327.

Matsuo, M., Wang, T. T., and Kwei, T. K. (1972), *J. Polym. Sci. A-2* **10**, 1085.

Matzner, M., Schober, D. L., and McGrath, J. E. (1973), *Eur. Polym. J.* **9**, 469.

Maxwell, J. C. (1904), *Electricity and Magnetism*, London, Clarendon Press, (reprint, 3rd ed., Vol. 1, New York, Dover).

McCrum, N. G., Read, B. E., and Williams, G. (1967), *Anelastic and Dielectric Effects in Polymeric Solids*, Wiley.

McGarry, F. J. (1956), Report P66-5, Materials Research Laboratory, Massachusetts Institute of Technology, October 1966.

McGarry, F. J., and Mandell, J. F. (1972), in *Proc. 29th Ann. Tech. Conf. Reinforced Plastics Div., SPI*, Section 5-B, p. 1.

McGrew, F. C. (1958), *J. Chem. Ed.* **35**, 178.

McIntyre, D., and Campos-Lopez, E. (1970), *Macromolecules* **3**, 322.

Meals, R. N., and Lewis, F. M. (1959), *Silicones*, Reinhold.

Meares, P. (1965), *Polymers: Structure and Bulk Properties*, Van Nostrand.

Medalia, A. I. (1970), *J. Colloid Interface Sci.* **32**, 115.

Mehta, H. C., Chen, W. F., Manson, J. A., and Vanderhoff, J. W. (1975a), paper given at Inter-Associations Colloquium, Behavior in Service of Concrete Structures, Liege, June, 1975.

Mehta, H. C., Chen, W. F., Manson, J. A., and Vanderhoff, J. W. (1975b), *Transp. Res. Rec.*, in press.

Mehta, H. C., Manson, J. A., Chen, W. F., and Vanderhoff, J. W. (1975c), *Transp. Eng. J.*, ASCE **101**, No. TE1, 1.

Meier, D. J. (1969), *J. Polym. Sci.* **26C**, 81.

Meier, D. J. (1970), *Polym. Preprints* **11**, 400.

Meier, D. J. (1971), private communication, December 1971.

Meier, D. J. (1972), *Polym. Lett.*, submitted.

Meissner, B. (1967), *J. Polym. Sci.* **16C**, 781.

Meissner, W., Berger, W., and Hoffman, H. (1968), *Faserforsch. Textiltech.* **19**, 407.

Merz, E. H., Claver, G. C., and Baier, M. (1956), *J. Polym. Sci.* **22**, 325.

Messing, R. A., and Weetall, H. H. (1970), U. S. patent 3,519,538.

Meyer, J. A. (1965), *For. Prod. J.* **15**, 362.

Meyer, J. A., and Loos, W. E. (1969), *For. Prod. J.* **19**(12), 32.

Michaels, A. S. (1965), *Off. Digest* **37**, 638.

Michaels, A. S., and Bixler, H. J. (1961), *J. Polym. Sci.* **50**, 413.

Millar, J. R. (1960), *J. Chem. Soc.* **1311**.

Miller, A. A. (1959*a*), *J. Phys. Chem.* **63**, 1755.

Miller, A. A. (1959*b*), *Ind. Eng. Chem.* **51**, 1271.

Miller, M. L. (1966), *The Structure of Polymers*, Reinhold.

Mizumachi, H. (1969), *J. Adhesion Soc. Japan* **5** (6), 370.

Mizumachi, H. (1970), *J. Adhesion* **2**, 292.

Moacanin, S. L., Holden, G., and Tschoegl, N. W., eds. (1969), *Block Copolymers* (*J. Polym. Sci.* **26C**), Interscience.

Modern Plastics Encyclopedia (1970–1971), McGraw-Hill.

Modern Plastics Encyclopedia (1974–1975), McGraw-Hill.

Moehlenpah, A. E., Ishai, O., and DiBenedetto, A. T. (1970), *Polym. Eng. Sci.* **10**, 170.

Moehlenpah, A. E., Ishai, O., and DiBenedetto, A. T. (1971), *Polym. Eng. Sci.* **11**, 129.

Molau, G. E. (1965), *J. Polym. Sci.* **3A**, 1267.

Molau, G. E., ed. (1971), *Colloidal and Morphological Behavior of Block and Graft Copolymers*, New York, Plenum.

Molau, G. E., and Keskkula, H. (1968), *Appl. Polym. Symp.* **7**, 35.

Molau, G. E., and Wittbrodt, W. M. (1968), *Macromolecules* **1**, 260.

Monte, S. J., and Bruins, P. F. (1974), *Mod. Plast.* **51**(12), 68.

Mooney, M. (1940), *J. Appl. Phys.* **11**, 582.

Mooney, M. (1948), *J. Appl. Phys.* **19**, 434.

Mooney, M. (1951), *J. Colloid Sci.* **6**, 162.

Morgan, P. W. (1961), U. S. patent 2,999,788.

Morgan, R. J. (1974), *J. Mater. Sci.* **9**(8), 1219.

Morton, M., and Healy, J. C. (1968), *Appl. Polym. Symp.* **7**, 155.

Morton, M., Healy, J. C., and Dencour, R. L. (1969), in *Proc. Int. Rubber Conf. 1967*, Gordon and Breach Science Publishers, p. 175.

Mostovoy, S., and Ripling, E. J. (1969), *J. Appl. Polym. Sci.* **1**, 1083.

Movsum-Zade, A. A. (1964), *Vysokomol. Soedin.* **6**, 1340.

Mullins, L. (1969), *Rubber Chem. Technol.* **42**, 339.

Mullins, L., and Tobin, N. R. (1956), in *Proc. 3rd Rubber Tech. Conf.* (London, 1954), p. 397.

Mumford, R. B., and Nevin, J. L. (1967), *Mod. Text. Mag.* **48**(4), 51.

Murphy, M. C., and Outwater, J. O. (1973), in *Proc. 28th Tech. Conf. Reinforced Plastics/ Composites Inst., SPI*, Section 17-A, p. 1.

Murray, J., and Hull, D. (1969), *Polymer* **10**, 451.

Nakamura, Y., Negishi, M., and Kalinuma, T. (1968), *J. Polym. Sci.* **23C**, 629.

Narkis, M., and Nicolais, L. (1971), *J. Appl. Polym. Sci.* **15**, 469.

National Academy of Sciences (1974), Committee on Structural Adhesives for Aerospace Use, Final Report, U.S. Government Printing Office, July, 1974.

National Commission on Materials Policy (1973), Materials Needs and the Environment Today and Tomorrow, Final report, U. S. Government Printing Office, June 1973.

Natta, G., Farina, M., and Peraldo, M. (1958), *Acc. Naz. Lincei Rend.* **25**, 424.

Natta, G., Bassi, I. W., and Allegra, G. (1961), *Acc. Naz. Lincei Rend.* **31**, 350.

Nawy, E. G., Sauer, J. A., and Sun, P. F. (1975), *Transp. Res. Rec.*, in press.

Newman, S., and Strella, S. (1965), *J. Appl. Polym. Sci.* **9**, 2297.

Newman, S. B., and Wolock, I. (1957), *J. Res. Nat. Bur. Stand.* **58**(6), 339.

Nicholas, T., and Freudenthal, A. M. (1966), ONR project No. NR064-446, Contract No. Nohr 266(78), Tech. report No. 36.

Nicholl, W. A. (1969), Technical paper, 18th Int. Wire and Cable Symp., Atlantic City, New Jersey, December 3–5, 1969, reprinted as AD 698604 (Clearinghouse for Federal Scientific and Technical Information).

Nicolais, L. (1975), *Polym. Eng. Sci.* **15**, 137.

Nicolais, L., and DiBenedetto, A. T. (1973), *Int. J. Polym. Met.* **2**, 251.

Nicolais, L., and Narkis, M. (1971), *Polym. Eng. Sci.* **11**, 194.

Nicolais, L., and Nicodemo, L. (1973), *Polym. Eng. Sci.* **13**, 469.

Niederhauser, W. D., and Bauer, W., Jr. (1972), U. S. patent 3,641,199.

Nielsen, L. E. (1953), *J. Am. Chem. Soc.* **75**, 1435.

Nielsen, L. E. (1962), *Mechanical Properties of Polymers*, Reinhold.

Nielsen, L. E. (1965), *Trans. Soc. Rheol.* **9**, 243.

Nielsen, L. E. (1966), *J. Appl. Polym. Sci.* **10**, 97.

Nielsen, L. E. (1967a), *J. Compos Mater.* **1**, 100.

Nielsen, L. E. (1967b), *J. Macromol. Sci.* **A1**, 929.

Nielsen, L. E. (1968), *J. Compos. Mater.* **2**, 120.

Nielsen, L. E. (1969a), *Polym. Eng. Sci.* **9**, 356.

Nielsen, L. E. (1969b), *Trans. Soc. Rheol.* **13**, 141.

Nielsen, L. E. (1970a), Soc. Automotive Eng. Pub. No. 700068.

Nielsen, L. E. (1970b), *J. Appl. Phys.* **41**, 4626.

Nielsen, L. E. (1974), *Mechanical Properties of Polymers and Composites*, New York, Marcel Dekker, 2 vols.

Nielsen, L. E., and Chen, P. E. (1968), *J. Materials* **3**, 352.

Nielsen, L. E., and Fitzgerald, W. E. (1964), *Proc. R. Soc. London* **A282**, 137.

Nielsen, L. E., and Lewis, T. B. (1969), *J. Polym. Sci. A-2* **7**, 1705.

Nielsen, L. E., Wall, R. A., and Richmond, P. G. (1955), *Soc. Plast. Eng. J.* **11**, 22.

Ninomiya, K., and Maekawa, E. (1966), *Nippon Gomu Kyokaishi* **39**, 601.

Noland, J. S., Hsu, N. N.-C., Saxon, R., and Schmitt, J. M. (1971), in *Multicomponent Polymer Systems* (Adv. Chem. Series 99), Platzer, N. A. J., ed., Chapter 2.

Oberst, H., and Schommer, A. (1965), *Kunststoffe* **55**, 634.

Oberst, H., Bohn, L., and Linhardt, F. (1961), *Kunststoffe* **51**, 495.

Oberst, H., *et al.* (1970), U. S. patents 3,547,757; 3,547,758; 3,547,759; 3,547,760; 3,547,755.

Oberst, H., *et al.* (1971), U. S. patents 3,553,072; 3,554,885.

Oberth, A. E. (1967), *Rubber Chem. Technol.* **40**, 1337.

Odian, G. (1970), *Principles of Polymerization*, New York, McGraw-Hill, p. 25.

Odian, G., and Bernstein, B. S. (1963), *Nucleonics* **21**, 80.

Odian, G., and Bernstein, B. S. (1964), *J. Polym. Sci. A-4* **2**, 2835.

Odian, G., and Kruse, R. L. (1968), *Polym. Preprints* **9**, 668.

Oertel, H. (1965), *Bayer Farbenrev.* **11**, 1.

O'Mahoney, J. F. (1970), *Rubber Age* **102**(March), 47.

O'Malley, J. J., Crystal, R. G., and Erhardt, P. F. (1969), *Polym. Preprints* **10**, 796.

O'Malley, J. J., Crystal, R. G., and Erhardt, P. F. (1970), in *Block Copolymers*, Aggarwal, S. L., ed., New York, Plenum, p. 163.

Ore, O. (1963), *Graphs and Their Uses*, New York, Random House.

Orowan, E. (1948), *Phys. Soc., Reports Prog. Phys.* **12**, 186.

Otocka, E. P., and Kwei, T. K. (1968), *Macromolecules* **1**, 214.

Outwater, J. O. (1970), in *Proc. 26th Ann. Conf. Soc. of the Plastics Industry, Reinforced Plastics Div.*

Outwater, J. O., and Matta, J. (1962), in *Proc. 17th Ann. Tech. Conf. Reinforced Plastics Div., SPI*, Section 14-B, p. 1.

Outwater, J. O., and Murphy, M. C. (1969), paper IIc, 24th Ann. Tech. Conf., Composites Div., Soc. of the Plastics Industry.

Outwater, J. O., and Murphy, M. C. (1970), *J. Adhesion* **2**, 242.

Owen, M. J. (1974a), in *Fracture and Fatigue* (*Composite Materials*, Vol. 5), Broutman, L. J., ed., Academic, p. 314.

Owen, M. J. (1974*b*), in *Fracture and Fatigue* (*Composite Materials*, Vol. 5), Broutman, L. J., ed., Academic, p. 342.

Oxford English Dictionary (1933), Oxford, The Clarendon Press, Vol. II, p. 1147.

Papero, P. V., Kuba, E., and Roldan, L. (1967), *Text. Res. J.* **37**, 823.

Parkyn, B., ed. (1970), *Glass Reinforced Plastics*, Cleveland, Ohio, CRC Press.

Patrick, R. L., ed. (1973), *Treatise on Adhesion and Adhesives*, Marcel Dekker, Vol. 3.

Patrick, R. L., Gehman, W. G., Dunbar, L., and Brown, J. A. (1971), *J. Adhesion* **3**, 165.

Paul, D. R., and Kemp, D. R. (1972), *Am. Chem. Soc., Div. Org. Coat. Plast. Prepr.* **32**(2), 156.

Payne, A. R. (1958), in *Rheology of Elastomers*, Mason, P., and Wookey, N., eds., London, Pergamon Press, p. 86.

Payne, A. R. (1960), *J. Appl. Polym. Sci.* **3**, 127.

Payne, A. R. (1961), *Rubber Plastics Age* **42**, 963.

Perera, D. Y., and Heertjes, P. M. (1971), *J. Oil Colour Chem. Assoc.* **54**, 313, 774.

Peretz, D., and DiBenedetto, A. J. (1972), *Int. J. Fract. Mech.* **4**, 979.

Peterson, C. M. (1968), *J. Appl. Polym. Sci.* **12**, 2649.

Peterson, J. M., and Hermans, J. J. (1969), *J. Compos. Mater.* **3**, 338.

Peterson, C. M., and Kwei, T. K. (1961), *J. Phys. Chem.* **65**, 1330.

Petrich, R. P. (1972), Impact Reinforcement of Poly(vinyl chloride), paper presented at SPE RETEC meeting, Cleveland, Ohio, March 7, 1972 (also presented at Polymer Conference Series, Mechanical Behavior of Polymers, University of Utah, June 1974).

Picot, C., Fukuda, M., Chou, C., and Stein, R. S. (1972), *J. Macromol. Sci. Phys.* **B6**, 263.

Pierzchala, H. (1969), *Tonind.-Ztg. Keram. Rundsch.* **93**(9), 337.

Piggott, M. R. (1970), *J. Mater. Sci.* **5**, 669.

Piggott, M. R. (1974), *J. Mater. Sci.* **9**, 494.

Piggott, M. R., and Leidner, J. (1974), *J. Appl. Polym. Sci.* **18**, 1619.

Pinner, S. H. (1960), *Plastics* **25**, 35.

Pinner, S. H. (1961), *A Practical Course in Polymer Chemistry*, Pergamon.

Pisarenko, G. S., and Shemegan, A. A. (1972), *Problemy Prochnosti* **4**(March), 8.

Platzer, N. A. J., ed. (1971), *Multicomponent Polymer Systems* (Adv. Chem. Series 99), ACS, Introduction.

Plueddemann, E. P. (1970), in *Proc. 25th Ann. Tech. Conf., Reinforced Plastics/Composites Div., SPI*, Section 13-D.

Plueddemann, E. P., ed. (1974*a*), *Interfaces in Polymer Matrix Composites* (*Composite Materials*, Vol. 6), Academic.

Plueddemann, E. P. (1974*b*), in *Interfaces in Polymer Matrix Composites* (*Composite Materials*, Vol. 6), Plueddemann, E. P., ed., Academic, p. 174.

Plueddemann, E. P., Clark, H. A., Nelson, L. E., and Hoffman, K. R. (1962), *Mod. Plast.* **39**, 135.

Plunkett, R. (1959), Measurement of Damping, in *Structural Damping*, Ruzicka, J. E., ed., New York, ASME, Section 5, p. 117.

Pochan, J. M. (1971), *Polym. Preprints* **12**, 212.

Pollack, H. (1971), U. S. patent 3,591,673.

Porai-Koshits, E. A., ed. (1973), *Phase Separation Phenomena in Glasses*, New York, Consultants Bureau.

Porter, R. S., Southern, J. H., and Weeks, N. (1975), *Polym. Eng. Sci.* **15**, 213.

Prevorsek, D. C. (1971), *J. Polym. Sci.* **32C**, 343.

Prevorsek, D. C., and Lyons, W. J. (1964), *J. Appl. Phys.* **35**, 3135.

Price, F. P. (1958), in *Growth and Perfection of Crystals*, Doremus, R. H., Roberts, B. W., and Turnbull, D., eds., New York, Wiley, p. 466.

Prud'homme, J., and Bywater, S. (1970), in *Block Polymers*, Aggarwal, S. L., ed., New York, Plenum, p. 11.

Rabinowitz, S., and Beardmore, P. (1973), *CRC Crit. Rev. Macromol. Sci.* **1**, 1.

Radon, J. C. (1972), *Polym. Eng. Sci.* **12**, 425.

Ramalingam, K. V., Werezak, G. N., and Hodgins, J. W. (1963), *J. Polym. Sci. C* **2**, 153.

Ranney, M. W., Berger, S. E., and Marsden, J. G. (1974), in *Interfaces in Polymer Matrix Composites (Composite Materials*, Vol. 6), Plueddemann, E. P., ed., Academic, p. 132.

Rao, P. N., and Hofer, K. E., Jr. (1972), Report AD 867-805.

Rao, P. N., and Hofer, K. E., Jr. (1972), Report AD 741-580.

Ravve, A. (1967), *Organic Chemistry of Macromolecules*, Marcel Dekker.

Rawson, H. (1967), *Inorganic Glass Forming Systems*, New York, Academic, Chapter 8.

Rebenfeld, L., ed. (1971), *Applied Polymer Symposium 18*, Part 1, Wiley.

Richards, D. H., and Szwarc, M. (1959), *Trans. Faraday Soc.* **56**, 1644.

Riddell, N. M., Koo, G. P., and O'Toole, J. L. (1966), *Polym. Sci. Eng.* **6**, 363.

Rideal, E. K. (1922), *Phil. Mag.* **44**, 1152.

Riess, G., Kohler, J., Tournut, C., and Banderet, A. (1967), *Macromol. Chem.* **101**, 58.

Ripling, E. J., Mostovoy, S., and Bersch, C. (1971), *J. Adhesion* **3**, 145.

Ritchie, P. D., ed. (1965), *Physics of Plastics*, Van Nostrand.

Rivin, D. (1971), *Rubber Chem. Technol.* **44**, 307.

Rivlin, R. S. (1948*a*), *Trans. R. Soc. London* **A240**, 459, 491, 509.

Rivlin, R. S. (1948*b*), *Trans. R. Soc. London* **A241**, 379.

Rivlin, R. S., and Thomas, A. G. (1953), *J. Polym. Sci.* **10**, 291.

Robinson, R. A., and White, E. F. T. (1970), in *Block Polymers*, Aggarwal, S. L., ed., New York, Plenum, p. 123.

Rodriguez, F. (1970), *Principles of Polymer Systems*, McGraw-Hill.

Roe, R. J., Davis, D. D., and Kwei, T. K. (1970), *Polym. Preprints* **11**(2), 1263.

Rogers, C. E., and Ostler, M. I. (1973), *Polym. Preprints* **14**(1), 587.

Rogers, C. E., Kupta, B., Yamada, S., and Ostler, M. I. (1974), *Am. Chem. Soc., Div. Org. Coat. Plast. Prepr.* **34**(1), 485.

Rosen, B., ed. (1964), *Fracture Processes in Polymeric Solids, Phenomena and Theory*, New York, Interscience.

Rosen, S. L. (1967), *Polym. Eng. Sci.* **7**, 115.

Rosen, S. L. (1971), *Fundamental Principles of Polymeric Materials for Practicing Engineers*, New York, Barnes and Noble.

Rosen, S. L. (1973), *J. Appl. Polym. Sci.* **17**, 1805.

Rosenfeld, A. R., and Kanninen, M. F. (1973), *J. Macromol. Sci. Phys.* **B7**, 609.

Rowland, F., Bulas, R., Rothstein, E., and Eirich, F. R. (1965), *Ind. Eng. Chem.* **57**(September), 46.

Ruffell, J. F. E. (1952), in *History of the Rubber Industry*, Schidrowitz, P., and Dawson, T. R., eds., Heffer, Chapter 5.

Rumsheidt, G. E., and Bruins, P. (1960), U. S. patent 2,939,859.

Ryan, C. F. (1972), U. S. patent 3,678,133.

Ryan, C. F., and Crochowski, R. J. (1969), U. S. patent 3,426,101.

Saam, J. C., Gordon, D. J., and Fearon, F. W. (1971*a*), in *Colloidal and Morphological Behavior of Block and Graft Copolymers*, Molau, G. E., ed., New York, Plenum, p. 75.

Saam, J. C., Gordon, D. J., and Fearon, F. W. (1971*b*), *Ind. Eng. Chem., Prod. Res. Dev.* **10**(March), 10.

Saam, J. C., Gordon, D. J., and Lindsey, S. (1970), *Macromolecules* **3**, 1.

Sadron, C. (1962*a*), French patent 1,295,524.

Sadron, C. (1962*b*), *Pure Appl. Chem.* **4**, 347.

Sadron, C. (1963), *Angew. Chem.* **75**, 472.

Sadron, C. (1966), *Chim. Ind. Genie Chim.* **96**, 507.

Sadron, C., and Gallot, B. (1973), *Makromol. Chem.* **164**, 301.

Sahu, S., and Broutman, L. J. (1972), *Polym. Eng. Sci.* **12**, 91.

Sambrook, R. W. (1970), *J. IRI* **4**, 210 (1970).

Sambrook, R. W. (1971), *Rubber Chem. Technol.* **44**, 728.

Sato, Y., and Furukawa, J. (1963), *Rubber Chem. Technol.* **36**, 1081.

Sauer, J. A., Marin, J., and Hsaio, C. C. (1949), *J. Appl. Phys.* **21**, 1071.

Schapery, R. A. (1968), *J. Compos. Mater.* **2**, 380.

Schlesinger, W., and Leeper, H. M. (1953), *J. Polym. Sci.* **11**, 203.

Schmieder, K. (1962), *Kunststoffe* **1**(5), 7.

Schmitt, J. A. (1968), *IUPAC Preprint*, Toronto, September 1968, p. A9.1.

Schmitt, J. A., and Keskkula, H. (1960), *J. Appl. Polym. Sci.* **3**, 132.

Schonhorn, H. (1964), *J. Polym. Sci. B* **2**, 465.

Schrader, M. E. (1974), in *Interfaces in Polymer Matrix Composites* (*Composite Materials*, Vol. 6), Plueddemann, E. P., ed., Academic, p. 110.

Schrenk, W. J. (1973), *Polym. Preprints* **14**, 241.

Schrenk, W. J., and Alfrey, Jr., T. (1969), *Polym. Eng. Sci.* **9**, 393.

Schrenk, W. J., and Alfrey, Jr., T. (1972), presented at the American Chemical Society Meeting, April 9–14, 1972, Boston, Massachusetts.

Schultz, J. (1974), *Polymer Materials Science*, New York, Prentice-Hall, Chapters 1 and 5.

Schwartz, R. J., and Schwartz, H. S. (1968), *Fundamental Aspects of Fiber-Reinforced Plastic Composites*, New York, Interscience.

Schwarzl, F. R. (1967), On Mechanical Properties of Unfilled and Filled Elastomers, in *Mechanics and Chemistry of Solid Propellants*, Eringen, Liebowitz, Koh, and Crawley, eds., Pergamon.

Schwarzl, F. R., Bree, H. W., and Nederveen, C. J. (1965), in *Proc. 4th Int. Cong. Rheology*, Lee, E. H., ed., New York, Interscience, Part 3, p. 241.

Scola, D. A. (1974), in *Interfaces in Polymer Matrix Composites* (*Composite Materials*, Vol. 6), Plueddemann, E. P., ed., Academic, p. 217.

Scott, K. W. (1967), *Polym. Eng. Sci.* **7**, 158.

Scott, R. L. (1949), *J. Chem. Phys.* **17**, 279.

Sears, D. A. (1965), in *Plastics and Surgical Implants*, ASTM Spec. Tech. Publ. 386.

Sellers, J. W., and Toonder, F. E. (1965), in *Reinforcement of Elastomers*, Kraus, G., ed., Interscience, Chapter 13.

Seward, R. J. (1970), *J. Polym. Sci.* **14**, 852.

Seymour, R. B. (1971), *Introduction to Polymer Chemistry*, McGraw-Hill.

Sharpe, L. H. (1971), Recent Advances in Adhesion, *162nd Meeting Am. Chem. Soc. Div. Org. Coat. Plast. Chem. Preprints* **31**(2), 201.

Sheffler, K. D., Kraft, R. W., and Hertzberg, R. W. (1969), *Am. Inst. Min., Metall. Pet. Eng.* **245**, 227.

Shen, M. C. (1969), *Macromolecules* **2**, 358.

Shen, M. C. (1970), *Rubber Chem. Technol.* **43**, 270.

Shen, M., and Bever, M. B. (1972), *J. Mater. Sci.* **7**, 741.

Shen, M., Hall, W. F., and DeWames, R. E. (1968), *J. Macromol. Sci. Rev. Macromol. Chem.* **C2**, 183.

Shen, M., Chen, T. Y., Cirlin, E. H., and Gebhard, H. M. (1971), in *Polymer Networks: Structure and Mechanical Properties*, Chompff, A. J., and Newman, S., eds., New York, Plenum, p. 47.

Shimada, J., and Kabuki, K. (1968), *J. Appl. Polym. Sci.* **12**, 671.

Shinohara, V. J. (1959), *J. Appl. Polym. Sci.* **1**, 251.

Shuler, M. L., Tsuchiya, H. M., and Aris, R. (1973), *J. Theor. Biol.* **41**(2), 347.

Shultz, A. R., and Gendron, B. M. (1972), *J. Appl. Polym. Sci.* **16**, 461.

Shuttleworth, R. (1968), *Eur. Polym. J.* **4**, 31.

Shuttleworth, R. (1969), *Rubber Chem. Technol.* **42**, 936.

Siau, J. F., and Meyer, J. A. (1966), *For. Prod. J.* **16**(8), 47.

Siau, F., Meyer, J. A., and Skaar, C. (1965a), *For. Prod. J.* **15**, 162.

Siau, F., Meyer, J. A., and Skaar, C. (1965b), *For. Prod. J.* **15**, 426.

Siau, J. F., Davidson, R. W., Meyer, J. A., and Skaar, C. (1968), *Wood Sci.* **1**(2), 116.

Sih, G. C. (1970), private communication.

Sih, G. C., and Irwin, G. R. (1970), *Eng. Fract. Mech.* **1**, 603.

Silman, I. H., and Katchalski, E. (1966), *Ann. Rev. Biochem.* **35**, 873.

Singer, K., Vinther, A., and Fördös, Z..(1971), Danish Atomic Energy Commission and Concrete Research Laboratory (Karlstrup), Concrete Research Laboratory Report BFL 264.

Skoulios, A., and Finaz, G. (1962), *J. Chim. Phys.* **59**, 473.

Skoulios, A., Tsouladze, G., and Franta, E. (1963), *J. Polym. Sci.* **4C**, 507.

Slonimskii, G. L. (1958), *J. Polym. Sci.* **30**, 625.

Smallwood, H. (1944), *J. Appl. Phys.* **15**, 758.

Smith, T. L. (1958), *J. Polym. Sci.* **32**, 99.

Smith, T. L. (1959), *Trans. Soc. Rheol.* **3**, 113.

Smith, T. L. (1964), *J. Appl. Phys.* **35**, 27.

Smith, T. L. (1970), in *Block Polymers*, Aggarwal, S. L., ed., New York, Plenum, p. 137.

Smith, T. L., and Rinde, J. A. (1969), *J. Polym. Sci. A-2* **7**, 675.

Smith, T. L., and Stedry, P. J. (1960), *J. Appl. Phys.* **31**, 1892.

Society of Plastics Engineers (1975), *Polym. Eng. Sci.* **15**(3), based on the United States–Italy Symposium on "Advanced Polymer Mechanics–Composite Materials and Ultra-High-Modulus Fibers," Santa Margharita Ligure, May 27–31, 1974.

Soen, T., Horino, H., Ogawa, Y., Kyuma, K., and Kawai, H. (1966), *J. Appl. Polym. Sci.* **10**, 1499.

Soldatos, J., and Burhans, R. (1970), *Ind. Eng. Chem., Prod. Res. Dev.* **9**, 296.

Solomotov, V. I. (1967), Atomic Energy Commission Translation Series, Tr. 7147 (transl. from Russian).

Sorensen, W. R., and Cambell, T. W. (1961), *Preparative Methods of Polymer Chemistry*, Interscience.

Souder, L. C., and Larson, B. E. (1966), U. S. patent 3,251,904.

Speri, W. M., and Jenkins, C. F. (1973), *Polym. Eng. Sci.* **13**, 409.

Sperling, L. H., ed. (1974a), *Recent Advances in Polymer Blends, Grafts, and Blocks*, New York, Plenum.

Sperling, L. H. (1974b), presented at the MARM meeting, ACS, Wilkes-Barre, Pennsylvania, April 1974.

Sperling, L. H. (1974c), in *Recent Advances in Polymer Blends, Grafts, and Blocks*, Sperling, L. H., ed., New York, Plenum, p. 93.

Sperling, L. H. (1974d), *Fibre Sci. Technol.* **7**, 199 [*Polym. Preprints* **14**, 431 (1973)].

Sperling, L. H. (1974e), presented at the American Chemical Society Meeting, Atlantic City, New Jersey, September 1974.

Sperling, L. H. (1974–1975), in *Encyclopedia of Polymer Science and Technology*, continuation series.

Sperling, L. H. (1975), to be published.

Sperling, L. H., and Arnts, R. R. (1971), *J. Appl. Polym. Sci.* **15**, 2317.

Sperling, L. H., and Friedman, D. W. (1969), *J. Polym. Sci. A-2* **7**, 425.

Sperling, L. H., and Mihalakis, E. N. (1973), *J. Appl. Polym. Sci.* **17**, 3811 (1973).

Sperling, L. H., and Sarge III, H. D. (1972), *J. Appl. Polym. Sci.* **16**, 3041.

Sperling, L. H., and Thomas, D. A. (1974), U. S. patent 3,833,404.

Sperling, L. H., and Tobolsky, A. V. (1966), *J. Macromol. Chem.* **1**, 799.

Sperling, L. H., and Tobolsky, A. V. (1968), *J. Polym. Sci. A-2* **6**, 25.

Sperling, L. H., Cooper, S. L., and Tobolsky, A. V. (1966), *J. Appl. Polym. Sci.* **10**, 1725.

Sperling, L. H., Taylor, D. W., Kirkpatrick, M. L., George, H. F., and Bardman, D. R. (1970a), *J. Appl. Polym. Sci.* **14**, 73.

Sperling, L. H., George, H. F., Huelck, V., and Thomas, D. A. (1970b), *J. Appl. Polym. Sci.* **14**, 2815.

Sperling, L. H., George, H. F., Huelck, V., and Thomas, D. A. (1970c), *Polym. Preprints* **11**, 477.

Sperling, L. H., Huelck, V., and Thomas, D. A. (1971), in *Polymer Networks: Structure and Mechanical Properties*, Chompff, A. J., and Newman, S., eds., New York, Plenum.

Sperling, L. H., Chiu, T. W., Hartman, C., and Thomas, D. A. (1972), *Int. J. Polym. Mater.* **1**, 331.

Sperling, L. H., Chiu, T. W., and Thomas, D. A. (1973), *J. Appl. Sci.* **17**, 2443.

Sperling, L. H., Chiu, T. W., Gramlich, R. G., and Thomas, D. A. (1974), *J. Paint Technol.* **46**, 47.

Spurr, O. K., and Niegisch, W. D. (1962), *J. Appl. Polym. Sci.* **6**, 585.

Stallings, R. L., Hopfenberg, H. B., and Stannett, V. (1972), *Am. Chem. Soc., Div. Org. Coat. Plast. Prepr.* **32**, 131.

Stamm, A. J. (1964), *Wood and Cellulose Science*, New York, Ronald Press.

Stannett, V. T., and Hopfenberg, H. P. (1971), in *Cellulose and Cellulose Derivatives*, Bikales, N. M., and Segal, L., eds., Wiley–Interscience, Part V, p. 907.

Stanton, G. W., and Traylor, T. G. (1962), U. S. patent 3,049,507.

Stein, R. S. (1966), *J. Polym. Sci.* **15C**, 185.

Stein, R. S., and Wilkes, G. L. (1969), *J. Polym. Sci. A-2* **7**, 1696.

Steinberg, M. (1973), Concrete Polymer Composite Materials and its Potential for Construction, Urban Waste Urilization, and Nuclear Waste Storage, Informal Report, Brookhaven National Laboratory, May 1973.

Steinberg, M., Kukacka, L. E., Colombo, P., Kelsch, J. J., Manowitz, B., Dikeou, J. T., Backstrom, J. E., and Rubinstein, S. (1968), Concrete-Polymer Materials, First Topical Report, Brookhaven National Laboratory, report BNL 50134, December 1968.

Steinberg, M., Dikeou, J. T., Kukacka, L. E., Backstrom, J. E., Colombo, P., Hickey, K. B., Auskern, A., Rubinstein, S., Manowitz, B., and Jones, C. W. (1969), Concrete-Polymer Materials, Second Topical Report, Brookhaven National Laboratory, report 50218 (T-560), December 1969.

Steinman, H. W. (1970), *Polym. Preprints* **11**, 285.

Sterman, S., and Marsden, J. G. (1966), Bonding Organic Polymers to Glass by Silane Coupling Agents, Glass Division Symposium, American Ceramic Society Meeting, Washington, D. C., May 10, 1966.

Stern, H. J. (1967), *Rubber, Natural and Synthetic*, Maclaren, 2nd ed.

Sternstein, S. S. (1972), *J. Macromol. Sci. Phys.* **B6**, 243.

Sternstein, S. S., Ongchin, L., and Silverman, A. (1968), *Appl. Polym. Symp.* **7**, 175.

St. John Manley, R. (1963), *J. Polym. Sci.* **1A**, 1875.

Stoelting, J., Karasz, F. E., and MacKnight, W. J. (1969), *Polym. Preprints* **10**, 628.

Stoelting, J., Karasz, F. E., and MacKnight, W. J. (1970), *Polym. Eng. Sci.* **10**, 133.

Stone, F. G. A., and Graham, W. A. G., eds. (1962), *Inorganic Polymers*, New York, Academic, Chapter 5.

Studebaker, M. L. (1965), in *Reinforcement of Elastomers*, G. Kraus, ed., Interscience.

Sultan, J. M., and McGarry, F. J. (1973), *Polym. Eng. Sci.* **13**, 29.

Sundstrom, D. W., and Chen, S. Y. (1970), *J. Compos. Mater.* **4**, 113.

Sussman, V. (1970), *Soc. Plast. Eng., Ann. Tech. Conf.*, Tech. paper No. 28, 72.

Sweitzer, C. W. (1961), *Rubber Age* **89**, 269.

Szwarc, M. (1956), *Nature* **178**, 1168.

Szwarc, M. (1973), *Polym. Eng. Sci.* **13**, 1.

Szwarc, M., Levy, M., and Milkovich, R. (1956), *J. Am. Chem. Soc.* **58**, 2656.

Szymczak, T. J. (1970), M.S. Thesis, Lehigh Univ., June 1970.

Szymczak, T. J., and Manson, J. A. (1974a), *West. Electr. Eng.* **18**(1), 26.

Szymczak, T. J., and Manson, J. A. (1974b), *Mod. Plast.* **51**(8), 66.

Takayanagi, M. (1963), *Mem. Fac. Eng., Kyushu Univ.* **23**, 11.

Takayanagi, M. (1972), private communication, February 8, 1972.

Takayanagi, M., Harima, H., and Iwata, Y. (1963), *Mem. Fac. Eng., Kyushi Univ.* **23**, 1.

Tarkow, H., Baker, A. J., Eickner, H. W., Eslyn, W. E., Hajny, G. J., Hann, R. A. Koeppen, R. C., Millet, M. A., and Moore, W. E. (1970), in *Kirk-Othmer Encyclopedia of Chemical Technology*, 2nd ed., Vol. 22, p. 358.

Taylor, G. B., Dietz, G. R., and Duff, R. M. (1972), *Nuclear Technol.* **13**, 72.

Tazawa, E., and Kobayashi, S. (1973), in *Polymers in Concrete*, American Concrete Institute Publication SP-40, paper No. 4, p. 57.

Tetelman, A. S., and McEvily, A. J. (1967), *Fracture of Structural Materials*, Wiley.

Theberge, J. E. (1969), trade literature, Liquid Nitrogen Processing Co., *Modern Plastics*.

Thirion, P., and Chasset, R. (1962), in *Proc. 4th Rubber Technol. Conf.*, London, p. 338.

Thomas, J. P. (1960), Report AD 287-826.

Thomka, L. M., and Schrenk, W. J. (1972), *Mod. Plast.* **49**(4), 62.

Thurn, H. (1960), *Kunststoffe* **50**, 606.

Tiffan, A. J., and Shank, R. S. (1967), U. S. patent 3,305,514.

Till, P. H., Jr. (1957), *J. Polym. Sci.* **24**, 301.

Timmons, T. K., Meyer, J. A., and Cote, W. A., Jr. (1971), *Wood Sci.* **4**(1), 13.

Tobolsky, A. V. (1960), *Properties and Structure of Polymers*, New York, Wiley.

Tobolsky, A. V., and Aklonis, J. J. (1964), *J. Phys. Chem.* **68**, 1970.

Tobolsky, A. V., and DuPre, D. B. (1968), *J. Polym. Sci. A-2* **6**, 1177.

Tobolsky, A. V., and MacKnight, W. J. (1965), *Polymeric Sulfur and Related Polymers*, Interscience.

Tobolsky, A. V., and Mark, H., eds. (1971), *Polymer Science and Materials*, New York, Wiley–Interscience.

Tobolsky, A. V., and Rembaum, A. (1964), *J. Appl. Polym. Sci.* **8**, 307.

Tobolsky, A. V., and Shen, M. C. (1965), *Advances in Chemistry Series, No. 48*, American Chemical Society.

Tobolsky, A. V., and Shen, M. C. (1966), *J. Appl. Phys.* **37**, 1952.

Tobolsky, A. V., and Sperling, L. H. (1968), *J. Phys. Chem.* **72**, 345.

Tobolsky, A. V., Carlson, D. W., and Indictor, N. (1961), *J. Polym. Sci.* **54**, 175.

Tobolsky, A. V., Lyons, P. F., and Hata, N. (1968), *Macromolecules* **1**, 515.

Tompa, H. (1956), *Polymer Solutions*, Butterworths, Chapter 7.

Torvik, P. J., (1971), Technical Report, AFIT, TR 71-2, Wright-Patterson Air Force Base, Ohio, May 1971.

Touhsaent, R. E., Thomas, D. A., and Sperling, L. H. (1974a), *J. Polym. Sci.* **46**, 175.

Touhsaent, R. E., Thomas, D. A., and Sperling, L. H. (1974b), ACS Organic Coating and Plastic preprint **34**(2), 309.

Trachte, K., and DiBenedetto, A. T. (1971), *Int. J. Polym. Mater.* **1**, 75.

Treloar, L. R. G. (1944), *Trans. Faraday Soc.* **40**, 59.

Treloar, L. R. G. (1958), *The Physics of Rubber Elasticity*, Clarendon.

Trommsdorf, E., Kohle, H., and Lagally, P. (1948), *Makromol. Chem.* **1**, 169.

Tsai, S. W. (1964), Technical Report NASA CR-71, July 1974.

Tsai, S. W. (1965), Technical Report NASA CR-224, April 1965.

Tsai, S. W., Halpin, J. C., and Pagano, N. J., eds. (1969), Composite Materials Workshop, Stamford, Connecticut, Technomic Publishing Co.

Tsao, G. T. (1961), *Ind. Eng. Chem.* **53**, 395.

Tschoegl, N. W. (1971a), private communication.

Tschoegl, N. W. (1971b), *J. Polym. Sci.* **32C**, 239.

Tsoumis, G. (1968), *Wood as a Raw Material*, Pergamon, p. 69.

Turbak, A. F., ed. (1970), *Membranes from Cellulose and Cellulose Derivatives* (Applied Polymer Symp. 13), Interscience.

Turley, S. G. (1963), *J. Polym. Sci.* **C-1**, 101.

Turley, S. G. (1968), *Appl. Polym. Symp.* **7**, 237.

Turner, P. S. (1946), *J. Res. Nat. Bur. Stand.* **37**, 239.

U. S. Bureau of Mines (1971), Preliminary Survey of Polymer-Impregnated Rock, U. S. Bureau of Mines Report of Investigations, No. 7542.

Van Amerongen, G. J. (1964), *Rubber Chem. Technol.* **37**, 1067.

Vanderhoff, J. W., Hoffman, J. D., and Manson, J. A. (1973), *Polym. Preprints* **14**(2), 1136.

Van der Poel, C. (1958), *Rheol. Acta* **1**, 198.

Van der Waal, C. W., Bree, H. W., and Schwarzl, F. R. (1965), *J. Appl. Polym. Sci.* **9**, 2143.

VanOene, H. (1972), *J. Colloid Interface Sci.* **40**, 448.

Van Vlack, L. H. (1964), *Elements of Materials Science*, Reading, Massachusetts, Addison-Wesley, Chapters 9–11.

Vermillion, J. L. (1968), in *Modern Plastics Encyclopedia*, McGraw-Hill, p. 187.

Vincent, P. I. (1967), in *Proc. Conf. on Polymer Structure and Mechanical Properties, Natick, Massachusetts, April 19–21, 1967* (NASA CR 100-141).

Vold, M. J. (1963), *J. Colloid Sci.* **18**, 684.

Vollmert, B. (1962), U. S. patent 3,055,859.

Voyutskii, S. S. (1963), *Autohesion and Adhesion of High Polymers*, Wiley.

Voyutskii, S. S., and Vakula, V. L. (1969), *Mekh. Polim.* **5**, 455.

Voyutskii, S. S., Kamenskii, A. N., and Fodiman, N. M. (1966), *Mekh. Polim., Akad. Nauk LatvSSR* **3**, 446.

Wagner, E. R., and Cotter, R. J. (1971), *J. Appl. Polym. Sci.* **15**, 3043.

Wagner, E. R., and Robeson, L. M. (1970), *Rubber Chem. Technol.* **43**, 1129.

Wagner, H. B. (1965), *Ind. Eng. Chem., Prod. Res. Dev.* **4**, 191.

Wagner, H. B. (1966), *Ind. Eng. Chem., Prod. Res. Dev.* **5**, 149.

Wagner, H. B. (1967), *Ind. Eng. Chem., Prod. Res. Dev.* **6**(4), 223.

Wagner, M. P., and Sellers, J. W. (1959), *Ind. Eng. Chem.* **51**, 961.

Waldrop, M. A., and Kraus, G. (1969), *Rubber Chem. Technol.* **42**, 1155.

Wambach, A., Trachte, K. L., and DiBenedetto, A. T. (1968), *J. Compos. Mater.* **2**, 266.

Wang, T. T., and Kwei, T. K. (1969), *J. Polym. Sci.* **A-2 7**, 889.

Ward, I. M. (1972), *Mechanical Properties of Solid Polymers*, Wiley.

Ward, T. C., and Tobolsky, A. V. (1967), *J. Appl. Polym. Sci.* **11**, 2403.

Warson, H. (1972), *The Application of Synthetic Resin Emulsions*, Ernest Benn, London.

Washburn, E. W. (1921), *Phys. Rev.* **17**, 243.

Watson, W. F. (1955), *Ind. Eng. Chem.* **47**, 1281.

Weinstein, J. N., Bunow, B. J., and Caplan, J. R. (1972), *Desalination* **11**, 341.

Weinstein, J. N., Misra, B. M., Kalif, D., and Caplan, S. R. (1973), *Desalination* **12**, 1.

Wellons, J. D., Williams, J. L., and Stannett, V. (1967), *J. Polym. Sci.* **A-1 5**, 1341.

Wells, H. (1967), in *Proc. 22nd Ann. Meeting Reinforced Plastics Division, Soc. Plastics Industry, Washington, D.C.*, Section 17-B, p. 1.

Western Electric (1974), advertisement, *Science*, July 5.

Weyers, R. E., Cady, P. D., Blankenhorn, P. R., and Kline, D. E. (1975), *Transp. Res. Rec.*, in press.

White, F. T., and Mann, R. (1967), Br. patent 1,066,610 (to Assoc. Electrical Industries Ltd.).

Whiting, D. A., Blankenhorn, P. R., and Kline, D. E. (1973), *Polym. Preprints* **14**(2), 1154.

Whiting, D. A., Blankenhorn, P. R., and Kline, D. E. (1974), *J. Test. Eval.* **2**, 44.

Williams, D. J. (1971), *Polymer Science and Engineering*, New York, Prentice-Hall.

Williams, M. L., and DeVries, K. L. (1970), in *Proc. 5th Int. Cong. Rheol.*, Vol. 3, p. 139.

Williams, M. L., Landel, R. F., and Ferry, J. D. (1955), *J. Am. Chem. Soc.* **77**, 3701.

Williams, T., Allen, G., and Kaufman, M. S. (1973), *J. Mater. Sci.* **8**, 1785.

Wippler, C. (1959), *Rev. Gen. Caoutch.* **36**, 369.

Wollek, R. F. (1965), U. S. patent 3,193,049.

Work, J. L. (1972*a*), presented at 30th ANTEC Soc. Plast. Eng. Meeting, Chicago, Illinois, May 1972.

Work, J. L. (1973), *Polym. Eng. Sci.* **13**, 46.

Wu, E. M. (1974), in *Fracture and Fatigue* (*Composite Materials*, Vol. 5), Broutman, L. J., ed., Academic, p. 191.

Wu, S. (1974), *J. Macromol. Sci. Rev. Macromol. Chem.* **C10**(1), 1.

Wu, T. T. (1966), *Int. J. Solids Struct.* **2**, 1.

Yakobson, F. I., Amerik, V. V., Peterson, V. F., Shteinbak, V. S. H., and Ivanyukoo, D. V. (1970), *Plast. Massy* No. 3, p. 11.

Yim, A., and St. Pierre, L. E. (1969), *Polym. Lett.* **B7**, 237.

Yim, A., Chahal, R. S., and St. Pierre, L. E. (1973), *J. Colloid Interface Sci.* **43**, 583.

Yoffe, E. (1951), *Phil. Mag.* **42**, 739.

Zhurkhov, Ş. N.; and Tomashevskii, E. E. (1966), in *Physical Basis of Yield and Fracture*, Oxford, Inst. Phys. and Phys. Soc.

Ziegel, K. D. (1969), *J. Colloid Interface Sci.* **29**, 72.

Ziegel, K. D., and Romanov, A. (1973), *J. Appl. Polym. Sci.* **17**, 1119, 1131.

Zimmerman, J. (1968), U. S. patent 3,393,252.

Zimmerman, J., Pearce, E. M., Miller, I. K., Muzzio, J. A., Epstein, I. G., and Hosegood, E. A. (1973), *J. Appl. Polym. Sci.* **17**, 849.

Polymer Index

507

Subject Index

509